Commodity Marketing

from a Producer's Perspective

730-Day Profitable Marketing
of Grain and Livestock

Second Edition

025

Commodity Marketing

from a Producer's Perspective

730-Day Profitable Marketing
of Grain and Livestock

Second Edition

Donald G. Chafin, Ph.D.

Professor of Agriculture
Wilmington College
Wilmington, Ohio

Paul H. Hoepner, Ph.D.

Professor Emeritus: Agricultural and Applied Economics
Virginia Polytechnic Institute and State University
Blacksburg, Virginia

INTERSTATE PUBLISHERS, INC.
Danville, Illinois

Commodity Marketing
from a Producer's Perspective
Second Edition

Prior edition: 1989

Library of Congress Control Number: 00-102810

ISBN 0-8134-3179-4

1 2 3 4 5 6 7 8 9 10   06 05 04 03 02 01

Order from

Interstate Publishers, Inc.

510 North Vermilion Street
P.O. Box 50
Danville, IL 61834-0050

Phone: (800) 843-4774
Fax: (217) 446-9706
Email: info-ipp@IPPINC.com

World Wide Web: www.interstatepublishers.com

Preface

In this book we develop a systematic approach to commodity marketing, specifically designed from a farmer and rancher perspective. Used as a manual, this book can transform producers from mere sellers to thoughtful merchandisers. But the book is more than a guide for farmers—it is for anyone who wants to study the market from a producer's point of view. The more commodity producers learn about how markets function, the better their marketing decisions.

This textbook grew out of experience in teaching agricultural marketing at Wilmington College in Ohio and in researching commodity prices at Virginia Polytechnic Institute and State University. Since the agriculture program at Wilmington stresses practicality, other marketing texts did not suit our purposes. Most are descriptive or theoretical and focus on intermediary activities in the agricultural economy. Wilmington students wanted a different perspective. They wanted to know how to make practical marketing decisions. Fluctuating commodity prices provided an opportunity to teach marketing principles from an applied perspective. That perspective is the thrust of this book. Though much of our material centers around the commodity futures markets, we also define and describe various marketing tools. We explain how to apply each tool to decision making in the market place.

We organized the book in 10 chapters for a 10-week or one-quarter college course, but it can also fit a semester course. It can be used in four-year college programs, two-year technical agricultural programs, high school courses, or farmer-rancher adult education courses.

We do not claim that all of our ideas are original. We too have been students, learning from others. We have selected and adapted concepts from various sources that we feel are most relevant to building an understanding of farmer marketing. From that foundation, we have built a new perspective for commodity marketing. In our organization, selection of topics, clarity of presentation, and teaching aids, we offer the reader a new approach to the study of commodity marketing—730-Day Marketing Program.

ACKNOWLEDGMENTS

Students at Wilmington College and the farmer-producers in the community inspired us to develop this book. Their interest and acceptance of its point of view encouraged us to exert the effort necessary to complete our study.

Every college teacher, seminar leader, lecturer, magazine writer, and marketing text author whom we have met has contributed directly or indirectly to this effort. It is impossible to mention all of them, but their contributions were important in our learning how to present this topic. However, omissions in the text are our own.

Special thanks are due to Ms. Patricia Campbell for typing the manuscript.

Contents

Chapter 1

Framework for Analyzing a Market Economy

Framework for Analyzing
a Market Economy

The World Changed in 1972

Agricultural prices changed dramatically in 1972 when the United States sold wheat to the Soviet Union. As soon as the public learned of the sale, many commodity prices, particularly grain prices, exploded.

Many other more subtle economic changes also contributed to this price explosion. Though the Soviet grain purchase was the watershed event marking the beginning of the agricultural commodity price level rise, four structural factors identified by Carroll Brunthaver appeared in the food demand equation.[1] These were world population growth, income growth, the protein principle, and monetary exchange rates. The Russian purchase in itself would not have had long-reaching effects had not these other significant changes also been occurring in the world.

In the late 1960's and early 1970's, farmers in Taiwan were buying tractors to increase their productivity. South Koreans were building a subway in their capital city of Seoul that would be in operation even before the subway in Washington, D.C. The once sleepy country of Spain had become the fastest growing economy in the industrialized world. The Soviet government gave top priority to satisfying consumer demands for an improved diet, a sharp departure from the traditional concern with heavy industrial development. The Chinese began buying western technology in the form of airplanes, petrochemical plants, and machine tools. In developing countries of the world, children were learning to read, and their parents were learning more productive skills. All over the world new medicines were helping people live longer, happier lives. All of these changes were part of a major economic phenomenon—a basic shift in the demand curve for commodities.

Never before had so many people lived on this planet. Never before had the world's industries possessed so much capacity to produce what consumers wanted. Never before had the demand for resources been so great. The changes affected not only oil and food, but also lumber, metals, energy, fertilizers, and a whole host of basic commodities. Demand was simply outstripping supply. The results were sharp price increases as shortfalls in supply were unable to satisfy the rapidly growing demand. It was not a change in supply conditions, but a fundamental shift in the demand for commodities that caused the headlines of leaping prices.

The change meant not only higher prices, but also more volatile prices. Prior to 1972, commodity prices were relatively stable. After 1972 and into 1980, commodity prices rose to record levels. More importantly, they fluctuated violently from month to month and from one year to the next. Figure 1–1 shows the dramatic doubling of the futures price index from 1972 to 1973, its further rise in 1980, its collapse in 1981, its rebound during the 1988 drought, its decline into the mid-1990's, and a recovery

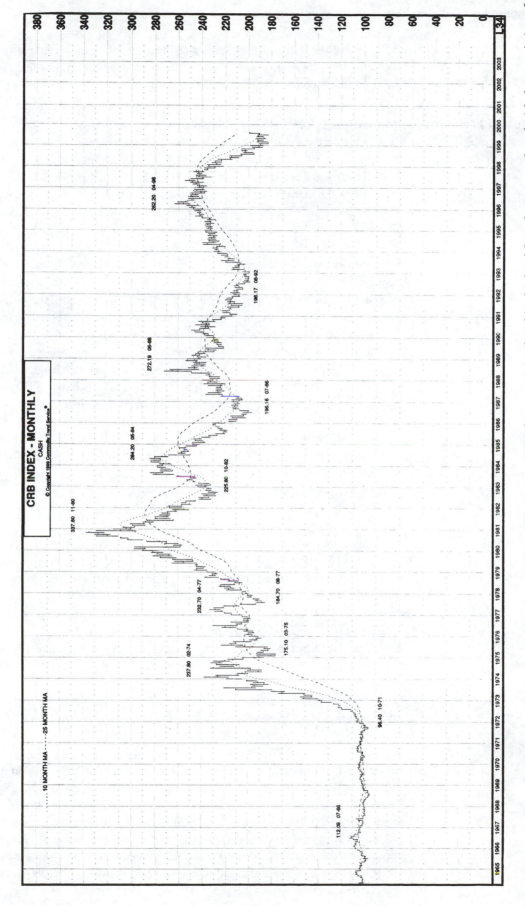

FIGURE 1–1: Futures Price Index, 27 Commodity Markets, 1965 to 1998. Source: CRB charts contained within this book have been reproduced with the permission of Commodity Trend Service, Inc. All rights reserved.

at the end of the century. This graph also illustrates the intra-year volatility that has characterized economic events since 1972. Price volatility significantly heightened the level of uncertainty for commodity producers.

Changes in Demand

Population. In the first place, there were more people to be fed. The population of the world first topped 1 billion people about 1850. It had taken from the dawn of time until almost 3,000 years after the birth of Christ for the population to build up to that level. It took less than 100 years, by 1930, to add the second billion people. In 30 more years, 1960, the world's population reached 3 billion. The fourth billion was reached 15 years later, in 1975. In October of 1999, the world's population exceeded seven billion souls.

Early in the 1970's changes in political relations between the United States, the Soviet Union, and China opened U.S. markets to a much larger segment of the world population. Policy changes in the Soviet Union significantly affected that country's overall food demand.[2] From the 1920's until Stalin's death, the Soviet Union exported grain to gain foreign exchange, even while some of its people starved. In the 1960's, the decision was made to import grain to provide needed food. The U.S.S.R. actually had a bigger relative shortfall in grain crops due to poor harvests in 1963 than in 1972. The difference was that in 1963, the Soviet people slaughtered livestock, tightened their collective belts, and imported as little as possible. In 1972, Soviet leaders made a basic change in policy and decided to feed their people better. The Soviets began buying large quantities of U.S. grain for food and livestock feed.[3]

The volume of sales to the Soviet Union was not as large as sales volume to other U.S. trading partners, but the *additional* demand had important price effects on commodities. A USDA study estimated that the expansion of the Soviet grain trade increased U.S. corn prices 50 to 60 cents per bushel, wheat and soybean prices almost a dollar a bushel. Soviet trade also added about $30 per ton to the price of soybean meal.[4]

China added another demand factor to U.S. agriculture. After nearly 30 years of isolation, U.S. relations with the People's Republic of China were reestablished in the early 1970's. With almost 1 billion people, or a fourth of the entire population of the world, China was seen as the "greatest potential market for the world's agricultural products," as one observer stated:[5] "Before normalized relations with the United States, the Chinese had purchased grain almost exclusively from Canada and Australia. The United States became a supplier only when these countries were unable to fulfill all of China's needs. The United States had been a supplier of last resort. From 1972 to 1974, China purchased wheat, corn, and cotton from the United States. However, the next three years they purchased practically nothing. Such trading tactics made China the 'wildcard' participant in the U.S. grain markets."[6]

Income. More important than sheer numbers of people was the fact that the world's population was living better than people had lived in the past. The rising standard of living was not just an America phenomenon; it was happening all over the world. From the 1960's through the early 1970's, the world economy recorded annual growth rates averaging 5 percent. After 1970, growth in world gross domestic products, adjusted for inflation, slowed to about 3.7 percent. The per-capita growth rate of GNP averaged 2.3 percent after adjusting for inflation during the 1980 to 1987 period and is expected to grow at a 1.5 percent from 1985 to 2000.[7]

Higher productivity brought a rising standard of living for the world's population and increased the purchasing power necessary to buy goods and services, thereby translating needs and desires into effective demand.

The Protein Principle. Agriculture has an additional market demand factor—the protein principle. As soon as people have enough calories in their diet, they spend extra income to improve the quality of their diet. This means increasing demand for protein in the form of livestock products—meat, milk, and eggs. A 1 percent increase in income is usually associated with a 1 percent increase in livestock demand, in a free market. Compared to the U.S. meat consumption of 243 pounds per person* in 1984, the Chinese consumed only 29 pounds, the Japanese 74 pounds, Brazilians 63 pounds, Mexicans 80 pounds, Filipinos 25 pounds, and Soviets 48 pounds per person.[8] The European Common Market held its demand slightly lower by pricing grain—and therefore meat—very dearly. Japan held its consumption at a very low level by imposing restrictive import quotas, but Japanese consumers registered their desire for meat by bidding the price of choice beef cuts to $25 per pound.[9]

Livestock protein is resource-intensive. A broiler chicken is one of the most efficient converts of feed grain into meat protein. Nonetheless, about 2.5 pounds of cereals and protein supplement are required to produce a pound of broiler meat. It takes additional resources to slaughter the chicken, process it, and keep it refrigerated as it moves through the marketing chain. Thus, three times the resources are required to deliver a pound of broiler meat, compared to a pound of cereals, to the dining table. At the other end of the livestock scale, grain-fed beef is about 8 to 10 times as resource-expensive as cereals. The feed-to-meat conversion ratios mean that the demand side of the world grain consumption equation is multiplied many fold by the protein principal.

The demand for farm products was affected not just by an increase in numbers of people but by their desire and ability to eat better. Considering all factors, the cereal demand equation equals more people, multiplied by more income, multiplied once more by the protein principle. The result was that 3 to 10 times as many resources were required to satisfy the growing world demand for food.

Devaluation and Demand. The final factor in the food demand equation for U.S. farm products was the devaluation of the dollar in 1971 and 1973. The demise of the 1946 Bretton Woods Agreement on a post-war monetary system ended the practice of fixed currency rates. The value of the dollar, as measured by the currencies of other major countries, declined steadily between 1970 and 1973 as the Untied States progressed from a fixed exchange rate system (and an over-valued dollar) to a system of freely floating rates. By 1973, the dollar had fallen 16 percent from its 1970 parity value.[10] Cheaper dollars reduced the cost of U.S. farm products to international consumers.

This dramatic drop in value of the 1973 dollar brought worldwide food demand to the U.S. farm gate. U. S. agricultural exports increased from $7 billion in 1970 to $40.5 billion in 1980. Total agricultural exports approached a record $60 billion in 1996. During that decade, annual world trade in wheat and coarse grains increased almost 90 million tons; the United States supplied almost 75 million tons, or more than 80 percent of the increase. During the 1990's, the United States exported half of

---

*Carcus weight on red meat and ready-to-cook weight on poultry.

its wheat and soybeans, a third of its cotton, and a fourth of its corn production. The United States continued to export the production from 1 out of 3 harvested acres throughout the last two decades of the twentieth century.[11]

Demand and Supply in 1980's and 1990's

Demand

Commodity price indices reached record high levels in 1980, as the demand factors interacted to increase prices for agricultural commodities. However, a series of events that year abruptly turned the price situation downward. In January 1980, President Carter suspended grain and food shipments to the Soviet Union in retaliation for its invasion of Afghanistan. Soviet imports were drastically reduced. This fourth embargo since 1972 may have branded the United States as an unreliable supplier and intensified its image as a marginal supplier, to be sought after only when other countries could not provide exportable supplies.

An increase in the value of the dollar further dampened export demand by foreign buyers. But the dollar's rise was only a symptom of the economic changes that dampened world demand for commodities.[12] The boost in oil prices resulted in worldwide demand for dollars. Had extra dollars needed to pay for oil come from monies that would otherwise have been spent on other goods and services, higher petroleum prices would have been offset by lower prices of other items. But most oil users resorted to bank borrowing, as OPEC recycled its petrodollars. The jump in money supplies pushed the prices of virtually all commodities sharply higher. Higher commodity prices encouraged less developed countries to go on a borrowing binge, adding further fuel to the inflation fire.

By 1980, the acute inflation of the late 1970's began to backfire. As prices rose faster than income, real earnings dropped and worldwide real demand for goods and services softened. Debt-laden borrowers found it difficult to service their loans. The bad debt problems of least developed countries forced lenders to turn off the credit spigot. "Credit contraction abroad was the prime reason for the 1980-84 disinflation."[13]

Just as higher incomes had worked through the protein principal in a multiplier effect to increase demand for U.S. farm products (and to increase prices), deflation acted as a decelerating multiplier, causing commodity demand and prices to tumble. Until the slump, agricultural exports set records for 12 straight years (Figure 1–2). The dramatic growth in volume of exports during the 1980's averaged 8 percent per year. From the 1981 peak of $43.8 billion, U.S. exports fell to $30 billion in 1986. Export volume tumbled from a high of 163.9 million metric tons in 1980 to a round 108 million tons in 1987.[14]

Two major factors came into play in the late 1990's. Agricultural legislation known as the "Freedom to Farm Act" lifted government restrictions on acreage controls and production quotas. There were no set-aside acreage requirements, but the Conservation Reserve continued to hold fragile and marginal land out of production.

Triggered by a regional financial crisis that began in Thailand in July 1997, exports of U.S. agricultural commodities sagged significantly toward the end of the century. U.S. agricultural exports had risen steadily throughout the 1990's, reaching $60 billion in 1996. However, financial problems in Asia and the former Soviet

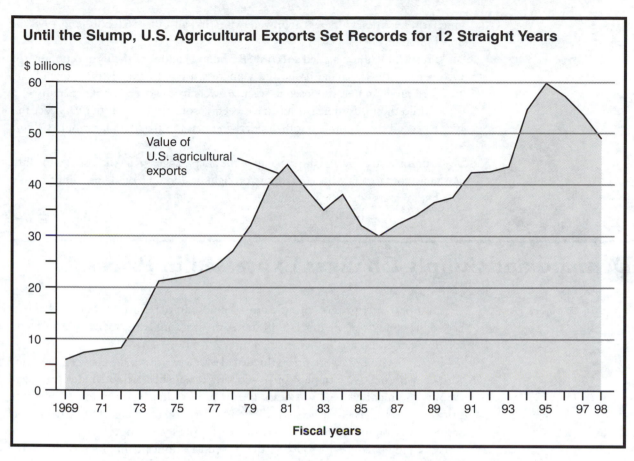

FIGURE 1-2: Value of U.S. Agricultural Exports, 1969 to 1998. Source: USDA, FAS.

Union weakened world demand and U.S. exports slipped below $50 billion at the end of the millennium. These trends are summarized in Figure 1–2.

Supply in the U.S.

In response to high commodity prices of the 1970's, Agriculture Secretary Earl Butz urged U.S. farmers to expand production and to "plant fence row to fence row." U.S. farmers responded by producing bin-bursting corn, wheat, and soybean crops. Under increased supply and reduced demand pressures, U.S. grain prices retreated to depression-era levels (on a deflated basis) in the mid-1980's. Government price support mechanisms were brought into play to set a floor under U.S. grain prices and support farmer income. Price supports, commodity loans, target prices, and payment-in-kind schemes played an important role throughout the decade, until 1988. Nature shut off the rain in June and July across most of the Corn Belt. Prospects for significantly lower yields in 1988 quickly turned price-depressing surpluses into strategic reserves.

Supply conditions changed by 1996. Coupled with the Freedom to Farm Act that eliminated acreage restrictions, diminished world demand caused corn, wheat, soybeans, and cotton inventories to pile to near record highs. Huge carry-over stocks forced grain prices to plunge to the low levels of the economically disastrous mid 1980's levels. Government farm program payments were increased to subsidize sagging farm incomes. Congress approved two farm bail-outs: $6.0 billion in 1998 and $8.9 billion in 1999.

"President Clinton signed a record $8.7 billion bailout of the farm economy [on October 22, 1999], saying it showed Congress needs to over-haul a Republican-authored program that supposed to wean growers from government programs.

"'While these additional funds have been absolutely critical, the very fact that we've needed them points out the underlying flaws in the 1996 farm bill,' Clinton said.

"Clinton also criticized the makeup of the aid package, saying it provides too little assistance for farmers who lost crops to drought or flooding, while providing money to other growers who don't need it."[15]

Demand and Supply Changes Expressed in Prices

Historical price patterns of agricultural commodities reflect the world's changing economic conditions. Figure 1–3 illustrates the gyrating corn, soybean, wheat, cotton, live cattle, and hog futures market prices.

From 1960 to 1973, corn prices fluctuated over a narrow range between $1.00 and $1.60 per bushel. Government price supports established a floor under price, and surplus commodities held in storage fixed a price ceiling. During the 1970's, corn prices rose to $4.00 per bushel and fell to $1.80. Intra-year swings of $1.00 per bushel were not uncommon into the 1980's. Soybean prices were more volatile than corn prices. In 1966, soybeans touched the previously unheard of price of $4.00 per bushel. Farmers were more accustomed to 50-cent-per-bushel fluctuations above and below $3.00. In 1973, soybean prices rose to $13.00 per bushel but promptly fell to about $5.00. Intra-year swings of $3.00 per bushel occurred throughout the 1970's and early 1980's and again in 1988. After years of trading in a band from $1.25 to $2.30 per bushel, wheat prices rose to almost $6.50 in 1973, then declined to $3.40 in 1974. Intra-year fluctuations of $1.50 continue. Cotton prices had a trading range of 20 to 40 cents per pound during the 1960's. Since the 1970's cotton has fluctuated that much or more within year. Hogs historically sold for between 15 and 30 cents per pound. They began to fluctuate from the mid-20-cent level to the mid-60-cent level, a range of 40 cents. Live cattle prices followed the rising price spiral and changed from a 5-cent-per-pound fluctuation to a 20-cent fluctuation.

Widely fluctuating commodity prices became the rule, not the exception, that U.S. farmers had to contend with. Table 1–1 shows the enormous effect on income created by fluctuating prices. By calculating the difference between annual high and low prices on nearby futures, average gross income changed $105 per acre on corn, $72 for soybeans, $53 for wheat, and $117 per acre of cotton. Livestock experienced gross value changes of $117 per head for fat cattle and $54 per head for hogs.

On a 500-acre crop farm, these price changes translated into about $50,000 fluctuations in gross income. In a feeding operation with 1,000 cattle, gross income could have fluctuated over $100,000 in a year due to varying prices. A pork producer finishing 1,000 head saw a $54,000 fluctuation in potential gross receipts. Effect on net income, or profit, was even more dramatic. The magnitude of income change emphasizes the importance of marketing knowledge for producers who want to reduce risk and increase profits.

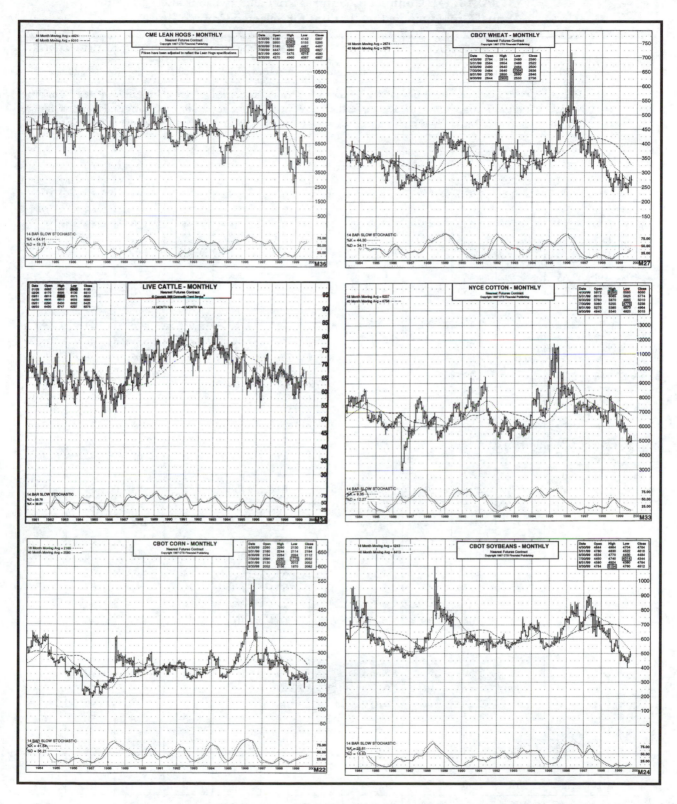

FIGURE 1–3: Monthly Futures Prices of Major Agricultural Commodities, 1990-2000.
Source: Charts reprinted by permission of Commodity Trend Service.

TABLE 1–1: Commodity Futures Price Fluctuations and Annual Gross Income Fluctuations for Selected Commodities*

Year		CORN 100 bu. yield Price Range ($/bu.)	CORN Income Change ($/a.)	WHEAT 40 bu. yield Price Range ($/bu.)	WHEAT Income Change ($/a.)	SOYBEANS 30 bu. yield Price Range ($/bu.)	SOYBEANS Income Change ($/a.)	COTTON 500 lb. yield Price Range ($/cwt.)	COTTON Income Change ($/a.)	LIVE CATTLE 600 lb. gain Price Range ($/cwt.)	LIVE CATTLE Income Change ($/hd.)	LIVE HOGS 220 lbs. Price Range ($/cwt.)	LIVE HOGS Income Change ($/hd.)
1980	H	3.96		5.45		9.55		98		75		54	
	L	2.57	139	3.78	67	5.70	116	72	130	59	96	27	59
1981	H	3.84		5.17		8.28		97		72		57	
	L	2.36	148	3.65	61	5.96	70	60	185	54	108	38	42
1982	H	2.84		4.05		6.78		72		74		68	
	L	2.13	71	3.00	42	5.20	47	61	55	54	120	43	55
1983	H	3.76		4.10		9.60		82		74		60	
	L	2.42	134	3.05	42	5.55	122	65	85	56	108	40	44
1984	H	3.65		4.35		9.00		82		73		58	
	L	2.52	113	3.55	32	5.60	102	63	95	60	78	44	31
1985	H	2.83		3.80		6.15		70		69		54	
	L	2.17	66	3.10	28	4.75	42	57	65	51	108	34	44
1986	H	2.60		3.66		5.65		70		63		64	
	L	1.50	110	2.50	46	4.65	30	29	205	52	66	38	57
1987	H	2.04		3.29		6.15		81		71		64	
	L	1.42	62	2.48	32	4.80	41	52	145	55	96	41	51
1988	H	3.59		4.42		11.00		70		76		74	
	L	1.84	175	2.88	62	6.01	150	50	100	62	84	51	51
1989	H	2.93		4.38		8.20		70		79		73	
	L	2.17	76	3.80	23	5.40	84	55	75	68	66	52	46
1990	H	3.02		4.12		6.55		94		81		91	
	L	2.15	87	2.38	70	5.52	31	64	150	72	54	64	59
1991	H	2.60		4.07		6.40		95		83		80	
	L	2.23	37	2.44	65	3.14	98	56	195	65	108	53	59
1992	H	2.74		4.63		6.37		67		80		70	
	L	2.04	70	3.02	64	3.24	94	51	80	70	60	53	37
1993	H	3.07		4.15		7.55		65		84		72	
	L	2.10	97	2.77	55	5.62	58	53	60	70	84	55	37
1994	H	3.12		3.51		7.32		87		78		72	
	L	2.10	102	3.04	19	5.27	61	66	105	62	96	40	70
1995	H	3.08		5.25		7.42		117		76		72	
	L	2.28	80	3.40	74	5.44	59	72	225	59	102	49	51
1996	H	5.54		7.50		8.56		89		74		90	
	L	2.58	296	3.77	149	6.60	59	69	100	54	120	60	66
1997	H	3.20		4.49		9.04		78		71		87	
	L	2.38	82	3.13	54	6.20	85	66	60	63	48	58	64
1998	H	2.84		3.48		6.94		83		70		63	
	L	1.85	99	2.34	46	5.09	56	59	120	57	78	21	92
1999	H	2.34		2.97		5.19		67		73		61	
	L	1.77	57	2.30	27	4.02	35	48	95	59	84	31	66
Average			105		53		72		116		88		54

*Interyear difference between high and low of nearby futures contract. Because of carry and seasonal prices the fluctuations may be exaggerated.

In the evolution of U.S. agricultural production, labor and machinery integration was the challenge of the 1940's and 1950's. Using chemical technology (fertilizers, herbicides, and pesticides) was the challenge of the 1960's. The ability to cope with price risk was the challenge created in the 1970's, and continued to the end of the century.

Production techniques have become necessary farm exercises, but marketing ability determines whether the farmer and rancher will be economically successful or bankrupt. Instead of being production oriented, tomorrow's successful farmers must understand markets and marketing. They must know:

HOW MARKETING SYSTEMS WORK AND WHY

—Knowing the Rules and Using Them to Advantage.

HOW AGRICULTURAL PRICES ARE AGREED UPON

—Making Judgements on Fundamental Demand and Supply Conditions As They Affect Prices;
—Understanding Technical Factors As They Express Emotions in the Market.

WHO OPERATES IN THE MARKET PLACE AND HOW

—Functions and Relationships Among Producers, Sellers, Processors, Users, and Various Investors;
—Knowledge of How Participants Use and Misuse Each Other.

HOW THE PRODUCER CAN OPERATE WISELY IN TODAY'S MARKET PLACE

—Tools of the Trade; How to Use Them and When to Use Them.

Marketing Concepts

We begin our investigation of markets and marketing by defining the concepts of marketing.

Agricultural marketing means many things to different people. Each person typically defines it from a partial perspective on the total process. The rancher may see it as loading livestock on trucks for delivery to a local auction. The farmer may see grain marketing as calling various elevators to determine which one is offering the highest price. An intermediary in the food chain may see marketing as transportation alternatives or processing activities. The retailer may see marketing as allocation of space on the grocery shelf. The consumer may see marketing as the weekly shopping trip to the grocery store. Each one sees an individual part of the total marketing process and defines the concept as it relates to his or her perspective.

"Markets" and "marketing" are broad and loosely used terms. We need to view them from various perspectives to gain a comprehensive understanding of marketing as a concept and to define its meaning as used in this book. We could begin with dictionary definitions. In another view, we could define jobs performed by marketing and who does them. Finally, we need to understand what marketing accomplishes as a process in creating consumer satisfaction.

Marketing Defined

Market and marketing are two different things; a market is a place, marketing is a process. The dictionary defines "market" as a region, a place, and "marketing" as buying, selling, and trading activities. When we refer to the "New York egg market," we refer to the millions of consumers of eggs in that city and the demand for eggs created by these consumers. Likewise, the "Cleveland milk shed" refers to the area surrounding the city that supplies fresh milk to the metropolitan area. Similar phrases with only the city or commodity changed may have significantly different meanings. However, a region of the country, whether seen from consumption or production perspective, is one description of a market.

A market is also a physical facility where trading occurs. The grocery trade uses the word "market" to designate a retail food store as distinct from a clothing store. Framers' markets are an institution in many communities where farmers congregate to sell their products directly to consumers. A market, from a livestock producer's point of view, could be a hog-buying station, an auction barn, or a livestock terminal market.

Marketing, as a process, occurs at a yard sale on someone's front lawn. A sign post along a busy thoroughfare, a cleaned-out garage, and a Saturday afternoon provide the requirements of such a market. Farmers utilize auctions as a means of disposing of equipment, livestock, and land. The sale barn brings a congregation of sellers and buyers together, usually on a weekly schedule, to market livestock. A meeting of people to buy and sell constitutes another definition of marketing.

A farmer listening to the noon market reports and a feedlot operator calling different buyers to check on prices being offered are both performing marketing activities. Any business day, grain elevators, livestock processors, and cotton merchants make offers and stand ready to purchase commodities from producers. Buying and selling activities associated with consummating the transactions are important components of the marketing process.

Dictionary definitions tend to emphasize the facilities and physical aspects of people getting together to trade. However, new technologies have de-emphasized many of the physical attributes of the market place. Telephones, video monitors, and satellite communication equipment have eliminated the necessity of people trading face to face. Grade standards permit buying and selling by description, rather than with the goods at hand. Business contracts assure the integrity of transactions. Technology and business procedures may have dispersed the market place, but they have not necessarily changed the marketing process.

Shepherd, Futrell, and Strain generalize these different descriptions of marketing into a single definition: "A market is a group of buyers and sellers with FACILITIES FOR TRADING [emphasis ours] with each other."[16] It is the "trading facilities," or business techniques, perspective of the market which we want to emphasize in this book. What are the tools available to farmer-rancher producers for marketing their products? How do they act and react with those business procedures in an endeavor to maximize profit?

Functions of Marketing

Marketing can also be characterized by the functions that are performed. A marketing function is a specialized activity, or job, performed in accomplishing the marketing process. Market functions bring together the products of thousands of scattered farm producers, convert them into usable forms, and distribute them to millions of consumers.

Kohls and Uhl classify the marketing process into three functional categories and nine activities: [17]

A. Exchange Functions
   1. Buying (assembly)
   2. Selling (dispersion)
B. Physical Functions
   3. Storage
   4. Transportation
   5. Processing
C. Facilitating Functions
   6. Standardization (grading)
   7. Financing
   8. Risk bearing
      a. physical
      b. monetary
   9. Market Information

Exchange functions are activities that transfer title to goods. They define who owns what at what point in time. Price determination enters the marketing process as a part of exchange. Buying includes activities for seeking out sources of supply, purchasing, and assembling products. Selling involves more than placing a price tag on items; it includes all the activities of merchandising: packaging, displaying, promoting, and advertising. The flow of commodities from producers to consumers requires an enormous number of sales transactions. For example, imagine how many ownership transfers are made during a year as 100 million hogs move off 100,000 farms to 1,000 meat packers, who sell millions of pounds of hams, bacon, pork chops, and sausages to thousands of retailers, who resell them to 50 million households. The inefficiency of 50 million consumers buying directly from hog farmers is staggering. Market intermediaries make exchange more efficient by reducing the number of buying and selling transactions.

Physical functions describe activities that involve handling, movement, and physical change of the commodity. Agricultural crops typically have short harvest periods, but they are consumed in a continuous year-round pattern. Storage makes goods available at the time consumers want them. It may mean holding raw products in storage bins or warehouses until they are needed for further processing, or it may mean holding supplies of finished products in inventories by processors, wholesalers, and retailers.

The transportation function makes goods available at the proper place. Agricultural production is widely dispersed across the United States and around the world. Consumers are concentrated in cities, often located across the continent from specialized production areas. Moving goods from one point to another requires evaluation of transportation alternatives, whether over land, on water, or in the air.

The processing function converts raw commodities into consumable products in the form desired by buyers. Processing frequently involves radical change in the appearance and characteristics of the product, such as butchering hogs, baking bread, or tailoring a wool suit. On the other hand, it may merely require washing and packaging eggs, trimming lettuce, or pasteurizing and bottling milk. Specialization and economics of scale in manufacturing generally eliminate extensive food processing from the farmer's marketing activities. However, grain farmers can choose to sell their production as meat, milk, or eggs and thus significantly change the form of their production.

Facilitating functions consist of business activities that allow smooth performance of the exchange and physical functions. Standardization, for example, establishes uniform measurements of quality and quantity. It makes sale by description possible. A bushel of U.S. No. 2 yellow soybeans or a load of Choice 3 beef carcasses means the same thing throughout the trade as any other lot so graded. Price differentials in a consumer-directed economic system reflect consumer preferences. Willingness or reluctance of consumers to pay different prices for quality differences registers their vote in the market place. The price differentials in turn relay to marketing agencies, and eventually to producers, what is wanted by the consumers. Only when the commodity is traded consistently in well-defined units of quality and quantity can the price quotation do an effective job of relaying that information. It is essential that an impartial judge make the grading decisions on biologically different commodities. Governmental agencies typically have the responsibility for providing grading personnel who make the quality designations and inspection decisions.

The financing function advances credit to carry out the activities of marketing. Large investments in facilities for processing and storage are required in the food chain. Furthermore, carrying inventories of raw or finished commodities from harvest until ultimate sale represents a large investment and incurs significant interest charges. The funds may be provided by lending institutions or by the owner's own resources. In either case, they constitute a significant cost item.

Risk bearing arises from a loss in product quality from deterioration or destruction by biological deterioration, insects, rodents, weather, fire, or accidents. Price risk refers to change in value of a commodity as market prices rise or fall. Producers may bear these risks themselves or transfer them, to the extent possible, to someone else. For example, in return for premium payments, the farmer can shift risk of loss by fire, wind, and other destructive elements to insurance companies that specialize in providing protection against many forms of destruction. Likewise, commodity futures markets provide vehicles to transfer price risk from producers, owners, or processors of commodities to professional speculators who specialize in this form of risk taking. A modest premium is required to transfer the risk of catastrophic loss. Risk bearing is closely associated with the storage and financing functions.

The market information or intelligence function involves collecting, interpreting, and disseminating facts and figures about commodities. Efficient and effective marketing requires reliable information about demand, supply, and prices. An effective pricing mechanism depends upon well-informed buyers and sellers. Furthermore, production planning, efficient transportation, and storage programs depend upon good information. Government reports, private consultants, market letters, radio and TV, and individuals themselves provide that information. But individuals operating in the food chain have the responsibility to interpret the material and make their own production and marketing decisions.

It is essential for agricultural producers to understand that all the functions discussed must be performed as raw commodities move from the field to the fork. Someone has to perform them. Someone must incur the costs associated with their performance; someone will reap the returns from satisfying consumer demand.

Eliminating Intermediaries. Quite often farmers and consumers talk about "eliminating the intermediaries" between producers and consumers and "abolishing costly marketing processes." The implicit assumption is that farmers would receive more for their products and consumers would pay less for food. While it may be possible to "eliminate the intermediaries," it is difficult or impossible to eliminate the marketing functions performed by them. Eliminating the intermediaries involves transferring the activity—its costs and returns—to someone else. Farmers may build storage facilities and assume storage, financing, risk-bearing, selling, and transportation functions. By taking on new roles, farmers eliminate brokers and commission merchants, but their functions cannot be eliminated from the marketing process. Whether the farmer-producer supplies the services or transfers the responsibility to intermediaries depends upon who can effectively do the job at the lowest cost. If a producer does them as an integral part of the production process, the returns accrue to the farm operation. If farmers cannot economically perform the service, a more efficient intermediary is necessary to move commodities through the channels of trade to the consumer. Surplus products stacked up at the farm gate and hungry city dwellers scrounging for a morsel of food would be the scenario without productive marketing activities..

Marketing functions that create place, time, and form utilities are as essential as the initial production of foodstuffs. We intend to develop methods of analysis so that commodity producers can decide when they should assume marketing functions to earn marketing profits.

Elements of Marketing: Consumer Utilities

Marketing has been likened to a bridge between producer and consumer. Wheat in the field of a Kansas farmer, milk in the cooler of a Minnesota dairy farmer, or cotton on the plantation of a Mississippi grower might as well not be produced unless a marketing system exists to transform that product into a *form* desired by the consumer, at the appropriate *time* and in the right *place*. Marketing, as a continuation of the production process, converts raw material into a consumable good by moving the material to where it can be used, making it ready for use at the right time, and converting it into a form suitable for consumption.

The production activities of the marketing process create consumer utilities called (1) place utility, (2) time utility, and (3) form utility. Together they constitute what have been called the **elements of marketing.**[18]

Place Utility. Consumers live primarily in cities. Producers are widely dispersed across the country. Because of various absolute and comparative advantages, areas of specialized production have evolved across the United States. For example, climatic conditions limit citrus production to Florida, Texas, and California. Land productivity leads to concentrating corn and soybean production in the Midwest. Rainfall dictates wheat production in the Plains states. Cattle feeding has become concentrated in Texas and Colorado. Wisconsin is identified as the dairy state. Sugar cane is grown in Florida, Hawaii, and Louisiana because it requires a long growing season . Figure 1–4 shows the major types of farming throughout the United States. Each region specializes in the production of commodities most appropriate to its resource base. But consumers' market baskets are filled every day with products from all regions.

Place utility is added to commodities by barge, railroad, airline, or truck companies that move goods from one point where they are in surplus to another point where they can have further utility added by processing or storing, and to their final

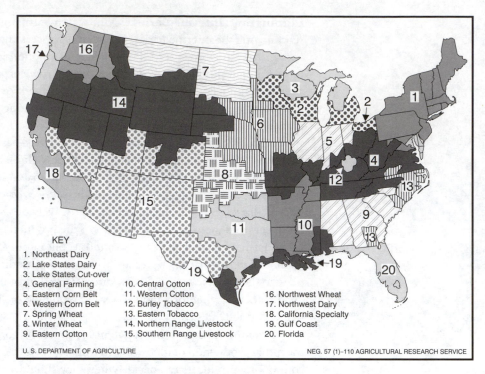

KEY

1. Northeast Dairy
2. Lake States Dairy
3. Lake States Cut-over
4. General Farming
5. Eastern Corn Belt
6. Western Corn Belt
7. Spring Wheat
8. Winter Wheat
9. Eastern Cotton

10. Central Cotton
11. Western Cotton
12. Burley Tobacco
13. Eastern Tobacco
14. Northern Range Livestock
15. Southern Range Livestock

16. Northwest Wheat
17. Northwest Dairy
18. California Specialty
19. Gulf Coast
20. Florida

U. S. DEPARTMENT OF AGRICULTURE   NEG. 57 (1)–110 AGRICULTURAL RESEARCH SERVICE

FIGURE 1–4: Major Types of Farming in the United States. Source: USDA.

destination where they are consumed. In order to move commodities from surplus areas (production concentration) to deficit areas (consumption concentration), price differentials must be sufficiently wide to pay the assembly costs on one end, dispersion costs on the other, and transportation costs across that space.

Transportation agencies are not the only ones who contribute place utility to farm products. Farmers and ranchers who incur transportation charges in transferring goods from point to point also contribute to place utility. Producers decide whether or not they can profitably produce on the basis of the price at the farm gate. The price at the farm gate is the price in the market place minus transportation charges to move commodities across space. When farmers can transport commodities economically, they add place utility to their production.

Place utility will be important when we investigate commodity prices in determining **WHERE** farmers should sell their product.

Time Utility. Agricultural production is a biological process and therefore highly seasonal. Harvest periods for crops such as wheat, cotton, soybeans, corn, fruits, and vegetables are concentrated into a relatively short period of time. Livestock production is not as dependent upon seasonal factors as plant production, because livestock housing facilities have significantly modified the natural environment. Nonetheless, biological production patterns, institutional factors, and, to a degree, seasonality produce livestock receipts that vary substantially throughout the year. Consumers, on the other hand, expect well-stocked grocery shelves every day of the year. Consequently, commodities must be stored in order to extend the time period of availability. Time utility is added when products are stored from harvest until they are desired for consumption.

Holding storable commodities from harvest until later periods of consumption incurs costs. These costs include buildings and facilities for storage, interest charges on ownership of inventory, and risk of loss from spoilage and price change. In order to enjoy the consumption of goods six months after production, consumers must be willing to pay the price at harvest *plus* the cost of holding those goods for six months. If consumers are unwilling to pay these added holding costs, goods will be consumed at harvest or left unharvested to rot in the field.

This time element is important in later understanding "carrying charges" on storable commodities. They dictate to the farmer **WHEN** to sell the product.

Form Utility. Corn and soybeans can become breakfast cereal, cooking oil, or live-stock feed products. Wheat at the farm gate can become bread in the supermarket or feed for hogs. Added production and processing creates form utility by transforming raw materials into more desirable consumer food products. The significance of marketing, when viewed as a production process, is that it contributes to making a product more useful to consumers. A shopper can buy flour and ingredients, cake mix, refrigerated dough, or bakery items when he or she decides to treat the family. The extent of preparation depends upon the price difference between what the consumer is willing to pay for the finished or intermediate product and what the farmer receives for the raw materials. If the price differential is sufficiently great, an intermediary will commit resources to convert the raw material into the "finished" product and the consumer will buy.

This element is important to farmers because it determines in **WHAT** form they sell their production. Will it be cash grain or hogs, grass-fat or grain-fed beef, bulk or bottled milk?

Economics of Agricultural Marketing

The economics of marketing are broader in scope than just the physical and financial aspects of marketing. Economic considerations deal with three separate, but related, problems, as outlined by Shepherd, Futrell, and Strain:[19]

1. *Consumer Demand* for farm products separately and as a group,

2. *The Pricing System* that reflects demand back to distributors, processors, and producers.

3. *The Costs* associated in exchanging title and getting the physical product from the producer to the consumer in the form wanted, at the time and in the place desired.

Consumer Demand: Farmer's View. The production manager's job is not merely producing a commodity. First, he or she must find out what potential customers want. Producing that product comes next. The payoff comes from seeing that the product reaches consumers, as they desire it. Too often, farmers have been insensitive to consumer demands. Their attitude has been: "Here is the stuff. We have produced it. You market it as best you can. Look at this lard: Sell it at some price or develop a new use for it. And look at all this butter: What are you going to do with it?" Clearly this is working the problem at the wrong end. We are taking hold of the pig after it is fully grown and hard to handle. A large share of marketing problems in agriculture arise because the right crop was not planted or because the right quality (choice beef, for example) was not produced. To begin the marketing process with a product at the farm gate is to begin too late. At that point, marketing problems are often beyond a profitable solution.

Wayne Purcell emphasized the role of marketing in a broader context. He states that marketing should be defined as "the set of economic and behavioral activities that are involved in coordinating the various stages of economic activity from production to consumption." He makes no effort to separate production from marketing. Instead, he sees production as a part of an interrelated set of economic activities. "Emphasis is placed on the workings of the marketing system as the means of achieving coordination between production activities and consumer demands."[20] He sees the role of marketing as system coordination.

Farmers and ranchers should accept the fact that market conditions dictate their operational plans. As price takers, they must interpret market conditions *before* they make production plans. If farmers overproduce an item, the surplus can be sold only at significant price discounts. On the other hand, if there is a shortfall in production, consumers will bid up the price of available supplies. Producers must attune their thinking to what the market is telling them. They must produce the right amount of what consumers desire. The market place is where the signal is given. Producers need to learn to read market signals, since these signals represent consumer demands that can be converted into profit.

Pricing System. Marketing provides a system of communication, conflict resolution, and coordination. The farmer wants a high price for the products of his or her labor and management; the consumer wants a low-cost supply of food for the family; and the intermediary seeks the maximum possible profit margin between what is paid for raw commodities and what is received for finished products. Marketing somehow must reconcile these conflicting economic objectives. It must transmit to sellers what is wanted; it must transmit to buyers what is available; and it must achieve compromise between producers' high price goals and consumers' low cost goals. Marketing fills these requirements by establishing price.

In a competitive economy, pricing is the master control system guiding production, marketing, and consumption. Fluctuating competitive prices perform three major economic jobs:

1. They guide and regulate production decisions.

2. They guide and regulate consumption decisions.

3. They guide and regulate marketing decisions over time, space, and form.

Prices serve as guides before decisions are made and as rewards for correct decisions. To some individuals or firms, high and rising prices mean increased profits, and they provide incentive to increase production. But to customers, high prices mean "slow down" or perhaps "do without." The opposite signal would be flashed by low or falling prices as they tell producers to reduce production and consumers to increase consumption.

Profitable production and marketing decisions involve selection among many possible alternatives, each one having its own relative price and profitability. Farmers may use land, labor, and capital to produce corn, soybeans, or wheat. They must choose between selling cash grain or marketing it through livestock. Farmers respond to relative prices as they make production decisions. They switch acreage from corn to soybeans when the price ratio favors soybean production. Or they switch back to corn if its relative price is higher. Next, prices guide the movement of commodities through the marketing system, automatically turning on the green light here and the red light there, signaling to intermediaries the location of needs and availability of supplies. Finally, consumers react to prices by choosing to buy or not buy, and thereby indicate what products, quantities, and qualities they want and are ready to pay for. Tradition, custom, and personal preference also play a part in production and consumption decisions. Over time, however, price signals and profit are better explanations of market choice than the non-price factors.

Thus, prices established in the marketing system play an important role in economic life. They must be low enough to clear the market of available supplies, or high enough to ration and distribute available supply among customers. They must induce further production of things wanted by customers, or discourage production when surpluses begin to pile up. And they must pay the costs of providing the services added by intermediaries. Too often, people seem dissatisfied with the role of prices in dictating decisions. But the prosperity of American farmers and the abundant products on grocery shelves attest to the fact that somehow the process works, even if at considerable discomfort to some.

Costs of Marketing. Costs are associated with transforming raw farm commodities to finished consumer products at the time, at the place, and in the form that consumers desire. These marketing costs are required to pay for the resources used in the marketing process.

How efficiently the marketing system performs its functions can be gauged by relating its operating costs to its accomplishment. One gauge of performance is profitability. Like farmers, intermediaries incur production costs that add value to commodities. Figure 1–5 shows what costs are paid by the markup between what inter-

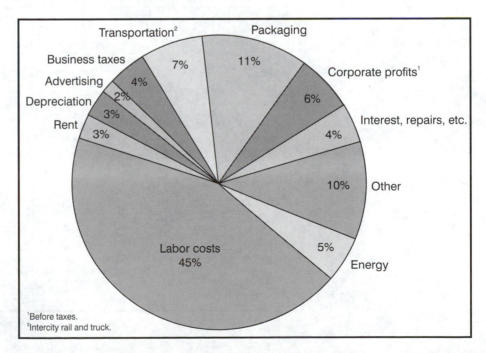

FIGURE 1-5: What Makes Up the Farm-Food Marketing Bill? Source: Kohls and Uhl 216.

mediaries pay for farm products and the price at which they sell finished goods to consumers. The largest single cost item is labor, accounting for almost one-half of the gross operating margin. Pre-tax profit accounts for about 6 percent of consumer expenditures for food.

Another way to view marketing costs is to look at the relative share of the consumer's food dollar that accrues to intermediaries and producers, as shown in Table 1–2. The farmer's share consistently averaged around 40 percent, until the 1980's when the farmer's share shrunk into the low 30 percent level and continued to erode to less than 25 percent at the end of the century.

The farmer's share has been widely used as a measure of the fairness of farm prices and the efficiency of food marketing. The fact that it held at about 40 percent for so long does not mean that it represents the correct balance of income for producers versus intermediaries. Marketing costs must be borne by consumers as they place demands upon the food chain to produce and provide an abundant supply of wholesome food when, where, and in the form desired. Consumers are indifferent, however, whether farmers or intermediaries incur the costs and reap the rewards of providing the functions. In this book, we will provide methods for farmers to evaluate the profitability of assuming some marketing costs and profiting by performing some of the marketing functions.

TABLE 1–2: The Market Basket of Farm Foods and the Farmer's Share of the Consumer's Food Dollar, 1915-1997★

Years	Market Basket Retail Cost	Marketing Charges	Farm Value	Farmer's Share of Consumer's Dollar
	($)			(%)
1915-19	339	210	189	47
1920-24	444	263	181	41
1925-29	439	256	183	42
1930-34	327	211	116	35
1935-39	341	204	137	40
1940-44	397	208	189	47
1945-49	762	381	384	51
1950-54	940	509	431	46
1955-59	957	576	381	40
1960-64	1004	625	379	38
1965-69	1101	686	435	39
1970-74	1416	829	586	41
1975-79	2053	1232	821	40
1980-84	2831	1799	1033	36
1985-89	3217	2208	1012	31
1990-94	3900	2839	1062	27
1997	4640	3564	1076	23

Source: Agricultural Statistics, various issues, USDA, Economic Research Service.

★The market basket contains the average quantities of 65 domestic farm-originated food products purchased in 1960-1961 by wage earners, clerical worker families, and workers living alone. The composition and weighting of the market basket were revised several times from 1915 to 1997.

Summary

World economic conditions changed significantly in 1972. The Russian Wheat Deal was the watershed event, but underlying population growth, personal income, quality of diet, OPEC, and currency values were basic demand changes that affected marketing. Commodity prices rose dramatically and fluctuated violently during the decade, then fell after 1980 but continued to fluctuate. Opportunities created by increased uncertainty added a new dimension to profitable farm production. Marketing and marketing decisions took center stage in the profit equation.

We covered several perspectives in defining markets and marketing in order to form a foundation for understanding the nature of marketing. The functional approach established what marketing jobs are accomplished and emphasized their necessity but left unanswered the question of who would perform them. The elements of marketing described time, form, and space utility in satisfying consumer demands. Finally, the economic perspective viewed marketing in its pricing role as "the invisible hand" that guides production, marketing, and consumption decisions. Each of these concepts of marketing will be applied to real-world conditions as we move through the marketing decision process for farmers and ranchers.

Questions for Study and Discussion:

1. What four economic conditions contributed to the dramatic increase in commodity prices in the early 1970's at the time of the Soviet wheat purchase?

2. What were the two distinguishing features of commodity prices in the 1970's, 1980's, and 1990's?

3. Distinguish between markets and marketing.

4. What is a marketing function?

5. List the functions of marketing and explain each of them.

6. List the three elements of marketing and explain their meanings.

7. Can the intermediary be eliminated from the food chain? Explain your answer.

8. Discuss the economic aspects of agricultural marketing.

Exercise:

Focus on actual marketing by maintaining a daily record of local prices and their effects on value of inventory (these price data will be used later in other exercises). Assume a sample farm with 50,000 bushels of corn and 20,000 bushels of soybeans in storage (or use another crop or livestock of particular importance in your area). Each day, using a photocopy of the following form, record the local cash price and multiply the quantity in inventory by that price. Calculate the daily change in value of inventory that results from the price change. The purpose of the exercise is to emphasize the daily change in economic well-being of the producer as market prices fluctuate. Some days your equity will increase; other days it will fall.

	Daily Price Record and Inventory Value Calculation					
	Corn—50,000 Bushels			Soybeans—20,000 Bushels		
Date	Price	Inventory Value	Change in Value	Price	Inventory Value	Change in Value

Suggested Complementary Reading:

Merchants of Grain, Dan Morgan, 1979, The Viking Press. Though this book is out of print it is available in many libraries. It contributes to the global view of commodity marketing and emphasizes the intricacies of the grain trade. It provides a broad perspective to grain merchandising.

Endnotes for Chapter 1:

1. Carroll G. Brunthaver, "What's Happening to Demand," speech delivered before American Farm Bureau Conference, Atlantic City, N.J., 1974.

2. Royal Fraedrich, "Get Set for the 1980's When World Food Needs Become World Food Demand," *Farm Futures*, 1979.

3. Clayton K. Yeutter, "Russians' Need for U.S. Wheat Proves a Point," *Cincinnati Enquirer,* 1974.

4. Lloyd D. Teigen and Abner W. Womack, *An Annual Econometric Model of the Livestock and Feed Sector*, Proceedings of Livestock Futures Research Symposium (Chicago: Chicago Mercantile Exchange, June, 1979).

5. Conrad Leslie, "China Revisited," *Thompson McKinnon Commodity Letter*, Vol. II, No. G SC-15, Nov. 13, 1978.

6. Leslie, "China Revisited," 3.

7. Gerald O. Barney, *The Global 2000 Report to the President,* Vol. 1 (Washington, D.C.: U.S. Govt. Printing Office), 1980.

8. USDA, *World Livestock and Poultry Situation* (Washington, D.C.: Foreign Agricultural Service, April, 1985).

9. Personal interview with Dr. Toshihiko Hayashi, Professor of Economics, Toyonaka, Osaka, Japan, Feb., 1985.

10. _____, "The Dollar in Perspective," *Barometer of Business,* Harris Bank, Chicago, July, 1982.

11. Bob Berglund, "Remarks Before Embassy Agricultural Attaches," *Selected Speeches and News Releases* (Washington, D.C.: USDA, Dec. 10, 1979).

12. T. J. Holt, "Newsletter for Decision Makers," *The Holt Executive Advisory,* Vol. X, No. 13 (T. J. Holt & Co., 290 Port Road West, Westport, Conn. 06880, July 12, 1985).

13. Holt, "Newsletter for Decision Makers."

14. USDA, "Revising U.S. Exports: Why Is It Taking So Long?" *Farmline,* Vol. VII, No. 9 (Washington, D.C.: Economic Research Service, Sept., 1986).

15. _____, "Clinton criticizes but signs $8.7 billion farm aid bill," *The Roanoke Times,* page A6, October 23, 1999.

16. Goeffrey S. Shepherd, Gene A. Futrell, and J. Robert Strain, *Marketing Farm Products,* 6th ed. (Ames: Iowa State University Press, 1976) 16.

17. Richard L. Kohls and Joseph N. Uhl, *Marketing of Agricultural Products,* 6th ed. (New York: Macmillan Publishing Co., Inc., 1985) 23.

18. Shepherd, Futrell, and Strain, *Marketing Farm Products* 18.

19. Shepherd, Futrell, and Strain, *Marketing Farm Products* 7-13.

20. Wayne Purcell, *Agricultural Marketing: Systems Coordination, Cash and Futures Prices* (Reston, Va.: Reston Publishing Co., 1979) 12.

Chapter 2

Marketing Tools

Marketing Tools

It takes 90 to 200 days to grow a grain crop or finish a pen of cattle. It can take less than two minutes to sell the product. But in order to maximize profit, farmers should take up to 24 months to market their products. This book builds toward a *730-Day Marketing Plan*.

Successful marketing is more than just selling. The key to profitable marketing requires recognition of profitable opportunities and use of the correct tool to capture those opportunities.

In large measure, farmers are price takers because of the competitive nature of their business. Any single producer acting alone cannot affect market price, either up or down. The producer takes what the market offers. Watching the panorama of changing prices, producers must decide WHEN and HOW to respond to market offers. WHEN they react to pricing opportunities forms the structure of their marketing strategy (discussed later). HOW they react depends upon the marketing tools they have at their disposal.

Different tools accomplish different objectives. Successful marketing requires knowing what marketing tools are available and knowing what they can accomplish. To better understand the tools, we begin our discussion by defining two components of pricing and the business events that occur in executing a sale. Later in this chapter, we will dissect these business events and describe various marketing tools for making profitable sales.

Market Price Level and Basis Price Level

The pricing transaction has two separate but related components. One concerns the market price that is discovered on the commodity exchanges and is referred to here as price level. The other concerns the difference between a cash market price and a selected futures contract price. This price spread is called "basis" and is referred to here as basis level.

Each price component has its own price-making forces. The market price level on futures exchanges is set by anticipated worldwide future demand and future supply conditions as interpreted by traders active on the exchanges. The price at a local market (spot market) is influenced by the general market price level, but it responds to local conditions, here and now, and therefore may be higher or lower than the futures price level. Most of the differences in prices between a central market and those available to a producer located in the country can be attributed to transportation and storage costs. However, market imperfections and price distortions account for a part of the difference also. These distortions provide profitable opportunities for astute marketers. An awareness that local price imperfections can and do exist is knowledge that farmers can use profitably.

To help distinguish between the two prices, let us examine the needs of a continuous-flow corn-oil processor in Dayton, Ohio. This processor draws most of its corn as raw material for processing from a 200-mile radius of the plant. If farmers in that region are reluctant sellers, the processor must increase the cash bid, or "spot" price, in order to convince sellers to part with their grain. There may be willing sellers at a lower price in Iowa, but frozen rivers and rail car shortages could prevent Iowa corn from satisfying the Ohio processor's current needs. Since it is costly to shut down and start up this operation, the processor is forced (within limits) to bid on premium relative to the price level that exists in other parts of the country. In market terms, local basis would be strong. Later, when rivers thaw and transportation problems are solved, there will be no reason for the processor to continue bidding a premium. The processor would lower its "spot" price relative to the market price level. The spread between local cash and central market cash price would deteriorate, and basis would weaken.

Farmers are concerned with capturing advantageous prices in both price level and basis level markets. They seek both a high price level and a strong local basis when they make pricing decisions. When the local cash price is strong relative to the futures price level, the basis portion of the pricing decision is good and should be captured with one of the marketing tools available. When price level is high, but basis is weak, another tool should be used. The discussion will explain how marketing tools fix either the price level or basis level component of price while leaving the other open to fluctuation. Some tools fix both price level and basis level, thereby eliminating all price fluctuation.

Key Business Events

The key business events of commodity transactions are:

DELIVERY—physical transfer of the commodity from buyer to seller.

PRICING—agreement on an acceptable price by buyer and seller.

TITLE TRANSFER—determining who owns the commodity at what instant in time.

SETTLEMENT—payment from buyer to seller.

All of these business events can occur together at a single instant, or these events can be separated by many months. Pricing need not occur in conjunction with actual delivery of the commodity; it can be done before planting or delayed until after harvest. Transferring title of ownership may occur when grain is physically exchanged, but it need not be simultaneous. Payment can occur at a time far removed from the other events. By separating the business events, the farmer has greater opportunity to execute each one at the most advantageous time.

Marketing Tools

Cash Sale

With cash sale, the seller delivers the commodity to a buyer and takes an immediate cash payment in exchange. All business events—*delivery, pricing, title transfer, and settlement*—occur in one transaction. The commodity must be physically in existence at the time of the transfer. Cash sales may be made directly from the field, immediately after harvest, or later from storage.

A cash sale removes all price fluctuation. Both price level and basis level are fixed. There is no chance to profit from additional price improvement; likewise, there is no chance to lose from future price decline. Picking the day to make a cash sale can be an excruciating experience because it finalizes a year of plowing, planting, and praying. Because cash sales so completely remove pricing options, many producers split their sales into several decisions, hoping that the average price achieved will be higher than a one-time sales decision.

Compared to the other market alternatives, this method is almost business risk-free for the seller, assuming the buyer's check does not bounce. The seller retains title, and often physical ownership, until the sale is made. To minimize the risk of nonpayment, producers should accept checks drawn on a local bank and cash them immediately. If the buyer declares bankruptcy in the interim between issuing and clearing the check, the seller will be only one of many unsecured creditors who claim title to the assets of the buyer.

The cash price quoted for a bushel of grain is based upon grading standards as defined by the United States Grain Standards Act. If the grain as delivered does not meet those standards, discounts will be calculated and the price received will be less than the quoted price. Producers should understand the discounts and reasons for them. They will thus be in a position to manage production and harvest practices to minimize or, better yet, optimize discounts.

Grading standards for corn are outlined in Table 2–1.[1] (Soybeans and wheat have similar standards, as shown in tables at the end of the chapter.) Grade differences depend upon test weight, moisture, broken grain and foreign material, and damaged kernels. In addition, mustiness, souring, heating, or insects lower quality. Elevators usually publish discount tables and reduce the price paid for any conditions that are substandard.

Moisture discounts are usually higher than other discounts and cause the most misunderstanding between farmers and elevator managers. Furthermore, there is little standardization of procedure among elevators in their methods of dealing with wet grain, drying, and shrinkage calculations. Though they may be quoting the same bid price, the net price received by the producer after deductions can vary considerably from elevator to elevator. The producer must determine which offer represents the best selling opportunity.[2]

On-Farm Storage

On-farm storage facilities permit delay of delivery, pricing, title transfer, and settlement transactions. Rather than forcing the farmer to deliver grain at harvest, storage provides greater flexibility in fulfilling business events. Storage bins are typically used to **speculate** on the possibility of rising prices. Certainly, no one would store at harvest if he or she expected prices to be lower in the future.

TABLE 2–1: Grades and Grade Requirements for Corn

| Grade | Minimum Test Weight per Bushel | Maximum Limits of — | | | |
| | | Moisture | Broken Corn and Foreign Material | Damaged Kernels | |
				Total	Heat-Damaged Kernels
	(lb.)	------------------- (%) -------------------			
U.S. No. 1	56.0	14.0	2.0	3.0	0.1
U.S. No. 2	54.0	15.5	3.0	5.0	0.2
U.S. No. 3	52.0	17.5	4.0	7.0	0.5
U.S. No. 4	49.0	20.0	5.0	10.0	1.0
U.S. No. 5	46.0	23.0	7.0	15.0	3.0
U.S. Sample grade	U.S. Sample grade shall be corn which does not meet the requirements for any of the grades from U.S. No. 1 to U.S. No. 5, inclusive; or which contains stones; or which is musty, or sour, or heating; or which has any commercially objectionable foreign odor; or which is otherwise of distinctly low quality.				

Source: USDA, *The Official United States Standards for Grain* (Washington: GPO, 1978) 2.3.

Unpriced grain held in storage is subject to both price level and basis level fluctuations. Price level may rise or fall; basis level may strengthen or weaken over time while grain is held in storage.

Using grain-drying, -handling, and -holding facilities as extensions of the production process can help facilitate timely harvest. Harvest losses increase substantially after corn dries down to 22 percent moisture in the field. At a price level of $2.85, the loss during the time corn drops only 2 percent in moisture (from 22 to 20 percent) could cost 7 cents per bushel on an average yield. This drop represents a large portion of the annual cost of owning grain-drying and -handling facilities.

However, the production process ends the moment dry grain is deposited into a storage bin. At that point, a producer with on-farm storage is performing a marketing function and should view this activity as that of a storage space merchandiser. The farmer-storer incurs the investment and ownership costs of the facility. Farmers also incur quality risk through the chance of physical deterioration, and price risk due to price fluctuations. However, they minimize business risk by retaining title and physical ownership. Full storage bins produce a sense of satisfaction and financial security when winter snows begin to blow, and they provide a commonly-accepted method for speculating on price change. Unfortunately, the results of this speculative activity are often not properly evaluated. When 10,000 bushels of soybeans worth $9.00 at harvest are sold six months later for $7.00, a loss of $20,000 (plus holding costs) has been incurred. Some would argue that the farmer never had the $20,000 to begin with. In reality the bypassed opportunity represents a real loss when it comes time to pay bills.

Hired Storage

Hired storage is *delivery today*; title transfer, pricing, and payment are to finalized at some time in the future. The commodity changes hands physically, but business transactions are delayed. This method allows sellers to speculate on the cash market without the capital investment required by on-farm storage facilities; instead they pay a space rental fee to the elevator operator. Neither price level nor basis level is fixed by storing unpriced grain.

The producer usually deposits grain with the commercial elevator and draws a *warehouse receipt*, as evidence that the holder of the receipt is entitled to hold, receive, or dispose of the goods represented by the receipt (Figure 2–1). Negotiable warehouse receipts are documents of title; the transfer of the warehouse receipts will transfer the title to the grain they represent. Loss or theft of warehouse receipts represents loss of title to the goods. Warehouse receipts are generally accepted by lenders as

OHIO DEPARTMENT OF AGRICULTURE—GRAIN WAREHOUSE DIVISION—NEGOTIABLE WAREHOUSE RECIEPT

Licensed Handler
Location of Warehouse
Date Receipt Issued ________
Name & Address of Depositor ________

Receipt No.
Ohio Code No.
CCC Code No.

Commodity	Grade	Gross Lbs.	Doc	Net Lbs.	Net Bushels	Rate of Storage Charges
						Elevation Ppd. In ____ Out ____ Storage paid to

SUPPLEMENTAL INFORMATION FOR U.G.S.A.—NOT PART OF WAREHOUSE RECIEPT

T.W.	%Mst	%Pro-Spl-Snd	%TD	%HD	%FM	%SBK	Def	Date Rec'd	RR	Bg	Tk	CCCChgs	Type Stg	Area

ADVANCES MADE OR LIABILITY INCURRED FOR WHICH THE HANDLER CLAIMS A LIEN ________

The agricultural commodity identified hereon is stored in a warehouse under the control of the licensed Agricultural Commodity Handler named hereon and is insured against loss or damage by fire, internal explosion, lightning, or windstorm. The commodity has been graded according to the United State Grain Standards Act. The licensed Handler is not the owner of the commodity covered by this receipt, either solely, jointly, or in common with others unless the facts of such ownership are stated hereon. The person to whom this receipt is issued, or to whom it is negotiated by endorsement, is protected against loss of the commodity, or the value thereof, according to the terms of Section 926.18 of the Ohio Revised Code. Upon return of this receipt properly endorsed and payment of the Handler's lien claimed hereon, said grain or grain of the same or better grade will be delivered to the above named depositor or his order.

REMARKS:

____
Authorized Signature

OHIO DEPARTMENT OF AGRICULTURE—GRAIN WAREHOUSE DIVISION—NEGOTIABLE WAREHOUSE RECEIPT

AGR 1253 GOLDENROD—ORIGINAL CANARY—EXAMINER'S PINK AND GREEN—WAREHOUSEMAN'S BLUE—DEPOSITOR'S

SIGN BOTH THE STATEMENT AND THE ENDORSEMENT

STATEMENT OF OWNERSHIP AND ENCUMBRANCES

The undersigned hereby certifies on the date stated that he is the owner, or authorized agent of the owner, of the commodity covered by this receipt and that, other than the Handler's lien evidenced on the face of this receipt and the following, there are no liens, mortgages or other encumbrances against said commodity.

Date ________, 20____

Signed ________

ENDORSEMENTS

FIGURE 2–1: Grain Warehouse Receipt. (Top–Front; Bottom–Back)

gilt-edged collateral against which they will lend a high percentage of market value of the grain represented. Thus, these receipts can be used to generate a cash flow while retaining title (subject to lien) of the commodity.

Most all states have grain warehouse licensing requirements for commercial grain warehousers. They make it unlawful to accept grain for storage and to issue warehouse receipts without being licensed by either the state or federal government. Federally licensed warehouses are inspected twice annually to verify the presence of grain in the warehouses. Inspections are a measure of protection for the producer. To qualify for licensing, the grain warehouse must be bonded to guarantee financial integrity. However, producers should not be misled by an elevator's poster that says "Licensed and Bonded," until they know the amount of the bond. In many stages, the bond is only a small fraction of the elevators capacity in terms of value. In the event of a shortage, claims against the bond can quickly exceed its value. At that point, the warehouse receipt is no more valuable than the financial integrity of the warehouser himself or herself.

As evidence of title, the warehouse receipt clearly distinguishes the storage transaction from a "delayed pricing" sale. The receipt indicates that the grain still belongs to the farmer or holder of the warehouse receipt.

Delayed Pricing

Delayed pricing, or priced later sales, involves *delivery and title transfer today,* while pricing and payment are delayed until a time in the future. When farmers deliver grain to the elevator and choose the delayed pricing option, they are given a delayed price contract, which specifies the quantity and quality of grain delivered and outlines the charges assessed to the seller. The business events of delivery and title transfer occur simultaneously; pricing and payment are delayed until later.

Many producers deposit grain in the elevator, accept a scale ticket that is marked *D.P.* (delayed pricing), and assume that they have stored the grain to be priced later. In fact, they have *not* stored the grain; they *have sold it.* Title to the grain has been transferred to the elevator. Except for pricing and payment, the sale is complete. An example of a D.P. agreement, shown in Figure 2–2, clearly states that title is being transferred.

The delayed price agreement gives farmers the privilege of choosing a price that is being paid by the elevator any day in the future, generally up to the end of the marketing year for the crop. If the local cash price is higher later in the marketing season than when farmers delivered the grain, they gain from the higher level. If it is lower, they receive less. Thus, farmers can use D.P. to speculate on local cash price. For this privilege, the elevator assesses a service charge, a delayed pricing fee. The price farmers receive is equal to local cash price less the service fees. Farmers receive payment for the grain at the time they establish the price.

Delayed pricing permits producers to speculate on the cash market without investing in on-farm facilities or paying elevator storage charges. Neither price level nor basis is fixed. However, delayed pricing fees often equal the basis gain and thereby limit the farmers' opportunity to capture basis improvement.

For the elevator operator, delayed pricing can mean larger operating volume. Because it is a sale and not a storage transaction, the elevator operator need not keep the grain on hand. It can be shipped out immediately to allow elevator operators to move more grain through their facility during the season and therefore not limit their business volume to their storage capacity. Some elevators ration storage space by

DELAYED PRICE AGREEMENT

Date: _____ Agreement No. _____

Commodities under agreement: _____

It is hereby agreed that I, the undersigned seller, may from time to time by my own choice, sell and deliver to the undersigned buyer agricultural commodities as listed above, on which the price is to be established at a later date. I pledge that the commodities delivered pursuant to this agreement shall be free of any lien or encumbrance.

In selling commodities under this agreement, I, the seller, fully understand that *I am transferring title* to the undersigned buyer upon delivery, and that after delivery I am a creditor of the buyer for the market value of commodities so delivered until the price is established and settlement is completed. If the buyer defaults in his/her obligation for settlement I may be a common (unsecured) creditor of the buyer for the value of the commodities not settled for.

Upon demand by the seller, the buyer is obligated to pay his/her regular bid price on the date of demand for the commodities being priced by the seller which have been delivered under this agreement, less any service charge which is due and payable to the buyer.

A separate contract may be entered into between buyer and seller to set forth the terms and conditions of establishing the price. This contract shall also list the charges to be assessed against the seller for services rendered by the buyer from the time of delivery of the commodity until the price is established by the seller. Said contract may be printed on and be a part of the inbound scale ticket, or may be contained on a separate form which shall be supplied to the seller either before or at the time of delivery of the commodity under contract.

If a separate contract is not used, the condition and charges for services under this agreement for which the seller shall be liable to the buyer are as follows:

_____ ¢ per bushel from the date of delivery through _____, 20_____, then

_____ ¢ per bushel per (day) (month) after the above date until the date of pricing.

If, after the signing of this agreement and the delivery of commodities thereon, the buyer causes a change to be made in the method of pricing or the service charge as listed hereon which would apply to the delivery of commodities after said change, a new agreement shall be signed between the buyer and the seller showing the date of the new agreement and the new service charges which are then in effect.

Section 926.29(C) of the Ohio Revised Code states that the seller under a delayed price agreement *may* demand as security for payment an amount that, at the time of delivery, is equal to the government loan rate for the commodity or 75% of the current bid price, whichever is less.

I am ☐ I am not ☐ demanding a payment guarantee from the buyer.

_____ | _____
Seller | *Buyer*

_____ | _____
Address | *Address*

_____ | _____
Authorized Signature | *Authorized Signature*

UNDER THE REQUIREMENTS OF ODA RULE 901:7–2–17(D): Agricultural commodities may be delivered on a delayed price agreement for a period of one year after its effective date. Commodities delivered thereafter shall be covered by a new agreement which must be executed not later than the first delivery under the new agreement.

FIGURE 2–2: Delayed Pricing Agreement.

accepting only part of the producer's deliveries for storage, thus forcing the rest of the delivered crop into the market as a cash sale or delayed price transaction.

Since title to the grain passes to the elevator at the time of delivery, elevator operators are free to dispose of the grain in any manner and at any time that it is most profitable to do so. There are three avenues they may follow. First, elevators may keep the delayed price grain on hand and treat it the same as stored grain. When the farmer decides to fix a price, the elevator sells the grain and passes the proceeds to the farmer. Second, since title has been transferred, the elevator operator can sell the D.P. grain and have interest-free use of the money until the farmer prices the grain. Finally, elevator operators can store the grain in either their own facilities or with terminal elevators, cause warehouse receipts to be issued against the grain, and pledge them as collateral for loans to the elevator.

Delayed pricing can be a business quagmire for farmers. Because title has been transferred to the elevator operator, the sale is often not covered by the grain warehouse licensing and handling requirements of state and federal agencies. Having relinquished title to the grain, the farmer is an unsecured creditor of the elevator until paid. The scale ticket and delayed price agreement are no more than a promise to pay when a price is established. If the elevator becomes insolvent, the D.P. promise is secondary to those holding warehouse receipts and mortgage interests in the available grain and other assets. In bankruptcy, after those other liens have been satisfied, the holders of delayed price agreements will share whatever remains with other general creditors. Bankrupt elevators, by definition, do not have sufficient assets to cover the claims of all creditors; D. P. obligations are seldom paid.

Delayed pricing service fees charged by elevators often are more flexible than their published storage rates. Their fees are highest during harvest and drop in late winter or early spring; they may be zero at certain times of the marketing year. D. P. service charges generally reflect the market's need for cash grain, as determined by the basis market. Free delayed pricing is most commonly offered late in the marketing year, when small harvests are in prospect. Farmers need to clear out storage bins for the new crop and typically feel there is price strength in the market. Under these conditions, basis is usually strong. Elevators entice grain movement from farmers to themselves with low or zero D. P. fees. Later on, basis typically weakens with the harvest movement of grain, irrespective of whether the general price level has increased or not. If elevator operators protect themselves with a hedge against price level change, delayed pricing is a method for them to profit by speculating the basis. If they receive grain when the basis is strong, but the farmer prices the crop when basis is weak, the elevator, rather than the farmer, profits from basis change. On the other hand, if basis strengthens between delivery and pricing, elevator operators stand to lose from their basis speculation.

In order to calculate whether delayed pricing is a profitable marketing alternative, a farmer needs to look at potential costs and returns. Soybeans might be selling at $5.00 per bushel at a local elevator during October. The price of a July futures contract might be trading at $5.80 that same day. The farmer could capture the 80-cent difference by delivering D.P. grain and simultaneously selling a July futures to hedge away price level change.* The farmer's costs are the D.P. service charge, the interest

---

*Assuming basis converges to zero by July 1.

that would have been received on money in the bank instead of keeping it invested in grain, the brokerage fee on the futures contract, and interest on funds required as margin on the futures transaction. A typical D.P. service charge on soybeans for central Ohio is 24 cents per bushel for 120 days and 3 cents per month thereafter. The charge would total 36 cents by July. The farmer's additional carrying charge is the $5.00 per bushel tied up for 8 months—money that could be used to pay off loans or be placed in interest-bearing investments. Using an annual interest rate of 12 percent, the interest cost of storage for 8 months would be 40 cents. To hedge in the return, broker fees on futures transactions cost about 1 cent per bushel, and interest on the margin deposit costs 1.6 cents. Total carrying costs would, therefore, be 78.6 cents, leaving a 1.4-cent profit if the farmer were to use D.P. and hedge in a return. With such a low locked-in profit margin, most producers do not place the hedge in the futures. Instead, they deliver D.P. grain in order to SPECULATE that local price level will rise and thereby afford them higher pricing opportunities.

Farmers accept the dual risks of lower prices and the financial integrity of the buyer in using this marketing tool as a method to speculate. There are other ways that farmers can speculate on price without speculating on the financial strength of the elevator operator as well. These other methods permit farmers to capture advantageous basis markets when they exist, basis appreciation that is not limited by D.P. fees.

As a marketing tool, delayed pricing can be a very risky and expensive way to speculate.

Deferred Payment Contracts

For sales with deferred payment provisions, *delivery, title exchange, and pricing* transactions are completed at the same instant. Only settlement is deferred until a later date, usually in the next tax year. As with a cash sale, price level and basis level are both fixed at the time delayed payment contracts are executed.

Deferred payments can provide a tax advantage. With proper stipulations in the agreement, the Internal Revenue Service has accepted the fact that the producer did not have right of payment in the year of the sale. As a result, the IRS has concluded that a producer could be taxed only on income from the sale in the year when payment was actually received. Tax rulings have made it clear that to achieve the tax deferral objective, there must be a valid contract that calls for payment during the following year. Otherwise, the IRS will consider the income "constructively received" in the year of delivery. The contract cannot be supported by a note or security that the seller could redeem for cash. Or if the farmer has the right to collect payment at any time after delivery of the commodity and all that is lacking is demand for payment, the farmer would have to include the sale in the tax year of delivery, regardless of when actual payment is received.[3]

Deferred payments can also be risky. Since delivery and transfer of title precede payment, deferred payment contractual arrangements carry the same business risk as delayed pricing sales. Sellers are unsecured creditors until cash is received. They should have confidence in the financial position of the buyers before agreeing to defer payment.

An example of the Deferred Payment contract and its provisions are detailed in Figure 2–3.

DEFERRED PAYMENT CONTRACT

a partnership
John Doe Grain Co., Anytown, Ohio, a corporation, hereinafter known as Buyer, and _____, hereinafter
an association
known as Seller, covenant and agree as follows.

Buyer agrees to purchase from Seller approximately _____ bushels of _____ at the agreed price of $_____ per bushel under the following terms and conditions to wit:

Seller shall deliver the above described grain within _____ days to Buyer's elevator in Anytown, Ohio.

Seller hereby warrants that title to the above grain is free from encumbrance of any nature whatsoever and that Seller has good and merchantable title, with the right to dispose of the same. Seller hereby authorizes the Buyer to contract for the sale or disposition of said grain for its own account either before or after delivery to the elevator above designated without further authorization from Seller except this contract, and buyer shall not be required to account for or pay the balance of said purchase price until the _____ day of _____, 20_____.

Seller hereby acknowledges receipt of the sum of $_____ to apply on this contract, and the balance shall not be due until the _____ day of _____, 20_____, it being the intent of all parties hereto that said sale shall not be completed until final payment by the Buyer to the Seller, and it is specifically agreed that Seller shall, under no circumstances, be entitled to the balance due herein until the _____ day of _____, 20_____.

Buyer agrees that it will pay the balance due without interest on the due date as fixed herein to the Seller.

It is further agreed by and between the Buyer and Seller that this contract shall be binding upon the heirs, administrators, and executors of the respective parties and that this contract cannot be assigned.

Dated this _____ day of _____, 20_____.

JOHN DOE GRAIN COMPANY

By _____

_____
Witness

_____
Buyer

_____
Witness

_____
Seller

FIGURE 2–3: Deferred Payment Contract.

Basis Contracts

Executing a sale with a basis contract allows *delivery, title transfer, partial settlement of payment,* and a *basis price* fixed relative to a futures contract price. This sale method allows a farmer to establish a price level at a later date, hopefully when the price level has risen. It thus allows speculation on price level, but not basis level.

The basis contract in Figure 2–4 specifies delivery date, quality, amount, and delivery point. The difference between it and a cash contract is in the price notation. Price on a basis contract is stated relative to a specified futures contract, rather than as a fixed price. For example, the price may be "30 cents under March Soybeans," rather than a $7.52 flat price, as could be in the case in a cash contract. Consequently, if March futures contract price rises, the seller will get a higher price, equal to 30 cents less than the March contract, however high it rises. On the other hand, if the March futures contract price declines, the seller will receive a lower price, still 30 cents below the March futures price.

Farmers have the option of selecting the day they wish to fix the price level for their grain. They may choose any day following execution of the contract until about the first day of the month of maturity of the futures contract specified. Any time before maturity, they simply notify the buyer to complete the transaction and forward remaining payment.

Or, farmers can continue to delay the pricing decision. They can roll forward the basis contract to another futures contract month to continue to hold the open specu-

---

BASIS PURCHASE CONTRACT

DATE _____   CONTRACT NO. _____

UNDERSIGNED SELLER hereby sells _____ bushels of _____ to BUYER subject to the following terms and conditions:

grade and grain

1.   SELLER has, as of the date of this contract, delivered to BUYER the quantity and type of grain indicated above. The delivery of grain is documented on scale tickets at BUYER'S scale.

Title and all other rights of ownership and control over the grain under this contract are transferred to BUYER.

2.   BUYER hereby grants the SELLER option to establish final cash price in accordance with the following terms and conditions:

(a) Basis portion of final cash price is hereby established at _____ ¢ under the _____ Future Option on the Chicago Board of Trade.

(b) SELLER may price grain between 9:40 a.m. and 1:00 p.m. on any day the Chicago Board of Trade is open between today and the _____ day of _____.

(c) SELLER may, at his/her option and expense, spread to a basis to another Chicago Board of Trade contract on or prior to date in section (b) above.

(d) if SELLER does not establish final cash price on or before above date, the BUYER will establish price based on the closing range of the first trading day following the above date on the Chicago Board of Trade.

(e) All bushels shall be priced or spread forward at one time.

3.   Final price shall be subject to grade discounts at market scale of BUYER at time of delivery. Shrinkage, drying, and any other charges that may be applicable shall be deducted before determining the final quantity and price of grain sold under this contract.

4.   The BUYER agrees to pay the SELLER _____ dollars per bushel at the time this contract is entered into between the BUYER and SELLER, and to pay the final remaining balance on the day this contract is priced as outlined in Item 2, section (b).

5.   SELLER hereby warrants that title to the above grain is free from encumbrance of any nature whatsoever, and the SELLER has good and merchantable title, with the right to dispose of same. It is further agreed by and between the BUYER and SELLER that this contract shall be binding upon the heirs, administrators, and executors of the respective parties and that this contract cannot be assigned without given consent of the other party.

_____   By_____
                Seller   Representing Buyer

_____   _____
          Phone Number   Dated

FIGURE 2–4: Basis Purchase Contract.

lative position. For example, they would roll forward to the July contract, if the price level increase they were expecting had not occurred by March, and if they still believed it would do so by July. The elevator typically assesses the seller the brokerage costs of the rollover.

The most important advantage of the basis contract method of sale is that it permits the producer to split pricing decisions into a basis level sale and a price level sale. When the basis is strong, but price is low and the producer feels it will rise, execution of a basis contract captures the advantageous basis level and allows continued speculation on price level. After the price level has risen, a producer can make the price level commitment. The producer thus has the opportunity to capture high prices in both markets if the price level does, in fact, rise as expected.

With basis contracting, farmers can select a delivery date during periods when few competing jobs need to be done on the farm. The risks of quality deterioration and shrinkage are eliminated by making delivery to the buyer. This may be especially advantageous with low-quality grain. Farmers can move the grain without having to fix price level.

Risks of nonpayment are also reduced with a basis contract, because farmers can draw a large percentage (70 to 80 percent) of the payment at the time of delivery of the grain. They are paid the remainder of the crop value when they price the grain by notifying the grain company that they want to complete the transaction. If the price level drops to the point that the advance payment exceeds the remaining value of the crop, the farmer can continue to speculate on price by returning part of the advance to the grain company. Otherwise the grain company can cancel the basis contract and settle the account.

Cash Sale and "Buying the Board"

Producers can achieve the same results as basis contracting by making a cash sale *and* substituting a paper position in the futures market for the physical grain. The cash sale portion has the same business and legal implications noted earlier; delivery, title transfer, pricing, and settlement are completed. Both price level and basis level are fixed by the cash sale. To maintain a speculative position, (hoping to profit from a price level increase) the producer "buys the Board" equal to the number of bushels in his or her cash sale.★ When the futures price level rises to the price objective, the farmer liquidates the futures position and profits from the price level rise. Should the price level decline, the farmer can liquidate the futures contract and absorb the loss; or the farmer can continue holding the position by making deposits to cover the declining value. The futures contract has a finite life. At some point in time, the contract expires. The farmer must eventually close out the position and withdraw the profits or absorb the losses.

With this marketing tool, the farmer replaces the physical grain with a paper position in the futures market. The cash sale fixes the basis level and requires delivery; the futures contract retains an open price level position, subject to fluctuation.

Business risk is minimized, since full value payment is made on the cash sale. But the futures transaction requires deposits and brokerage fees to execute the futures position. It is also subject to margin calls to cover adverse price movements.

★"Buying the Board" is defined as executing long futures positions, which will be explained later in this chapter.

Cash Contracts, Forward Sale, Booking

Cash contracting, forward sale, and booking are synonymous terms for a marketing tool that completes the sale by fixing *price, title exchange,* and *delivery date(s)* on the day of the execution; only physical delivery and settlement remain to be completed in the future. Since the latter two typically occur simultaneously, payment risks are minimized. Nevertheless, an insolvent buyer at time of delivery can make the contract worthless, and a favorable contracted price is no longer enforceable.

The advantage of this tool is that the producer can establish a favorable price without having the commodity on hand. Farmers can contract to sell intended production many months in advance of physical delivery. The delivery time need not even be during harvest; delivery can occur from storage anytime. By removing price uncertainty with a forward contract, the producer can more accurately project net returns from an enterprise before a crop is planted or livestock placed in a feedlot.

With cash contracting, both price level and basis level are fixed by the commitment. The price stated on the contract is what the producer will receive when delivery is made. All price speculation is eliminated with the cash forward contract. However, both the price level and the basis level are seldom favorable at the same time. That is the major disadvantage to this tool; it does not allow splitting the price decision between price level and basis level sales.

One of the risks of executing a cash contract is the possibility that prices will rise after contracting. Farmers bemoan their "loss in profit" when prices rise and they cannot cancel their delivery obligation. They should not cash contract until they are satisfied with the price being offered. They must give up the opportunity to reap higher prices in return for protection against the risk of lower price.

Production risks are magnified by the fact that farmers are obligated to deliver a specified quantity and quality. Should drought, hail, or other disaster occur, farmers are legally bound to purchase the commodity from someone else to deliver against the contract, or make payment to compensate the contract buyer for any loss that might be incurred. For this reason, producers are often reluctant to sell what they do not have, or else they cash contract only a portion of their expected production.

Figure 2–5 shows production and price changes that can occur after forward pricing commitments are made. Prices may rise or decline; actual production may exceed or fall short of the contracted amount. The example assumes 100 bushels contracted at $3.40 per bushel, for $340 gross receipts. Higher or lower production, or price, changes the gross actually realized.

Outcome I has both higher price ($3.60) than contracted and higher than expected production (120 bu.). If a producer contracted 100 bushels but harvests 120, the extra 20 bushels can be sold at the higher price, but the 100-bushel contracted amount sells at the contracted price. Producers bemoan this result because they could have received $24.00 more per acre had they not contracted. The result rates Good, however, because it is typically profitable. In the aggregate, bumper production seldom accompanies high price, so the probability of this combination of events occurring is low.

Outcome II presents the strongest argument for cash contracting; contract price is higher than harvest time market price ($2.80), and production is higher than expected (120 bu.). The extra 20 bushels must be sold at the lower market price and elicit the response, "I should have contracted more." The sale of the 100-bushel base amount produced the happy result of protecting income against lower price. In the aggregate, higher production usually accompanies lower price, as this combination

	Price		
	Higher	Contracted at 3.40/bu	**Lower**

<table>
<tr><th colspan="2">I</th><th></th><th colspan="2">II</th></tr>
<tr><td>Production, bu</td><td>120</td><td></td><td>Production, bu</td><td>120</td></tr>
<tr><td>Price, bu</td><td>$3.60</td><td></td><td>Price, bu</td><td>$2.80</td></tr>
<tr><td>Gross/Acre
Contracted
100 × $3.40 =
plus
20 × $3.60 =
Total</td><td>

$340

72
$412</td><td></td><td>Gross/Acre
Contracted
100 × $3.40 =
plus
20 × $2.80 =
Total</td><td>

$340

56
$396</td></tr>
<tr><td>Naked
120 × $3.60 =</td><td>
$432</td><td></td><td>Naked
120 × $2.80 =</td><td>
$336</td></tr>
<tr><td>RESULTS: GOOD</td><td></td><td>100 bu.
× 3.40
$340</td><td>RESULTS: EXCELLENT</td><td></td></tr>
</table>

Higher (row label, Production, Contracted at 100 bu/A)

<table>
<tr><th colspan="2">III</th><th></th><th colspan="2">IV</th></tr>
<tr><td>Production, bu</td><td>80</td><td></td><td>Production, bu</td><td>80</td></tr>
<tr><td>Price, bu</td><td>$3.60</td><td></td><td>Price, bu</td><td>$2.80</td></tr>
<tr><td>Gross/Acre
Contracted
80 × $3.40 =
less
20 × $3.60 =
Total</td><td>

$272

−$72
$200</td><td></td><td>Gross/Acre
Contracted
100 × $3.40 =
less
20 × $2.80 =
Total</td><td>

$340

−$56
$284</td></tr>
<tr><td>Naked
80 × $3.60 =</td><td>
$288</td><td></td><td>Naked
80 × $2.80 =</td><td>
$224</td></tr>
<tr><td>RESULTS: BAD</td><td></td><td></td><td>RESULTS: GOOD</td><td></td></tr>
</table>

Left axis label: **Production** — Higher / Lower

FIGURE 2–5: Effects on Income of Price Change and Production Differences from Contracted Amount.

depicts. The results rate Excellent because of the high probability of this combination of events and the income protection afforded by it.

Outcome III illustrates what many farmers fear the most—higher market price ($3.60) and low yields (80 bu.). Production shortfalls must be offset by purchases at higher price to fill the contract. There is a double loss from not having the commodity and having sold what little there is at the lower contracted price. These results must be rated as Bad.

One solution to guard against production shortfalls is to contract less than full production, such as 30 to 50 percent of expectations. Another is to buy crop insurance. It is available across the United States for practically all crops. For anyone who cannot bear risk, both production insurance and price protection can mean the difference between remaining in the business or not.

Outcome IV details the combination of lower price ($2.80) and lower production (80 bu.). The results are Good for contracting, because price protection compensates for production problems. Producers enter the open market to replace the shortfall in production at a lower price than they get by delivering against their contract. They buy low-cost grain from neighbors and sell it on contract at a higher price.

The extent of forward pricing anticipated production from a growing commodity depends upon the individual producer. An individual may be highly speculative or ultra conservative, financially stretched or free of debt. The farmer may be located in a marginally productive area or in one that never has had a crop failure. All these factors come into play in determining how much of expected production one is willing to commit by forward pricing. In some years, the highest price for the next production season occurs shortly after harvesting the old crop. Instead of weighing the likelihood of successful versus disastrous production and the relative profits compared with possible losses, many farmers avoid "selling what they do not have" and thereby forego profitable opportunities. They demonstrate their faith in a successful crop by borrowing heavily to buy seed, fertilizers, pesticides, and fuel, but they will not forward contract production even though prices may be at historically high levels. Fear of Outcome III and the quest for Outcome I should not deter farmers from making profitable sales when the market offers profitable opportunities, even though the crop is not on hand. Forward selling of intended production is common practice in other business enterprises.

Farmers must accept the legal implications of contract pricing. Forward contracts exchange title and are legally enforceable. The Uniform Commercial Code (UCC), now adopted in every state except Louisiana, provides that contracts for the sale of goods such as grain and livestock, which involve $500 or more, are legally not enforceable unless they are in writing. To protect themselves, elevators use the standard form of contract, or confirmation, illustrated in Figure 2–6. The confirmation normally includes the amount, quality, agreed price, and terms of delivery and thereby helps to avoid misunderstandings. The UCC also specifies that if the transaction is between two "merchants," the seller need not sign and return the confirmation in order to be bound by its terms. Most courts have looked at the market sophistication of the farmers involved and concluded that they should be called "merchants." If there are second thoughts or if the confirmation does not sound like the trade made verbally, the seller should contact the buyer immediately in an attempt to resolve any discrepancies. If the difference cannot be resolved, the seller has 10 days after receiving the confirmation to reject the offer in writing.

The terms of the contract should be reviewed by an attorney. The omitted provisions may cause more problems than the stipulated ones. For example, what if harvest is delayed by weather? Generally, the contracts will allow the buyer to accept or refuse later delivery at the buyer's option. What if the elevator is full and the elevator cannot accept delivery as called for? Who pays storage charges and who pays for yield loss or deterioration in quality while the crop stands in the field until the buyer can take delivery? It is unlikely that legal risks of marketing can be total eliminated, but they should be recognized and evaluated by the producer-seller *before* a problem occurs.

Futures Contracts

A futures contract is an agreement entered into today to complete a business transaction on some future date. It specifies the date and place of the transaction, and the quantity and quality of the goods to be delivered, along with the price at which the

GRAIN SALES CONTRACT

CONTRACT NO. _____

THIS AGREEMENT, made and entered into this _____ day of _____, 20 ____, by and between _____ of (city and state) _____ (hereinafter called Seller), and _____of (city and state) _____ (hereinafter called Buyer), wherein each of the parties agrees to the following terms and conditions of the sale of Seller's _____ (hereinafter called grain).

    1. Seller agrees to sell and deliver to Buyer at Buyer's elevator or to Buyer's designated recipient for Buyer's account on or before the _____day of _____, 20 ____, _____bushels of _____, Grade No. _____, said grain to be clean, sound, and merchantable according to the established grading standards of the USDA, and subject to Buyer's discounts applicable at the day of delivery, for the agreed-on price of _____Dollars ($ _____) per bushel, less any pre-payment, discount, or dockage arising from the condition of the grain. All grain to meet Pure Food and Drug Act requirements.

    2. Seller agrees that in event Buyer or its designated recipient is incapable of receiving all or part of said grain on the date it is offered for delivery by Seller, Seller will hold all or so much of the grain as cannot be received and deliver it to Buyer or its designated recipient upon receipt of notice that the grain can be delivered.

    3. Seller and Buyer agree that the selling price quotation for the grain herein is binding upon both parties and that market fluctuations after the date of this contract shall have no effect on either party's obligation hereunder and will not excuse non-performance of this contract by either party.

    4. Seller and Buyer agree that this is a contract for delivery of Seller's grain and is not a futures contract which can be purchased back by Seller. Seller's failure or inability to deliver said grain in no way relieves him/her or his/her obligation of delivery of said grain, and Seller agrees Buyer may enforce this contract by an action for specific performance, except should delivery of grain be made impossible solely because of a natural disaster. Buyer agrees that upon proper notification from Seller, prior to the delivery date or within the extended delivery period, if Buyer exercises his/her option to extend the contract, to permit Seller to reimburse Buyer for any and all losses Buyer may suffer from Seller's partial performance of this contract in lieu of delivery of all or part of said grain.

    5. In the event of default by either party, all sums due shall be payable without relief from valuation and appraisement laws and with reasonable attorney fees, and with interest at the rate of eight per cent (8%) per year, calculated monthly from the date of default.

    6. Seller agrees that Buyer may extend the due date for delivery of the grain beyond the aforementioned date at Buyer's sole option on written notification of such extension to Seller mailed to Seller's address as shown on this contract.

    7. Seller agrees that Buyer or its designated recipient shall have the sole option of accepting or rejecting any grain in excess of the contract amount and if accepted, it is agreed Buyer will pay Seller Buyer's closing market price on the date of delivery if delivered to Buyer's elevator, and Buyer's closing market price on the date of receipt of notification of such excess delivery from either Seller or Buyer's designated recipient if delivered to designated recipient.

    8. Seller warrants that all grain delivered to Buyer will be free and clear of all liens and encumbrances and that Seller has and will pass merchantable title.

    9. The word "bushels" as used herein means the equivalent number of pounds of said grain as established by the Bureau of Weights and Measures.

    10. This Agreement constitutes the entire understanding of the parties and may be modified only by a signed writing of both parties, excepting Buyer's right to extend this contract, and is binding on the heirs, assigns, and successors of the parties.

SELLER: _____
                      NAME

BUYER: _____

_____
                ADDRESS

BY: _____
                    TITLE

FIGURE 2–6: Cash Contract.

transaction is to be consummated. Neither title nor money changes hands until the date specified. A futures contract represents an obligation made on a particular day to enter into the cash transaction at a later date.

Contracting for future delivery is not an uncommon business practice, despite the fact that people often ask: "How can you sell something you do not have, or buy something that does not exist?" For example, a futures contract is established between a building contractor and a home buyer for the construction of a house. Contractors are selling something they do not have—the buyer is buying something that does not exist. Contractors enter into a contractual agreement with a customer that at some future date they will have erected a house on the customer's lot. The customer agrees to accept delivery on that date. Plans, specifications, and payment provisions are detailed in the contract. Upon entering the agreement, the contractor has neither a house nor the concrete, lumber, paint, or other materials to build the house. The builder knows, however, where to acquire the materials and hire the labor. The contractor knows how long past construction projects have taken and anticipates the cost likely to be incurred in the building process. A contractor enters into a futures contract to build a house based upon expectations about events that will transpire in due time. The lot owner wishing to own a home enters this agreement in anticipation of taking delivery and making payment on some future date. Any futures market, whether in houses or commodities, is an intentions and expectations market.

A farmer or rancher should have no more trepidation in executing a commodity futures contract than in signing a contract with a house builder. There is an intention to produce the commodity by some future date. There is the expectation that the cash transaction will be completed, as specified, in due time.

Commodity futures contracting is the process of setting a price in the present for a business transaction that will be completed in the future. *Price, quantity, delivery date, and delivery point* are fixed as part of the contract. Actual delivery and payment remain to be completed.

This marketing tool provides price protection similar to cash contracting but differs in its mechanics and pricing effects. In cash contracting, the farmer deals with a local elevator. With futures contracts, a farmer deals with an account executive or a broker directly through a commodity exchange. Cash contracting specifies a flat price the producer is to receive; both price level and basis level are fixed. A futures contract fixes the price level but permits speculation on basis level. With cash contracting, the farmer knows exactly what price is to be received; with futures contracts, the farmer knows only approximately what will be received, since basis level remains to be established.

As a commodity futures contract approaches maturity, final settlement must be made. Settlement of a futures contract differs from settlement of a cash contract, however. With the latter, the buyer expects delivery of the commodity from the seller and is usually inflexible in that requirement. Futures contracts can be settled by one of two methods:

1. Delivery Against the Futures Contract by completing the transaction as specified in the contract.

2. Offsetting Trades by making an opposite future transaction to cancel the original contract and thus eliminate the obligation for fulfilling it.

The flexibility of settlement provided by these two alternatives provides futures contracts a distinct advantage over cash contracts.

Delivery Against Contracts: Cash Settlement. Cash settlement of a futures contract typically requires delivery by the seller and payment by the buyer. The producer physically delivers the commodity to the delivery point on the date specified on the contract. The buyer accepts title and pays full value on the transaction.

Location of delivery points is critically important to the seller. Delivery points are usually located in commercial centers, far removed from the farm in both distance and mechanics of operation. Chicago and Illinois river terminals are the major grain delivery points[4]. Live cattle delivery is at key livestock commission firms and slaughter plants from Texas to Iowa, since this region produces the vast proportion of the nation's livestock[5,6]. The cotton futures market specifies delivery at the key warehousing and marketing points for the largest U.S. cotton regions: Memphis, Tennessee; Greenville, South Carolina; Houston and Galveston, Texas; and New Orleans, Louisiana are major delivery points. Kansas City, Missouri and Hutchinson, Kansas are delivery points for hard red winter wheat. Minneapolis, Duluth, and Red Wing, Minnesota accept spring wheat. The frozen orange juice contract requires delivery at regulated warehouses at or near Florida processing plants.

Feeder cattle and hogs are cash settled according to an index calculation of value published by the Chicago Mercantile Exchange. Feeder cattle and hog futures contracts are not settled by physical delivery of live animals.

The seller must pay transportation charges from the farm or ranch to the delivery point. In addition, the seller must pay handling charges to the terminal operator and brokerage fees for services rendered. The net price realized at the farm gate is the contracted futures price less these charges. (This difference between futures price and realized price is the subject of basis discussion in Chapter 7.)

A trader who buys a grain futures contract and holds it for delivery does not expect trucks to back up to his or her home or office and unload the actual commodity. Even if the buyer wants to accept delivery against the futures contract, the buyer is not given the commodity. Instead, a shipping certificate is issued from the facility that accepted the seller's delivery at the delivery point.

Commodity buyers rarely take delivery. Since many types and grades are deliverable against the contract and since the option to deliver or not remains with the seller, the buyer does not know in advance exactly what would be delivered. Processors require a careful selection of specific grades that may not be matched by the goods being delivered.

Those who do accept delivery on futures contracts are typically merchants or dealers. These traders find it practical to accept delivery, because they deal in many grades and types of the actual commodity and feel confident that they can find a buyer for whatever grade the seller delivers.

In actual practice, less than 1 percent of most futures contracts are settled by delivery, as shown in Table 2–2. The fact that delivery of the actual commodity can be made to satisfy futures obligation links together a price relationship between cash markets and futures markets. If cash and futures prices get too far out of line with each other, there are traders (arbiters, practicing arbitrage) who earn their livelihood by buying one and selling the other to bring them back into proper economic relationship.

The purpose of futures trading is to shift price risk rather than provide a mechanism for cash sale. Actual delivery and acceptance of a physical commodity to satisfy a futures contract is not essential for futures to fulfill that purpose; the mere possibility of physical exchange insures the proper price relationship.

**TABLE 2–2: Delivery of Selected Commodities
to Satisfy Futures Contracts 1998**

Contract	Volume of Trading	Total Contracts Settled by Delivery or Cash Settlement	Percentage Settled by Delivery
Live Cattle (40,000 lbs.)	9,857,000	6,232	0.06%
Lean Hogs (40,000 lbs.)	3,196,000	696	0.02%
Wheat (5,000 bu.)	5,682,000	22,837	0.41%
Corn (5,000 bu.)	15,795,000	54,659	0.35%
Soybeans (5,000 bu.)	12,431,000	347,166	2.79%
Cotton (50,000 lbs.)	876,000	2,342	0.13%
Orange Juice	472,000	5,110	1.08%

Source: Various Futures Exchanges.

Offsetting Trades. Producers generally do not find it advantageous to make delivery. Their products may not match the grade or quantity specified on futures contracts, and the costs of making delivery to a distant terminal may be prohibitive. It is usually more advantageous for them to simply offset the transaction in the futures market and deliver their product locally. Settlement, by the offsetting trades, eliminates the mechanical problems of delivery, without impinging on the effectiveness of transferring risk.

Producers who SELL a futures contract in essence OWE delivery of the commodity. At any time before maturity of the contract (even the next day), they can BUY back the futures contract and thereby offset the original commitment. This is called "covering" the position with an offsetting trade. Producers buy back the same number of contracts they sold. They are then free to sell the cash commodity to anyone, since its futures delivery commitment has been canceled.

The flexibility of canceling futures contract commitments with offsetting trades fosters hedging. Commodity producers or owners transfer the economic effects of price change from themselves to others by hedging. To understand the mechanics of price risk transfer afforded by hedging, we need to understand the parallel relationship between cash prices and futures prices.

Parallelism of Cash and Futures Prices. Futures trading creates neither a loss nor a profit for a producer who has a commodity to sell. It merely fixes a price level that will eventually be realized. This truism exists because of the parallel relationship between the cash price and the futures contract price of a commodity. The two prices

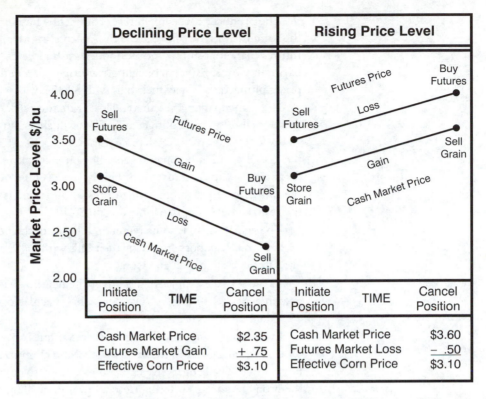

FIGURE 2–7: Parallel Movement of Futures and Cash Prices, Illustrating Offsetting Gains and Losses in a Selling (Short) Hedge.

tend to fluctuate together. When cash prices increase, futures prices increase; when cash prices decline, futures prices decline, as long as basis is constant.★

Cash price determines the inventory value of the growing or stored crops or livestock on hand. Rising cash price increases inventory value; falling price lowers inventory value. It is the erosion in value that a producer seeks to avoid by **hedging** to fix a price level in the futures market.★★

Hedging involves **selling** futures contracts equal to the amount of grain or number of livestock that a producer owns or expects to produce. The farmer-hedger OWNS grain, "knee high on the Fourth of July," and he or she OWES delivery of it against futures contracts. When the price level declines, as shown in Figure 2–7, two values change. First, the inventory value of grain in the field decreases. A drop in the cash price from $3.10 per bushel to $2.35 results in a loss in value of 75 cents per bushel. With 100,000 expected bushels of corn in the field, its total value decreases $75,000. Over the same time, however, the hedge account gained in value. The futures contract was sold (to establish the hedge) when the price level was $3.50 per bushel. It can be offset (bought back) at $2.75 per bushel. Over the period, $75,000 was credited to the brokerage account. The 75-cent loss in inventory value was offset by the 75-cent gain on the futures contracts.

★Parallelism assumes a perfect hedge without gain or loss from basis change. The assumption of a perfect hedge will be relaxed and explained in Chapter 7.

★★This is an example of a selling hedge; a buying hedge works by the same principle, but in reverse. It will be illustrated later.

The producer covers the futures contract by buying it back and sells the grain on the local cash market for $2.35. The 75 cents in price protection realized from the futures market position is added to local cash price of $2.35 per bushel; the hedger pays bills with $3.10 corn. Neighbors who *speculated* in the cash market and lost when prices plunged try to pay their bills with $2.35 corn.

In a rising market, the gains and losses are reversed. Anticipating lower prices, the farmer sells a futures contract at $3.50 per bushel. But the futures market rallies to $4.00 per bushel and the cash price rises from $3.10 to $3.60. When the hedger covers the hedge position by buying back the contract sold, there is a 50-cent-per-bushel deficit in the futures transaction. This is, however, recovered by a 50-cent increase in the cash price. By deducting the futures account deficit from the cash price received, the realized net cash price is once again $3.10.

To be sure, there is some regret on the part of the hedger who pays the bills with $3.10 corn. Neighbors are paying their bills with $3.60 corn. In this case, they speculated in the cash market and won.

In these two examples of a "perfect" hedge (no change in basis), the hedger was guaranteed a $3.10 price. Neighbors who speculated in both instances "enjoyed" $2.35 and $3.60 corn.

In both the rising- and falling-market examples, the farmer received a net price of $3.10. Hedging guaranteed that price level and eliminated the possibility of higher or lower price. The hedger gave up the opportunity to enjoy a higher price in return for the protection from suffering from a lower price. The price was fixed at a level that the producer decided was satisfactory when he or she set the hedge.

Farmer reluctance to hedge planned or growing production arises from production risk. Drought during the growing season could cut yields, reducing total production. Even though price per bushel increases, as illustrated in Figure 2–7, the hedger might not have enough bushels (multiplied by the higher price) to offset the futures position deficit. Total inventory value gain would be less than total futures deficit. Farmers hedging a growing enterprise do not have the luxury of assurance that production goals will be met. This problem can be remedied, as we will explain later.

Buying Hedge. Another type of hedge is a Buying Hedge. Buyers of raw materials utilize futures contracts to fix the cost of the resources they plan to use. For example, a High Plains cattle feeder can hedge the costs of feeder animals, feed corn, and soybean meal that will be used to produce slaughter-weight cattle. A **buying hedge** involves "buying" futures contracts equal to the amount of feed ingredients that a feeder expects to use. Having determined feed corn requirements, the livestock feeder would establish an equal, but opposite, position in the corn futures market. The feeder is "short" cash corn needed to feed the cattle. In order to fix feed cost before placing cattle in the lot, the hedger "buys" corn futures contracts. The feeder "OWES" feed to the cattle and "OWNS" the right to receive corn via futures contracts. The parallel movement of prices in the futures and cash markets, as shown in Figure 2–8, illustrates how hedging fixes the cost of feed corn.

Suppose the corn market quoted a futures price of $3.50 per bushel and a cash price of $3.00 when the feeder sets the hedge. In a rising market, both prices would be higher when buying cash corn from the market. The feeder might have to pay $3.60 for feed corn, incurring a $0.60-per-bushel higher cost. However, the futures contract price would also have risen. The hedger offsets the $3.50 "buy" position with a "sell" transaction at $4.10, producing a $0.60 gain in the futures account. Subtracting

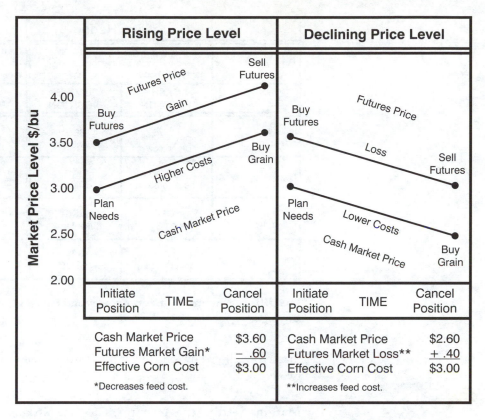

FIGURE 2–8: Parallel Movement of Futures and Cash Prices, Illustrating Offsetting Gains and Losses in a Buying Hedge.

the futures gain from cash purchase cost produces a $3.00 net cost of the feed ingredient.

In a declining market, the gains and losses would be reversed. The feeder buys cash corn at a price of $2.60, or $0.40 less than planned. Upon covering the futures contract at $3.10, however, there is a $0.40 loss in the position. Adding the futures loss to the lower purchase price of corn produces a net cost of $3.00 for the feed ingredient.

In a declining market, the gains and losses would be reversed. The feeder buys cash corn at a price of $2.60, or $0.40 less than planned. Upon covering the futures contract at $3.10, however, there is a $0.40 loss in the position. Adding the futures loss to the lower purchase price of corn produces a net cost of $3.00 for the feed ingredient.

In both market situations, the feed has a net cost of $3.00 per bushel. That is the cost that the feeder decided would be satisfactory when establishing the hedge position. The feeder gave up the opportunity to buy corn at a lower cost in return for protection against having to pay more for it. Gains in one market were offset by losses in the other; the price was fixed by the buying hedge just the same as in the example of the selling hedge.

The various positions and terminology of Selling and Buying Hedges are summarized in Figure 2–9.

Cash market prices and futures market prices **tend** to rise and fall together in a parallel manner. In reality, they do not move in perfect harmony over time. Later on we will remove the assumption of parallel movement of prices in futures and cash

	Selling Hedge		Buying Hedge	
Participants:	Farmers growing corn, wheat, cotton, soybeans. Livestock feeders selling cattle and hogs.		Cattle feeders buying feeder animals. Livestock producers buying corn and meal as feed ingredients.	
	CASH POSITION	**FUTURES POSITION**	**CASH POSITION**	**FUTURES POSITION**
Initiate position	Plant crop Buy feeders Breed sows	"SELL" contract	Plan production	"BUY" contract
Ownership	OWN the commodity	OWE delivery	OWE input item to operation	OWN the right to receive input
Terminology	LONG the cash	SHORT the futures	SHORT the cash	LONG the futures
Offset to initial position	Cash sale of commodity	"BUY" contract	Buy cash commodity	"SELL" contract

FIGURE 2–9: Summary of Terminology and Futures Positions for Hedge Selling and Buying.

markets to show how one might profit from the convergence or divergence of the two prices.

Margins: Settling the Accounts. There are two kinds of margins: initial and maintenance. Before a producer can enter into a futures contract, he or she must make a deposit of "good faith" money to establish financial integrity and guarantee performance. This deposit is similar to the house buyer's down payment when he or she signs a contract with a builder. In commodity trading, the payment is called the "initial margin deposit."

The second type of margin is maintenance margin. When the market moves against a futures position, a deficit occurs in the brokerage account. The broker makes a margin call to the producer, requesting an additional deposit of maintenance margin to cover the withdrawals and to maintain a minimum balance. In order to continue holding the contract position, the producer must send the deposit within a short time or else the broker has the authority to cancel the contract on the farmer's behalf.

With a perfect hedge, the parallel fluctuation of cash inventory value and futures account balance produces offsetting gains and losses that fix a preestablished price at the termination of the hedge. Day by day, however, from contract initiation until the hedge is covered, the inventory fluctuation is a non-cash change in value, but the futures account is settled in cash every day. Price change after initiating a futures position produces a cash surplus or cash deficit at the end of each day in the brokerage account. Daily settlement is made by the commodity brokerage firm in the producer's margin account. If the market price goes down after a selling hedge is made, the margin account will receive deposits each day equal to the price change from the previous day. The opposite situation occurs when the market price goes up after a futures contract is sold. In this case, the hedger must deposit payments to maintain a minimum balance in the margin account.

A cash surplus in a brokerage account is often seen as "profit" for the farmer-hedger. This is "found money"; it came so easily. The hedger did not have to work for it. All the hedger had to do was sell a futures contract and watch prices go down. The whole experience creates a false euphoria about what has actually happened.

Conversely, cash deficits or "margin calls" generate personal agony. Obviously, hindsight now tells the hedger that he or she sold crops or livestock at too low a price. Second, cash needed to meet margin calls can exceed available funds. Due to disgust over hedging at "low" prices, or inadequate financing, the farmer cancels the futures position at a loss. Invariably, price will turn the other way and begin to fall; but it is too late. The price protection the farmer sought is gone. Sold-out hedgers end the experience with a realized low price for their production and a loss in their futures account. They forever tell the story of "losing their shirt" in the futures market and repeat a vow never to "gamble" again by "speculating" in commodity futures. They are actually saying that they do not understand the purpose of futures trading nor the mechanics of the system. To use futures effectively, farmers need to understand the offsetting gains and losses that are occurring in two markets due to parallel movement in cash and futures prices.

Hedged-to-Arrive Contracts

Hedged-to-arrive contracting achieves the opposite result of a basis contract. With a basis contract, the producer locks in the basis level and continues to be exposed to price level fluctuation in the futures market. By contrast, the hedged-to-arrive contract fixes the price level, but the producer is exposed to basis fluctuation. Only the *price level is fixed* by hedged-to-arrive contracts at the time the contract is made. Delivery, title transfer, settlement, and fixing the basis are delayed until later.

Hedged-to-arrive contracts produce similar results as hedging with futures; the difference between the two arises from who handles the details. With hedged-to-arrive contracting, the elevator assumes responsibility for placing the futures position and absorbs the costs normally associated with hedging. These costs include margin calls, interest on margin deposits, brokerage fees, and associated management expenses. A farmer-hedger using futures would incur these costs and obligations personally.

The primary advantage of this method for the elevator operator is the potential for increased volume of business. While the elevator operator may give up some income from basis appreciation, increasing business volume and spreading fixed costs are definite advantages to the operator. For the grain seller, the hedged-to-arrive contract offers the price protection advantages of hedging without sacrificing the potential benefit from basis improvement. Because the hedged-to-arrive contract is settled in cash at time of delivery, there is little business risk. And the seller avoids margin costs and commission fees associated with hedging.

Elevator operators may allow the seller to roll forward the contract obligation. For example, the elevator may allow the farmer to sell July futures contracts to cover a harvest time delivery. This allows the seller to take advantage of present price spreads bid into the market. The seller would roll the futures position into another futures month as many times as desired, until delivery time is reached.

Serious controversies, leading to lawsuits, arose in the fall of 1996, as the corn market **did** what "it could not do"—exceeded $5.00 per bushel. Sellers' regret settled in when those who had sold futures or traded options at $3.50 saw neighbors selling a $500,000 corn crop when they had bargained for a $350,000 crop. Some farmers wanted to cancel their lower-priced commitments and claimed that they did not know they would be eventually held responsible for futures margin calls or the cost of additional options traded in a ballooning market. They claimed that they had been forced to trade in illegal off-exchange options and futures. The courts tended to side with the elevators and held producers accountable. The social consequences of these actions reverberated throughout the heartland as farmers who had invested substantial sums saw life savings evaporate in elevator bankruptcies as neighbors tried to avoid their financial obligations. Others argued that they were perfectly within their rights to delay delivery until the "basis was set" and that could be delayed for several crop years. The wreckage is still being sorted out and the Commodity Futures Trading Commission, as usual, has not been helpful.

Since the hedged-to-arrive contract is predicated on the farmer's planned delivery period, a future's price may be established up to 18 months in advance of actual delivery. This greatly increases the producer's opportunity to select a profitable price by extending the market season far into the future. By understanding month-to-month spreads in futures contracts and basis appreciation, a producer can use this marketing tool to capture basis returns as well as price level protection.

Agricultural Options[7]

In October 1984, trading of agricultural options was reinstated after a 50-year ban. They provide another marketing tool for farmers.

In contrast to a futures contract, an option is a contract that conveys the right, *but not the obligation*, to the option buyer to buy or sell the underlying commodity (either the physical commodity or the futures contract) at a specified price (the strike price) on or before a specified date of expiration. The person who grants the option buyer this right to buy or sell is known as the option writer. The writer receives a nonrefundable lump-sum payment, called a premium, as compensation.

There are two types of options: a PUT option and a CALL option. A PUT option conveys to the holder the right to *sell* at the strike price on or prior to the expiration date. A farmer can "put" the crop on the market at the stated price. A CALL option conveys the right to *buy* at the strike price on or prior to the expiration date. A feeder can "call" feed corn, for example, from the market at a stated price.

An alternative to hedging in the futures market is to hedge with a PUT option. A PUT option allows short hedgers, such as corn producers, to establish a floor, or minimum selling price, but it does not eliminate the upside price potential. A short hedger buys a PUT option by paying a premium, much like an insurance premium. There is no margin account or margin money, per se, required to hold an option. For the premium paid, the buyer has the right to receive a short futures contract at a predetermined price, referred to as a "strike price." The short hedger would exercise the right if price declined. The hedger thus sets a price floor.

There are also options for buyers of commodities known as CALL options. CALL options are used by long hedgers, or those needing feed supplies or feeder livestock. For the premium paid, they have the right to receive a long futures contract at the pre-stated exercise price. They exercise that right if input costs rise. They thus establish a price ceiling for the cost of feed and feeder livestock inputs.

To illustrate the components common to all options, consider a Call option that conveys the right to purchase a tract of farmland: Mr. Sellit offers Mr. Buyit the right to buy 100 acres at $1,000 an acre any time between now and July 1. For this right, Mr. Buyit pays Mr. Sellit an option fee of $2,500.

In options terminology, the specific 100-acre tract is the underlying commodity. The $1,000-per-acre purchase price is the strike price. July 1 is the expiration date. And the $2,500 paid by Mr. Buyit is the premium. At any time prior to the expiration date, Mr. Buyit can buy the underlying commodity at the option strike price, but Mr.

Buyit is not obligated to do so. That is, Mr. Buyit can exercise the option if he chooses. If Mr. Buyit decides not to exercise, he forfeits his premium and allows the option to expire worthless. Or Mr. Buyit might sell his option right to someone else. In any event, Mr. Sellit keeps the $2,500 premium.

In using options in farmer marketing, none of the business events of delivery, title exchange, or settlement are fixed by the transaction. Merely the right to receive either a short or long futures position is conveyed, should the buyer choose to exercise the option. As the name suggests, that right does not impose an obligation. Hence, options allow a farmer to fix a favorable price without being obligated to accept it. If price improves, the buyer can walk away from the original option and capture the better price being offered by the cash market. In that event, the maximum cost to the option holder is limited to the original option premium paid.

A farmer who wishes to protect the value of the growing soybean crop can purchase a PUT option as price insurance. A soybean option designated as a "Nov. 750 PUT" gives the option buyer the right to sell (go short) a November soybean futures contract (5,000 bushels) at a strike price of $7.50 a bushel at any time prior to expiration, regardless of what the futures price is at the time. Purchasing PUT options provides a means of obtaining protection against declining prices without giving up the opportunity to profit from a price increase. In essence, PUT options establish a price floor. The expected price is calculated by subtracting the option premium and expected basis from the strike price ($7.50 strike, − $0.75 premium, − $0.25 anticipated basis, = $6.50 anticipated cash price floor). If the futures price level declines to $7.00 just prior to expiration, the option could have a value of $0.50. Conversely, if the futures price rises to $8.50 prior to expiration, the option would be worthless and the farmer would let it expire without exercising it. The right to sell soybeans for $7.50 when the market is offering $8.50 makes the option worthless.

The buyer of a futures option, unlike the buyer of a futures contract, does not have to worry about daily margin calls. The buyer must instead pay an option premium at the time of purchase. Thus the option holder can weather a period of adverse price moves in the expectation that the position will eventually become profitable. If this expectation remains unfulfilled at the time of expiration, the buyer has the alternative to buy another option by paying another premium and thereby effectively "*meet another margin call.*" Options are, therefore, not cost-free. Using them to hedge production over a period of time may result in higher costs than meeting margin calls on futures contracts.

There are some additional elements of options that must be considered. Options are available for each of the specific futures contract traded. For example: corn has futures and options contracts for January, March, May, July, September, and December. However, options expire about a month before the underlying futures contract.

Options strike prices are listed in specific price increments. Corn and wheat are in 10 cent increments, while soybeans are in 25 cent increments. Options are traded in increments above and below the current futures level. Additional levels, up or down, will be added as market fluctuations warrant. If the July corn future is currently trading at $2.53, PUT and CALL options will be listed at $2.60, $2.70, $2.80, and above. Additional options are available at $2.50, $2.40, $2.30, and lower.

The cost of purchasing an option, called the option premium, consists of two components: **intrinsic value** and **time value.** Intrinsic value is the "built-in" value of the option. It is the difference between the option strike price and the underlying futures price ($2.50 strike price – $2.40 futures). The holder of the right to sell corn at

$2.50, while the market is trading at $2.40 enjoys a "built-in" value of 10 cents. This situation is referred to as an "In-the-Money" PUT. If December corn future were trading at $2.50, the $2.50 PUT would have **no** intrinsic value. Then it is referred to as an "At-the-Money" PUT. Were the market price of corn at $2.60, the $2.50 PUT would have no intrinsic value and would be dubbed an "Out-of-the- Money" PUT.

Time value is equal to the option premium minus the intrinsic value. If in July, a December $2.50 corn PUT sold for 34 cents while the December corn futures traded at $2.40, the PUT would have a 24-cent time value (34 cent premium – 10 cent intrinsic value = 24 cent time value). At or out-of-the-money options have no intrinsic value, but they **always have some time value.** For options with no intrinsic value, the entire premium equals its time value. The structure of strike prices and terminology is summarized in Figure 2–10.

Underlying December Corn Futures at $2.40

Strike Price	Puts		Calls	
$2.20	0.04 ⎫	Out-of-the-money	0.27 ⎫	In-the-money,
2.30	0.07 ⎭	No intrinsic value	0.20 ⎭	Intrinsic value
2.40	0.11 ⎫	At-the-money	0.15 ⎫	At-the-money
		No intrinsic value		No intrinsic value
2.50	0.34 ⎫		0.11 ⎫	
2.60	0.48 ⎬	In-the-money	0.08 ⎬	Out-of-the-money
2.70	0.59 ⎭	Intrinsic value	0.06 ⎭	No intrinsic value

FIGURE 2–10: Structure of Options, Strike Prices, Premiums, and Terminology.

Options are traded at the exchange via open outcry in the pits in the same manner that futures are traded. The value of the premium is determined in the open market by the willingness or reluctance of buyers and sellers to enter into the agreement, just as a mutually satisfactory price is arrived at in the adjacent futures pits. In a volatile market, trading at high or low price extremes, premiums will be higher. In a quiet, steady market at mid-price levels, premiums are lower. Furthermore, premiums are higher for more distant trading months, compared to nearby options; more events can occur in the long-run.

Summary

Various marketing tools allow producers to effect different marketing decisions. Different tools enable them to transact each business event at the most opportune time. Pricing, delivery, title transfer, and settlement need not occur simultaneously. Through a judicious use of these tools, each of the business events can be transacted at the most opportune moment over a broad span of time.

Time Frame	Tool	Pricing		Delivery	Title Exchange	Payment	Speculation			
		Price Level	Basis Level				Price	Basis	Quality	Quantity
Prior to harvest	Cash Contract[1]	Fixed	Fixed	Later	Later	Later	No	No	Yes	Yes
	Futures Contract[1]	Fixed	Open	Later	Later	Later	No	Yes	Yes	Yes
Harvest and post-harvest	Cash sale[1]	Fixed	Fixed	Now	Now	Now	No	No	No	No
	On-farm Storage	Open	Open	Later	Later	Later	Yes	Yes	Yes	No
	Commercial Storage	Open	Open	Now	Later	Later	Yes	Yes	No	No
	Delayed Pricing	Open	Open	Now	Now	Later	Yes	Yes	No	Yes[2]
	Delayed Payment[1]	Fixed	Fixed	Now	Now	Later	No	No	No	Yes[2]
	Basis Contract	Open	Fixed	Now	Now	Partial now; remainder later	Yes	No	No	No
	Cash Sale Plus Buy Futures[1]	Open	Fixed	Now	Now	Now on cash; later on futures	Yes	No	No	No

[1]Alternative tools also available to livestock producers to sell feeder cattle, fat cattle, and hogs.
[2]Payment is at risk.

FIGURE 2–11: Summary of Marketing Tools and Their Respective Business Transactions.

Distinguishing between the market price level and basis price level is important in determining which tool to use at the appropriate time.

The market pricing alternatives available to farmers and their respective pricing commitments are summarized in Figure 2–11. Grouping marketing tools under two kinds of strategies should help explain what they accomplish. For example, to fix price level, a farmer might use a cash sale, a forward contract, or a hedge. To speculate on price level, a farmer could store unpriced grain, use the delayed pricing contract, use the basis contract, or sell cash/buy futures. These four alternatives will hold grain for speculation. The proper time to use each tool will be discussed as part of strategy considerations in Chapter 6.

Questions for Study and Discussion:

1. Name and discuss the four business events of commodity transactions.

2. Discuss the difference between market price level and basis price level.

3. List the marketing tools of agricultural marketing. What price is fixed by each tool and what is left open to fluctuation? What speculation is removed with each tool?

4. Why are grain moisture considerations important in cash sales?

5. Why is delayed pricing a risky marketing tool?

6. Explain the reluctance of farmers to forward price production prior to harvest. Is the reluctance justified?

7. How does futures contracting differ from other tools as a marketing device?

8. What two methods may be used to satisfy a futures contract obligation?

9. What is the purpose of margins in futures trading?

10. Explain parallelism of cash and futures markets. Assume various prices and calculate gains and losses from cash and futures positions.

11. What marketing tool(s) would apply to each of these conditions:

    a.) Drought reduced crop, high prices, strong basis before harvest?

    b.) Harvest rolling in, huge crop, long lines at elevator, very weak basis, low July futures price?

    c.) Normal crop, weak basis, somewhat profitable July futures price?

    d.) Bonanza high price levels that occur once or twice in a decade?

    e.) Early January snowstorm slows railroads and clogs highway traffic. Basis is very strong, low to moderate July futures.

        1) Not using futures?

        2) Fluent using futures?

    f.) Prior to harvest, sufficient profitable futures, weak basis?

    g.) At harvest, no storage facilities, low carry in the futures contracts, possible strong export demand later in the year?

    h.) Prefer cash receipts next tax year.

        1.) Financially strong buyer?

        2.) Questionable buyer?

    i.) Cost of producing corn $2.50 per bushel. July futures $2.80. Banker insists on risk protection, build-up in livestock herds indicate strong feed demand next year?

    j.) As a high-volume poultry integrator, you wish to base next year's production planning on current low feed prices?

Exercises for Using a Standard Grain Shrinkage Table:

1. Compare value of shrink (grain price times percent shrinkage, from Table 2–3) against the discount charged by your local elevator. For example, 21 percent corn dried to 15.5 percent has 7.01 percent true shrinkage worth 21 cents per bushel on $3.00 corn ($3.00 x .070). By comparison, an elevator might discount 21 percent moisture grain by 44 cents.

2. Calculate the cost of over-drying corn. How much value loss occurs by taking grain from 15.5 percent to 13 percent? Add to the value loss the extra fuel cost required for extra drying. The sum of both costs is the cost of over-drying.

Endnotes for Chapter 2:

1. USDA, *The Official United States Standards for Grain* (Washington, D.C.:USDA, Jan., 1978).

2. A complete analysis of the economics of drying corn and soybeans is presented in Gene A. Futrell and Robert N. Wisner, eds., *Marketing for Farmers* (Doane Information Services, 11701 Borman Drive, St. Louis, Mo. 63146), 1982.

3. Standard Federal Tax Reports, "Tax Accounting Installment Method Prepaid Income," Vol. 4 (Chicago: Commercial Clearing House, Inc., 1985), paragraphs 2887L.025, 2887P.01.

4. _____, *Corn and Soybean Delivery Terms*, [After January 1, 2000], Chicago Board of Trade, 141 Jackson Blvd., Suite 2210, Chicago, IL 60604-2994, 1998.

5. _____, *CME Rulebook*, Chapter 16: Lean Hogs, Chapter 23 Feeder Cattle, Chicago Mercantile Exchange, Chicago, IL, November 1999.

6. _____, *Live Cattle Approved Stockyards, Commission Firms, and Slaughter Plants*, Chicago Mercantile Exchange, Chicago, IL, November 1999.

7. _____, *Selling Strategies for Crops with Options*, Chicago Board of Trade, 141 Jackson Blvd., Suite 2210, Chicago, IL, January 1997.

TABLE 2–3: Grain Shrinkage Chart (Standard Moisture Table)[1]

Initial Moisture	13.0%	13.5%	14.0%	14.5%	15.0%	15.5%	16.0%	16.5%	17.0%	17.5%	18.0%	18.5%	19.0%
Percent						*(percentage of shrinkage)*							
15.5	3.37	2.81	2.24	1.67	1.09	—	—	—	—	—	—	—	—
16.0	3.95	3.39	2.83	2.25	1.68	1.09	—	—	—	—	—	—	—
16.5	4.52	3.97	3.41	2.84	2.26	1.68	1.10	—	—	—	—	—	—
17.0	5.10	4.55	3.99	3.42	2.85	2.28	1.70	1.10	—	—	—	—	—
17.5	5.67	5.12	4.57	4.01	3.44	2.87	2.29	1.70	1.11	—	—	—	—
18.0	6.25	5.70	5.15	4.59	4.03	3.46	2.88	2.30	1.71	1.11	—	—	—
18.5	6.82	6.28	5.73	5.18	4.62	4.05	3.48	2.90	2.31	1.72	1.11	—	—
19.0	7.40	6.86	6.31	5.76	5.21	4.64	4.08	3.50	2.91	2.32	1.72	1.12	—
19.5	7.97	7.44	6.90	6.35	5.79	5.23	4.67	4.10	3.52	2.93	2.33	1.73	1.12
20.0	8.55	8.01	7.48	6.93	6.38	5.83	5.27	4.70	4.12	3.54	2.94	2.35	1.74
20.5	9.12	8.59	8.06	7.52	6.97	6.42	5.86	5.30	4.72	4.14	3.55	2.96	2.36
21.0	9.70	9.17	8.64	8.10	7.56	7.01	6.46	5.89	5.32	4.75	4.16	3.57	2.97
21.5	10.27	9.75	9.22	8.69	8.15	7.60	7.05	6.49	5.93	5.35	4.77	4.19	3.59
22.0	10.84	10.33	9.80	9.27	8.74	8.19	7.65	7.09	6.53	5.96	5.38	4.80	4.21
22.5	11.42	10.90	10.38	9.86	9.32	8.78	8.24	7.69	7.13	6.57	5.99	5.40	4.83
23.0	11.99	11.48	10.97	10.44	9.91	9.38	8.84	8.29	7.73	7.17	6.60	6.03	5.44
23.5	12.57	12.06	11.55	11.03	10.50	9.97	9.43	8.89	8.34	7.78	7.21	6.64	6.06
24.0	13.14	12.64	12.13	11.61	11.09	10.56	10.03	9.49	8.94	8.38	7.82	7.25	6.68
24.5	13.72	13.22	12.71	12.20	11.68	11.15	10.62	10.09	9.54	8.99	8.43	7.87	7.30
25.0	14.29	13.79	13.29	12.78	12.26	11.74	11.22	10.68	10.14	9.60	9.04	8.48	7.91
25.5	14.87	14.37	13.87	13.37	12.85	12.33	11.81	11.28	10.75	10.20	9.65	9.09	8.53
26.0	15.44	14.95	14.45	13.95	13.44	12.93	12.41	11.88	11.35	10.81	10.26	9.71	9.15
26.5	16.02	15.53	15.03	14.54	14.03	13.52	13.00	12.48	11.95	11.41	10.87	10.32	9.76
27.0	16.60	16.11	15.62	15.12	14.62	14.11	13.60	13.08	12.55	12.02	11.48	10.93	10.38
27.5	17.17	16.69	16.20	15.71	15.21	14.71	14.20	13.68	13.16	12.63	12.09	11.55	11.00

[1]Formulas: (1) $\text{Shrinkage} = 100.0\% - \left(\dfrac{\% \text{ dry matter in wet grain}}{\% \text{ dry matter in dry grain}} \times 100 \right) + 0.5\% \text{ handling shrink.}$

(2) Value of shrink = Price basis grade × shrinkage.

(3) Returns to drying = discount – value of shrinkage.

TABLE 2–4: Grades, Grade Requirements, and Grade Designations for Soybeans

The following grades, grade requirements, and grade designations are applicable under these standards:

Grade	Minimum Test Weight per Bushel	Moisture	Splits	Damaged Kernels		Foreign Material	Brown, Black, and/or Bicolored Soybeans in Yellow or Green Soybeans
				Total	Heat-Damaged		
	(lb.)	------------------------------				------ *(%)* ------	
U.S. No. 1	56.0	13.0	10.0	2.0	0.2	1.0	1.0
U.S. No. 2	54.0	14.0	20.0	3.0	0.5	2.0	2.0
U.S. No. 3[1]	52.0	16.0	30.0	5.0	1.0	3.0	5.0
U.S. No. 4[2]	49.0	18.0	40.0	8.0	3.0	5.0	10.0
U.S. Sample grade	U.S. Sample grade shall be soybeans which do not meet the requirements for any of the grades from U.S. No. 1 to U.S. No. 4, inclusive; or which are musty, sour, or heating; or which have any commercially objectionable foreign odor; or which contain stones; or which are otherwise of distinctly low quality.						

Source: USDA, *The Official United States Standards for Grain* 8.3.

[1]Soybeans which are purple mottled or stained shall be graded not higher than U.S. No. 3.

[2]Soybeans which are materially weathered shall be graded not higher than U.S. No. 4.

TABLE 2–5: Grades, Grade Requirements and Grade Designations for Wheat

The following grades, grade requirements, and grade designations are applicable under these standards:

Grade	Minimum Test Weight Per Bushel (pounds)		Percent Maximum Limits of—					Wheat of Other Classes[3]	
	Hard Red Spring Wheat or White Club Wheat	All Other Classes and Subclasses	Heat-Damaged Kernels	Damaged Kernels (Total)[1]	Foreign Material	Shrunken and Broken Kernels	Defects (Total)[2]	Contrasting Classes	Wheat of Other Classes (Total)[4]
U.S. No. 1	58.0	60.0	0.2	2.0	0.5	3.0	3.0	1.0	3.0
U.S. No. 2	57.0	58.0	0.2	4.0	1.0	5.0	5.0	2.0	5.0
U.S. No. 3	55.0	56.0	0.5	7.0	2.0	8.0	8.0	3.0	10.0
U.S. No. 4	53.0	54.0	1.0	10.0	3.0	12.0	12.0	10.0	10.0
U.S. No. 5	50.0	51.0	3.0	15.0	5.0	20.0	20.0	10.0	10.0

U.S. Sample grade   U.S. Sample grade shall be wheat which:

(1) Does not meet the requirements for the grades U.S. Nos 1, 2, 3, 4, or 5; or

(2) Contains a quantity of smut so great that 1 or more of the grade requirements cannot be determined accurately; or

(3) Contains 8 or more stones, 2 or more pieces of glass, 3 or more crotalaria seeds (*Crotalaria* spp.), 3 or more castor beans *(Ricinus communis)*, 4 or more particles of an unknown foreign substance(s), or 2 or more rodent pellets, bird droppings, or an equivalent quantity of other animal filth per 1,000 g of wheat; or

(4) Has a musty, sour, or commercially objectionable foreign odor (except smut or garlic odor); or

(5) Is heating or otherwise of distinctly low quality.

Source: USDA, *The Official United States Standards for Grain* 1.7.

[1]Includes heat-damaged kernels.

[2]Defects (total) include damaged kernels (total), foreign material, and shrunken and broken kernels. The sum of these 3 factors may not exceed the limit for defects.

[3]Unclassed wheat of any grade may contain not more than 10 percent of wheat of other classes.

[4]Includes contrasting classes.

Chapter 3

The Futures Markets: Focal Point in Commodity Marketing

The Futures Markets: Focal Point in Commodity Marketing

Futures Markets

Historical Evolution of Grain Marketing[1]

Agricultural commodity marketing, as we know it, did not just suddenly happen, it evolved over a considerable period of time along with the industrial and financial development of the nation. Removal of the Indian threat after the War of 1812 opened the Northwest Territory to rapid settlement. Pioneers found vast and rich resources that yielded bountiful production beyond the needs of the pioneer family. As they sought to sell the surplus, farmers brought their wagonloads of grain across the prairie on plank roads to the log cabin village of Chicago. Canal boats and barges loaded with grain converged on the town via the river system. When navigating the Great Lakes was possible, vessels crowded the Chicago harbor to take grain east for milling and processing, or to connect with boats on the Erie Canal for export.

After reaching Chicago, the crossroads of supply and demand, the farmer still had to haul the grain from store to store until a buyer could be found. Sometimes the farmer found willing buyers, sometimes not. During harvest, wagon lines were often 20 miles long. Excess grain was often dumped on the ground or in inadequate storage facilities. Spoiled grain was hauled onto Lake Michigan and dumped. Yet before the next harvest, supplies were short again. The farmer was at the mercy of severely fluctuating prices; either too much grain was received, dropping prices drastically for latecomers, or deliveries were too small, causing prices to skyrocket. Prices of consumer goods, such as bread, flour, and meal, rose and fell with the tide of raw materials. Sometimes products disappeared from the market entirely. Merchants safeguarded themselves against price change by offering less and less for raw commodities and charging more and more for the finished product.

With an ever-increasing volume of grain pouring into Chicago, old trade practices become inadequate. Standard grades had not been established. Lack of standard bushel weights fostered unfair practices in measuring. Storage facilities were inadequate. The only contracts were receipts from warehouses for grain of dubious quantity and quality. Bitter disputes often flared between buyer and seller.

Out of this chaos, one thing became clear—expanding production required a large, liquid, year-round central market for sellers and buyers of the new wealth of the prairie. The Chicago Board of Trade was founded in 1848 by 82 businessmen as a cash market. It was set up as a central meeting point in town for buyers and sellers. Trades could be made on the spot, hence the name "spot market." The stated purposes of the Board of Trade were "to maintain a commercial exchange; to promote uniformity in the customs and usages of merchants; to inculcate principles of justice and equity to trade; to facilitate the speedy adjustment of business disputes; to acquire and disseminate valuable commercial and economic information; and generally to secure to its members the benefits of cooperation in furtherance of their legitimate pursuits."[2]

One of the first activities of the Board of Trade was to appoint a committee to determine the quality of commodities being handled by members. Standardized definitions of grade quality, accepted everywhere in the world, were developed. Inspectors were appointed to grade incoming grain in conformity with Exchange standards. The method of buying grain by measured bushels was replaced by a weight system, and a weighing department was inaugurated to assess accuracy and integrity. In 1859, the state of Illinois legislatively set quality and quantity standards for grain trading. A governmental agency employed unbiased inspectors to make judgements on quality and enforce the standards. At this point, a grade designation such as U.S. No. 2 yellow corn meant the same throughout the trade, and grain could be bought and sold by description.

By 1860, forward contracting for future delivery of grain came into existence. Forward contracts were a vehicle for the seller and buyer to agree upon quality, quantity, and price for delivery of a commodity at a later date. Merchants located downstate along the Illinois River would buy surplus grain from local farmers at harvest, store it in their facilities, and forward contract its later sale to Chicago merchants. Forward contracts became the security instrument for financing grain held in storage.

Later, when river merchants brought grain to town, there was often a short supply and high spot price and sellers would not honor their earlier agreement. On the other hand , if there was a large supply and low price, buyers would default. The story is told of grain merchants going to Chicago with grain on the back of a wagon and a shotgun on the front. They would find the buyer and ask, "Which do I unload first?" Solution to the default problem was achieved with earnest money deposits by both buyer and seller, placed with a disinterested third party until delivery was made. These were the first margin deposits. Forward contracting, as a method of pricing, developed rapidly after adoption of this practice. Grain merchants were the first hedgers. They locked in a price to be received at time of delivery irrespective of whether cash market price rose or fell in the interim. By knowing their selling price, grain merchants carried less risk. They were willing to bid a relatively higher price to farmers for their grain.

As the practice of contracting for forward sales expanded, some buyers found that they had overbooked, while others were short of commitments. An active market for trading the forward sale contracts developed in order to match needs with availability; this practice laid the cornerstone of a futures contract market.

Trading in forward contracts led to two other practices. First, a clearinghouse was established to match buyers and sellers. Previously, merchants bringing grain into town to satisfy forward commitments had difficulty locating the buyer, since the forward contract might have been sold two or three times. The seller would be sent from place to place to find the holder of the contract. The clearinghouse provided a centralized clearing point to accept grain from the seller and match deliveries with contract holders. Warehouse receipts, as evidence of title, also came into being. With them, the grain merchant could deliver the warehouse receipt to the clearing corporation. In turn, the clearinghouse personnel knew who was to receive the grain; they matched the supplier with the buyer.

The second practice that evolved was speculative buying and selling of contracts. Individuals found they could buy excess contracts from an overextended buyer and later resell them to someone who was short of grain. Also, speculators soon learned they could sell contracts to people who needed grain at a later time and subsequently buy the cash commodity at, hopefully, a lower price to fulfill their contract commitment.

Maturity dates of the forward contracts and futures contracts changed gradually. Instead of contracting for delivery in 10, 30, or 90 days in the future, more distant delivery months were gradually agreed upon. Maturity dates for grain contracts generally corresponded with important production or marketing months. July was the first month for cash trading newly harvested winter wheat. September was the harvest month for spring wheat. By December, final crop size was known and most of the demand factors for the year were known. December was also the last month of shipping via the Great Lakes until the spring thaw in March. December was the first month that the new corn crop was available at the terminals. March related to the reopening of Great Lakes shipping. During May, traders could estimate carryover and assess spring planting intentions. July figures reflected rate of use of old crop and provided a view of summer growing conditions. September was the last old crop month.

Soybean trading months evolved somewhat differently. At one time, November was the month the new crop was available, and September was the last old crop month. Harvest patterns changed to make October the time when the bulk of the new crop was available. Consequently, September became a combination of old and new crop trading forces, and July became the end of the old contract year. July proved to be too far from harvest and allowed for extensive supply fluctuation due to summer growing conditions. Finally, August was added as a soybean trade month.[3]

The business practices of forward contracting a future cash transaction, margin deposits to insure financial performance, clearinghouse matching of buyers and sellers, speculative trading, and fixed months of contract delivery paved the way for trading of futures contracts. By 1865, and the end of the Civil War, trading in futures contracts had become standard business practice. With forward contracts, every detail of delivery, quality, quantity, and price was negotiable. With futures contracts, all but one variable, price, was standardized. It was negotiated in open auction.

Two points can be gleaned from this historical perspective. First, it took many years for the futures market and its business operating practices to evolve. Implementation of one practice led to development of another. Second, futures markets became important to farmer-sellers on one end of the food chain—and consumer-buyers on the other end. Futures pricing tended to minimize price fluctuation by shifting the business practice of matching producers' supplies to the needs of users from trade people into the hands of specialists. Trading in grain movement had much less risk of financial loss. Farmers received higher prices; consumers enjoyed lower costs. The system worked more efficiently.

Exchanges

By 1865, the basic structure of the Chicago Board of Trade was firmly established as we know it today. Out of the use of forward sales, the futures market, with its specific futures contracts, developed early in the history of the Chicago Board of Trade. The CBOT provided futures markets in wheat, corn, oats, soybeans, soybean oil and meal, iced broilers, plywood, silver, and gold. In 1975, the CBOT launched the first financial instruments futures contract in Government National Mortgage Association (GNMA) mortgaged-backed certificates. In 1977, two additional interest rate-

Information on Contributors and Exchange Affiliates

Amex Commodities Corporation (ACC)
86 Trinity Place
New York, NY 10006
(212) 306–1419

Cantor Financial Futures Exchange (CFFE or CX)
One World Trade Center
101st Floor
New York, NY 10048
(212) 938–3548
cx.cantor.com

Chicago Board of Trade (CBOT)
141 West Jackson Boulevard
Chicago, IL 60604
(312) 435–3500
comments@cbot.com
www.cbot.com

Chicago Mercantile Exchange (CME)
30 South Wacker Drive
Chicago, IL 60606
(312) 930–1000
info@cme.com
www.cme.com

Citrus Associates of the New York Cotton
    Exchange, Inc. (CACE)
Four World Trade Center
New York, NY 10048
(212) 742–5054

Commodity Exchange, Inc. (CEI or Comex)
1 North End Avenue
New York, NY 10282
(212) 299–2000
marketing@nymex.com
www.nymex.com

Kansas City Board of Trade (KCBT)
4800 Main Street
Kansas City, MO 64112
(816) 753–7500
(800) 821–5228
kcbt@kcbt.com
www.kcbt.com

MidAmerica Commodity Exchange
    (MACE or MidAm)
141 West Jackson Boulevard
Chicago, IL 60604
(312) 341–3000
dbed50@cbot.com
www.midam.com

Minneapolis Grain Exchange (MGE)
130 Grain Exchange Building
400 South 4th Street
Minneapolis, MN 55415
(612) 338–6212
mgex@ixnetcom.com
www.mgex.com

New York Board of Trade (NYBOT)
Four World Trade Center #9150
New York, NY 10048
(212) 742–6000

New York Futures Exchange (NYFE)
Four World Trade Center
New York, NY 10048
(212) 742–5054
marketing@nycc.com
www.nycc.com

New York Mercantile Exchange
    (NYME or NYMEX)
1 North End Avenue
New York, NY 10282-1101
(212) 299–2000
marketing@nymex.com
www.nymex.com

Philadelphia Board of Trade (PBOT)
1900 Market Street
Philadelphia, PA 19103
(215) 496–5000
(800) 843–7459
info@phlx.com
www.phlx.com

FIGURE 3–1: Futures Markets in the United States.

sensitive markets, U.S. Treasury bond futures and 90-day commercial paper loan futures, were launched.

Other exchanges developed in other cities representing major transportation centers through which a substantial portion of commodities moved. At the beginning of the century, eleven organized commodity exchanges were trading commodity futures contracts in the United States. Figure 3–1 lists the exchanges and their locations. The Chicago Board of Trade and the Chicago Mercantile Exchange were the largest, accounting for approximately 75 percent of the volume of all commodity futures trading.

The MidAmerica Commodity Exchange differs from the other exchanges and offers particular advantages to the farmer. Traders on the MidAmerica floor watch the price quote boards reporting price action from other exchanges. Their trading among themselves is at premiums or discounts to the prices on the other exchanges, and their contract sizes are smaller. Their activity is not concerned with price discovery since they reference their trades to a quoted price, but they provide market flexibility in contract size. MidAmerica offers 1,000-bushel contracts in corn, soybeans, and wheat, which are one-fifth the size of grain contracts on CBOT. It also offers a 15,000-pound lean hog contract (about 70 head) and a 20,000-pound cattle contract (about 20 head). These are one-half the size of similar contacts traded on the Mercantile Exchange. The advantage for producers is that the mini-contracts offer the opportunity to trade in smaller units. A smaller farmer is not eliminated from market participation. Larger operators can more closely tailor market positions with actual quantities of production in 1,000-bushel units compared to 5,000-bushel units. Finally, farmers can execute their pricing decisions in more separate transactions with smaller units.

Commodity Exchanges conform to the definition of a market: They provide a place for trading, and they facilitate exchange. The exchanges are organized, for the most part, as not-for-profit corporations of member-traders authorized to buy and sell futures contracts for the public. Each exchange has a trading floor where buyers and sellers meet; a governing board sets and enforces rules of orderly trading; and a clearinghouse facilitates trading and delivering commodities.

The organization of exchanges has been compared to a football game by T. A. Hieronymus: "The commodity exchanges provide the playing field and equipment; they write the rules; and they act as referee, head linesman and field judge."[4] But they do not handle the football. The players in the action are the hedgers and speculators.

What Is a Commodity?

The list of commodities traded on exchanges is extensive, diverse, and subject to change. Commodities are natural products, agricultural and manufactured products, currencies and financial instruments. A list at the end of this chapter provides detailed information on actively traded commodities and their futures contract specifications.

In order to qualify for futures trading, a commodity must possess the following characteristics:[5]

1. It must be a homogeneous product such that every lot is the same and units are interchangeable.

2. It must be susceptible to standardization and grading so that well defined, uniform grades become essentially homogeneous.

3. Supply and demand must be large enough so that no single person can control flow in and out of the market.

4. Supply to the market must flow naturally and freely without artificial restraints from government or private agencies.

5. Supply and demand must be uncertain and constantly changing. If supply and demand were both certain, prices could be adjusted without an organized market. However, if supply and demand are large and both are uncertain and subject to wide fluctuations from year to year and season to season, then the forces of supply and demand will be in a constant state of flux. This interplay of uncertain economic forces produces the constant price fluctuations which must exist in a successful futures market.

6. The commodity must be storable in order to allow for delivery on futures contracts that are due months after harvest or production. Any item which deteriorates rapidly would not meet this condition.

Commodities that are not adaptable to futures trading include (1) perishable commodities, such as fresh fruits and vegetables, (2) manufactured goods whose supply can be controlled and whose styles change, and (3) goods where supply has been monopolized by a few producers.

Obviously, livestock and livestock products do not satisfy all of these requirements. They are perishable and cannot be easily stored. Futures market trading in these commodities creates a different set of conditions and requires that a farmer or rancher gain a specialized understanding of the market for such goods. The peculiarities of futures trading in livestock will be discussed in Chapter 9.

Roles of Futures Markets

Commodity futures markets have three major roles to fulfill:

1. Price Discovery
2. Risk Transfer
3. Investment Medium

Exchanges provide the physical facilities and business framework for buyers and sellers to conduct trading. Worldwide demand and supply conditions flow into the trading pits and are interpreted by traders. Prices are discovered by the interaction of buyers and sellers.

Two types of people participate in the process—hedgers and speculators. Hedgers seek to transfer the effects of price change from their business, and speculators seek to profit from price change. Thus, roles of risk transfer and investment medium are fulfilled.

Futures Trading and Price Discovery[6]

What an item is worth is an age-old question of trade. Sellers obviously would like price to be high; buyers want it to be low.

Somehow a mutually agreeable compromise must be reached so that both are satisfied with the transaction. In an immediate cash trade, value of an item becomes what the buyer is willing to pay and the seller is willing to take. At the time, both the Indians and the Dutch were obviously satisfied with their trade of $24 worth of trinkets for Manhattan Island. But values do change.

In a barter economy, price is established by haggling. In a more complex economy it is established by the buyer's willingness or reluctance to part with his or her money in trade for the item and the seller's willingness to take the money in exchange for the goods.

Assessing a futures price is more complex than assessing a cash price. The cash price reflects "what is" today, but a futures price reflects market conditions as traders anticipate they will be. Because of the time lag between pricing and delivering the commodity, and the uncertainties of events in the interim, wide differences of opinion always exist about the "correct" futures price. The differences of opinion and unpredictable events insure that there will be plenty of buyers (who think price will go up) and sellers (who think it will go down) willing to participate in a futures market.

A trader in the futures market operates something like this: (1) The trader determines what he or she believes is the "correct" price for a commodity in some delivery month in the future, based upon knowledge and interpretation of expected demand and supply conditions. (2) If the current price of the futures contract the trader intends to trade is higher than the expected "correct" price, he or she "sells" (goes short) a contract obligation to deliver the commodity at the future date. The trader intends to purchase the commodity to satisfy the contract after the price has come down to the "true" supply-demand price and make delivery to fulfill the legal obligation. (3) If the current price is lower than the expected "correct" price, the trader "buys" (goes long) a futures obligation with intentions of selling it when the price rises to the "correct" level. If the trader's judgement is correct and price changes as expected, he or she profits from the difference between the original and final prices. This profit represents a loss to some other trader whose judgement was wrong.

The actions of traders lead to establishment of the "true" supply-demand price for the commodity as they express their collective judgements. The trader who sells the futures contract, believing that the current price is too high, tends to lower price in the direction that it should go if this judgement is correct. The action of the trader who buys the futures contract tends to raise the price toward what it should be if this judgement is correct. The buying and selling transactions do not balance each other out, even though every sale has a corresponding purchase. Obviously, if they did, the price would not change. For a trade to cause a change in price, that trade must be a reflection of the weighted judgement of market participants on their willingness to buy or sell. If buyers are more aggressive than sellers, the price will move upward; if sellers are more aggressive, the price will move downward. In terminology of the market, bear and bull markets develop in response to traders' actions. Bull markets are made by aggressive buyers ("bulls") who are willing to charge on to higher and higher prices. Bear markets are made by sellers ("bears") who want to retreat into hibernation away from stronger price action. An aggressive buyer must entice a reluctant seller to enter into a transaction by offering a slightly higher price than presently exists. This action thus bids up price. A seller entices a reluctant buyer into a trade by offering to sell at a slightly lower price. Thus, this action tends to lower price. Either up or down, a price direction develops from a trader's willingness to execute a trade at a price different from the previous one.

After the original "open commitments" of speculative traders have been made, some of them hold "long" positions; that is, they have bought futures commitments that have not been offset by a counterbalancing sale. They are committed to take delivery of the commodity. Other traders are "short" in that they have made commitments not yet offset by a counterbalancing buy. They owe delivery of the commodity. The circumstances under which the traders are willing to dispose of their current holdings intensify the direction of price action once a trend begins to develop. As supply and demand conditions unfold and point to higher prices, the natural reaction of the long-position traders is to maintain, or possibly increase, their open commitments. There are always traders willing to sell, but only at a higher price. How much higher depends upon judgements by both buyers and sellers of the eventual effects of a new development on supply and/or demand conditions. As a stronger market unfolds, the reaction of short sellers is to cancel their previous sale by purchasing a corresponding amount as quickly as possible. Since the short is more anxious to buy than someone else is to sell, the price of futures contract is pushed upward further by the decision of shorts to offset their original. Thus, as price direction is confirmed, it tends to feed upon itself.

More eager buyers than willing sellers will raise prices, transforming some sellers into buyers. More willing sellers than buyers will lower prices, changing some buyers to sellers. Judgements expressed by buyers and sellers cause the current market price of the futures contract to approach the price it should attain, based upon currently available information relating to expected supply and demand conditions in some future delivery month. Thus, a price at which trades are made is discovered.

In the process of price discovery, the commodity exchanges themselves do not buy or sell, nor do they set prices. The role of the exchanges is like a thermometer, which does not influence temperature but merely records it. As hot weather sends the thermometer reading up and cold weather sends it down, strong demand sends price up and heavy supply sends price down. "Through the commodity exchanges, supply, demand, and other price-making forces are translated into a single figure—into a price."[7] An exchange does not set price but simply provides facilities where collective judgements of buyers and sellers are expressed as price; hence: **the role of price discovery**.

How Trading Takes Place[8] The most visible aspects of exchanges are the exchange floor and the trading pits where trading takes place. Less visible, but as important, is the communication system that links brokers throughout the world with traders in the pits. The commodity trading floor is a physical place, but it represents a worldwide market. Up-to-the-second reports of exchange prices are available across the land on video equipment, and most newspapers print daily price quotations of futures market trading.

A picture tour of the Chicago Board of Trade can serve to illustrate the exchange as a market place. Figure 3–2 is a sketch of the trading floor where members of the exchange buy and sell futures contracts in open, competitive bidding. Trading takes place in "pits," which are raised octagonal platforms with descending steps. The shape permits all buyers and sellers to see all others. Traders stand in the pit for the commodity they wish to trade, on steps or in groups according to the delivery month in which they trade.

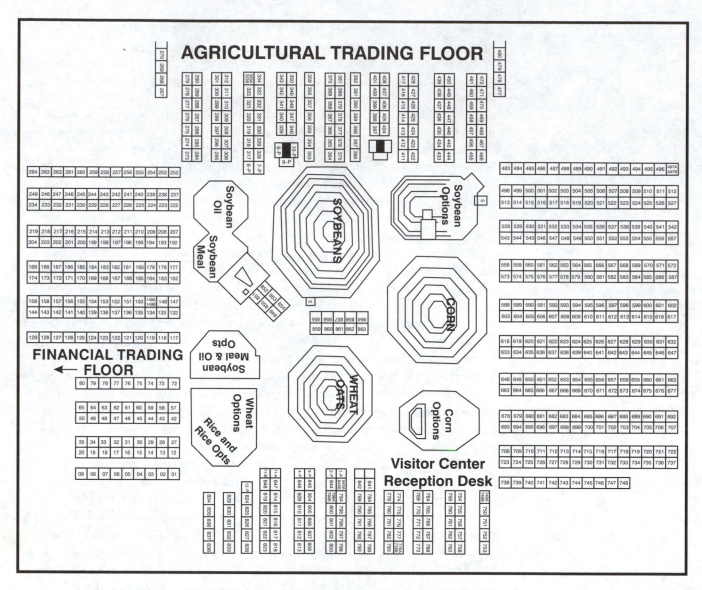

FIGURE 3–2: The Chicago Board of Trade, Trading Floor.

Trading is officially signaled each business day by a gong from the opening bell. No futures trades may be made before it sounds the official opening, nor can any be made after it rings the close. During trading hours, the trading floor is a beehive of activity as traders yell and signal their bids and offers.

At any moment there are thousands of buyers and sellers of futures contracts participating in the market, and there are even more potential participants. Anyone can participate. Contract orders from outside the exchange are relayed through brokers to member traders ("locals") on the floor. Batteries of telephone are located near the pits to handle the thousands of calls from commercial traders and brokerage firms. As orders are received by phone operators, they are time-stamped and rushed by messengers (runners) to traders in that pit. Exchange Rules and Regulations require pit traders to use open outcry to present bids and offers so that anyone who wishes to take the opposite side of the trade may do so. In addition, they use hand signals to clarify their verbal outcry. Position of the palm shows whether they are buying or selling. Traders hold the palm facing themselves when they want to buy and facing away when they wish to sell. For grains, each vertical finger represents one contract of 5,000 bushels. Finger signals with the hand held horizontally relay the fraction of a cent that the trader bids or offers relative to the last trade as posted on the electronic quote boards.

Figure 3–3 shows the one-quarter to full-cent designations of the finger positions in grain trades.

Once a trade has been confirmed with a nod, each trader lists the completed transaction on respective trading cards and the multi-part order form. A carded trade

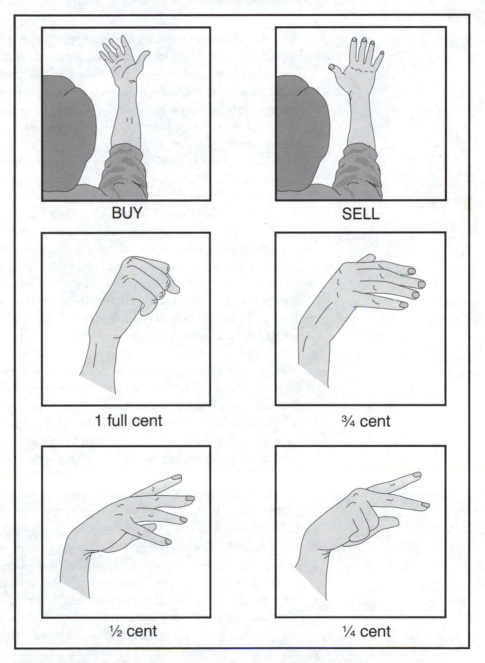

FIGURE 3–3: Hand Signals Used by Traders to Flash Bids and Offers. Source: Besant, *Action in the Marketplace*. Used by permission.

shows the number of contracts involved, the commodity, the maturity month of the contract, the price, the name of the clearing firm, the initials of the other broker, and an alphabetic symbol that indicates the time period in which the trade was executed. The broker jots down the price and other pertinent information on the order form that is tossed from the pit to the messenger for return to the phone desk. The information is then relayed to the office in which the order originated so the customer can be advised that the trade was completed. The whole process often takes less than two minutes between the time of placing the order until confirmation. The ease and speed of transaction reflect the liquidity of the market. Commodity markets lose their effectiveness if they cannot execute trades quickly and at slight price deviations.

Price Quotations. Above each trading pit is a raised platform where two or more Exchange employees (market reporters) constantly watch the changing price picture. As trades are executed at new prices, a reporter records the new price and transmits it to the central quotations computer. The computer flashes the new price on the quotation boards at the exchange and through the communications network to the world. Prices are posted immediately and constitute "the market" until another trade is made at a different price.

Newspapers expand the availability of market information by printing commodity quotations daily. A report from *The Wall Street Journal* in Figure 3–4 summarizes a day's trading activity in agricultural commodities. Similar commodities are grouped together under headings of grains and oil seeds, livestock and meat, food and fiber, metals and petroleum, interest rates, and foreign currencies. A description of each commodity indicates its name, where it is traded, the number of units, and the unit price designation. The month of maturity of the futures contract is noted on the left column. "Open" is the agreed price for the first transaction of the day. "High" is the highest price of any transaction during the day; "Low" is the lowest trading price of the day. "Settle" refers to the settlement price, which is the last trade, if there was only one. If frenzied trading at the closing bell creates a closing range of prices, officials of the exchange determine a settlement price, usually the midpoint of the trading range. "Change" refers to the increase or decrease in settlement price from the previous day's settlement price. "Lifetime high-low" reports the extremes in price at which a specific contract has been traded since its introduction. "Open interest" listed in the right column reports the number of contracts that have been initiated and are still in force for each contract month. "Est. Vol." indicates approximately how many contracts were traded during the trading day reported. "Vol. Tue." (for example) gives the accurate count of previous day's trades. "Open interest" reported at the bottom of the table is the total number of open contracts for all months. The last number, either positive or negative, indicates the change in open interest from the previous day.

The Clearing Corporation. The Clearing Corporation of the exchange handles the mechanics of matching trades between buyers and sellers.[9] Each trader turns endorsed orders of completed trades over to the clearing firm, which in turn reports the trades to the Clearing Corporation. The Clearing Corporation substitutes itself as the opposite party to each transaction. It becomes the buyer to every seller and the seller to every buyer and is, in effect, a party to every transaction. What was initially a sale by A to B becomes a sale by A to the Clearing Corporation. B's purchase becomes a transaction with the Clearing Corporation rather than with A. Because of this substitution of parties, it is not necessary for the buyer (or seller) to find the original

FUTURES PRICES

Monday, January 3, 2000

Open Interest Reflects Previous Trading Day.

GRAINS AND OILSEEDS

CORN (CBT) 5,000 bu.; cents per bu.

	Open	High	Low	Settle	Change	Lifetime High	Lifetime Low	Open Interest
Mar	204½	205¾	200½	200¾	-3¾	270	195¼	211,737
May	211¼	212¼	207¾	208	-3¼	261	202½	61,464
July	217¾	219	214½	214¾	-3¼	278½	209	58,065
Sept	225	225¼	221¾	222	-2	257	215¾	16,869
Nov	229	229	228½	228¾	-2	245	222½	559
Dec	233½	234½	231	231½	-2¼	279½	225¼	33,477
Mr01	242¾	242¾	239¾	239¾	-1¾	246½	233¾	869
Dec	253	253½	251½	252	-1½	263	246½	1,114

Est vol 72,000; vol Fri 0; open int 384,452, +306.

OATS (CBT) 5,000 bu.; cents per bu.

	Open	High	Low	Settle	Change	Lifetime High	Lifetime Low	Open Interest
Mar	110	110	107	108¼	-1	135	107	7,020
May	114¼	114¾	113½	113¾	-1	133	112¾	2,759
July	112¾	113	111¾	111¾	-¾	124¼	110½	1,466
Sept	117¼	117½	116¾	116¾	-¼	130	115¾	649
Dec	123¾	123¾	123¼	123¼	-¾	135	123	788

Est vol 1,200; vol Fri 0; open int 12,682, -15.

SOYBEANS (CBT) 5,000 bu.; cents per bu.

	Open	High	Low	Settle	Change	Lifetime High	Lifetime Low	Open Interest
Jan	461	466	455¾	456½	-5¼	632	415	13,916
Mar	472½	474¾	463¼	464½	-5¼	598	423½	58,472
May	478½	482	471	471¾	-6¼	554	432	24,780
July	485	489	477½	478½	-6½	647	440	23,882
Aug	488	488	478½	478½	-6½	541½	441	2,434
Sept	489½	489½	480	480½	-6¼	541	450	680
Nov	495	497	485½	486½	-6	631	453	10,227

Est vol 42,000; vol Fri 0; open int 134,425, -508.

SOYBEAN MEAL (CBT) 100 tons; $ per ton.

	Open	High	Low	Settle	Change	Lifetime High	Lifetime Low	Open Interest
Jan	148.00	148.00	145.50	146.40	-.30	164.00	123.30	10,410
Mar	146.80	148.70	145.70	146.70	-.10	162.00	126.30	31,427
May	148.50	148.50	146.00	146.30	-.90	162.60	127.30	20,929
July	150.00	150.00	147.20	147.50	-.80	164.20	130.00	16,676
Aug	148.10	148.10	147.20	147.20	-.80	164.00	131.00	5,439
Sept	150.00	150.00	147.20	147.20	-1.10	164.00	132.00	3,120
Dec	152.50	153.00	150.70	150.80	-1.00	169.00	135.50	6,748

Est vol 13,000; vol Fri 0; open int 96,632, -198.

SOYBEAN OIL (CBT) 60,000 lbs.; cents per lb.

	Open	High	Low	Settle	Change	Lifetime High	Lifetime Low	Open Interest
Jan	15.80	15.90	15.43	15.44	-.31	24.90	15.38	10,615
Mar	16.02	16.11	15.67	15.70	-.32	23.95	15.67	53,022
May	16.38	16.38	16.01	16.02	-.30	23.50	16.01	23,579
July	16.70	16.70	16.32	16.32	-.31	22.30	16.32	19,837
Aug	16.80	16.80	16.45	16.45	-.31	21.00	16.45	4,590
Sept	16.60	16.60	16.60	16.60	-.31	21.70	16.60	2,847
Oct	16.80	16.80	16.75	16.75	-.31	22.25	16.75	3,174
Dec	17.35	17.35	17.02	17.04	-.35	20.62	17.02	13,494

Est vol 15,000; vol Fri 0; open int 132,060, -776.

WHEAT (CBT) 5,000 bu.; cents per bu.

	Open	High	Low	Settle	Change	Lifetime High	Lifetime Low	Open Interest
Mar	249	251½	245¼	247½	-1	340	236½	85,073
May	260¾	263	256	258	...	322	246¾	10,091
July	273	273	265¾	267¾	-¾	347	256¾	21,326
Sept	281	281	276	277½	-1½	335	266½	1,395
Dec	293½	293½	290	292¼	-1	345	280½	2,983

Est vol 19,000; vol Fri 0; open int 120,912, -236.

WHEAT (KC) 5,000 bu.; cents per bu.

	Open	High	Low	Settle	Change	Lifetime High	Lifetime Low	Open Interest
Mar	277	277½	272½	274¼	-2	361½	262½	40,982
May	288	288	283½	284½	-1¾	340½	272¾	7,922
July	298	298¼	293½	294¾	-2¼	366	282	11,966
Sept	306¼	306¼	302½	304	-1	346	291	690
Dec	314	316	314	315½	-1½	354	302	638

Est vol 4,872; vol Fri 0; open int 62,198, +1.

WHEAT (MPLS) 5,000 bu.; cents per bu.

	Open	High	Low	Settle	Change	Lifetime High	Lifetime Low	Open Interest
Mar	321	321	315	316½	-1½	386	312½	12,319
May	327½	329	323½	325	-½	389	320	4,952
July	336½	336½	331¼	332½	-¾	380	327¼	2,296
Sept	...	344	339¼	339¼	-5¼	385	334½	1,018
Dec	...	...	...	348	+1	390	344	210

Est vol 2,470; vol Fri 0; open int 20,832, +480.

CANOLA (WPG) 20 metric tons; Can. $ per ton

	Open	High	Low	Settle	Change	Lifetime High	Lifetime Low	Open Interest
Jan	255.20	255.50	251.40	251.70	-2.30	353.50	250.00	4,159
Mar	260.20	260.50	256.50	256.90	-2.10	335.00	255.40	33,216
May	263.90	263.90	260.50	260.70	-2.30	306.50	260.00	8,269
July	266.70	267.00	264.70	265.00	-2.00	309.00	260.00	8,584
Aug	...	...	...	267.90	-2.10	302.50	264.00	450
Sept	...	...	...	270.90	-0.60	295.90	271.00	305
Nov	272.00	272.00	271.50	271.50	-0.50	315.90	269.0	854

Est vol 4,900; vol Thu 4,271; open int 55,837, -2,019.

WHEAT (WPG) 20 metric tons; Can. $ per ton

	Open	High	Low	Settle	Change	Lifetime High	Lifetime Low	Open Interest
Mar	129.00	129.00	128.30	128.30	-0.20	151.50	126.50	6,248
May	...	...	...	131.70	-0.20	141.00	129.50	626
July	134.10	134.10	134.10	134.10	...	146.30	132.80	315
Oct	...	...	...	138.10	...	141.50	137.00	531

Est vol 133; vol Thu 61; open int 7,720, -133.

BARLEY-WESTERN (WPG) 20 metric tons; Can. $ per ton

	Open	High	Low	Settle	Change	Lifetime High	Lifetime Low	Open Interest
Mar	113.50	113.50	112.40	112.40	-1.00	129.00	112.40	6,337
May	116.00	116.00	115.50	115.50	-0.90	125.40	115.10	3,525
July	...	...	...	117.30	-0.70	126.60	116.50	1,248
Oct	...	...	...	120.50	-0.70	127.90	118.50	4,339
Dec	122.50	122.50	122.50	122.50	-1.00	123.50	121.10	344

Est vol 156; vol Thu 539; open int 15,793, +15.

LIVESTOCK AND MEAT

CATTLE-FEEDER (CME) 50,000 lbs.; cents per lb.

	Open	High	Low	Settle	Change	Lifetime High	Lifetime Low	Open Interest
Jan	86.20	86.20	85.42	85.87	+.52	86.20	74.35	7,375
Mar	86.50	86.50	85.52	85.90	+.47	86.50	74.15	8,018
Apr	85.75	86.00	85.40	85.77	+.35	86.00	74.40	3,765
May	85.85	86.00	85.40	85.87	+.47	86.00	76.30	4,030
Aug	86.50	86.87	86.40	86.85	+.55	86.87	79.25	2,675
Sept	86.45	86.90	86.45	86.90	+.40	86.90	85.00	485

Est vol 3,540; vol Thu 42; open int 26,353, +25.

CATTLE-LIVE (CME) 40,000 lbs.; cents per lb.

	Open	High	Low	Settle	Change	Lifetime High	Lifetime Low	Open Interest
Feb	69.65	69.90	69.40	69.70	+.10	71.35	65.50	52,165
Apr	71.55	71.90	71.40	71.72	+.15	72.25	67.30	31,142
Jun	69.77	69.95	69.40	69.65	-.12	70.10	65.30	14,008
Aug	69.75	69.97	69.52	69.70	...	70.07	65.70	7,580
Oct	71.65	72.00	71.60	71.67	+.12	72.00	68.00	3,232
Dec	72.65	72.95	72.65	72.70	+.05	72.95	70.50	2,854

Est vol 13,578; vol Thu 302; open int 111,076, -180.

HOGS-LEAN (CME) 40,000 lbs.; cents per lb.

	Open	High	Low	Settle	Change	Lifetime High	Lifetime Low	Open Interest
Feb	54.95	56.30	54.95	55.97	+1.47	58.50	41.05	23,821
Apr	56.50	57.40	56.12	57.27	+1.67	59.05	42.90	10,037
June	63.70	64.47	63.60	64.32	+1.02	65.90	52.10	6,544
July	62.55	63.35	62.40	63.32	+1.22	65.40	51.80	2,125
Aug	61.50	61.70	61.05	61.70	+1.00	63.00	52.30	1,252
Oct	57.00	57.65	56.90	57.65	+1.00	57.65	49.00	817
Dec	55.40	55.95	55.40	55.95	+.95	55.95	49.50	242

Est vol 8,399; vol Thu 58; open int 44,860, -34.

PORK BELLIES (CME) 40,000 lbs.; cents per lb.

	Open	High	Low	Settle	Change	Lifetime High	Lifetime Low	Open Interest
Feb	80.50	81.15	79.65	81.15	+3.00	81.95	47.02	3,886
Mar	78.60	80.15	78.60	80.15	+3.00	80.85	47.80	444
May	79.00	79.60	78.00	79.60	+3.00	80.65	47.50	352
July	78.00	80.10	78.00	80.05	+2.95	80.40	54.20	158

Est vol 1,383; vol Thu 42; open int 4,883, -1.

FOOD AND FIBER

COCOA (NYBOT)-10 metric tons; $ per ton.

	Open	High	Low	Settle	Change	Lifetime High	Lifetime Low	Open Interest
Mar	840	846	820	830	-7	1,910	800	40,478
May	869	870	846	856	-7	1,697	827	17,484
July	893	896	873	883	-6	1,478	855	8,340
Sept	921	921	900	910	-8	1,494	886	9,446
Dec	957	957	940	946	-8	1,336	925	5,862
Mr01	...	...	...	984	-8	1,362	965	4,834
May	...	...	...	1,008	-8	1,222	1,002	2,932
July	...	...	...	1,034	-8	1,245	1,060	2,704
Sept	1,080	1,080	1,080	1,062	-8	1,246	1,080	993

Est vol 3,120; vol Thurs 3,217; open int 93,073, +724.

COFFEE (NYBOT)-37,500 lbs.; cents per lb.

	Open	High	Low	Settle	Change	Lifetime High	Lifetime Low	Open Interest
Mar	122.25	124.00	116.10	116.50	-9.40	149.00	83.50	29,132
May	125.00	125.50	119.00	119.50	-8.90	150.00	85.85	8,243
July	127.50	127.50	121.50	122.00	-8.75	149.00	87.50	3,229
Sept	128.50	128.50	125.00	124.10	-8.55	148.50	89.75	2,714
Dec	...	...	...	125.50	-8.75	150.50	92.00	2,073
Mr01	128.00	128.00	127.00	127.00	-8.75	153.85	94.25	431

Est vol 8,132; vol Thurs 5,695; open int 45,822, +758.

SUGAR-WORLD (NYBOT)-112,000 lbs.; cents per lb.

	Open	High	Low	Settle	Change	Lifetime High	Lifetime Low	Open Interest
Mar	6.15	6.18	6.05	6.10	-.02	10.00	5.06	81,338
May	6.30	6.31	6.21	6.24	-.07	9.05	5.26	47,581
July	6.42	6.43	6.30	6.34	-.10	9.15	5.40	28,371
Oct	6.53	6.55	6.42	6.48	-.10	9.00	5.68	9,631
Mr01	6.48	6.50	6.48	6.48	-.10	7.48	6.00	7,585
May	...	...	...	6.49	-.10	7.38	6.25	1,045

Est vol 12,427; vol Thurs 9,518; open int 175,618, -580.

SUGAR-DOMESTIC (NYBOT)-112,000 lbs.; cents per lb.

	Open	High	Low	Settle	Change	Lifetime High	Lifetime Low	Open Interest
Mar	18.20	18.20	18.20	18.20	-.04	22.30	17.50	2,215
May	18.47	18.50	18.47	18.50	+.01	22.33	18.40	2,178
July	19.18	19.18	19.18	19.18	+.18	22.39	18.80	2,292
Sept	...	...	...	19.75	-.10	22.38	19.16	1,516
Nov	...	...	...	19.17	+.26	22.30	18.75	819
Ja01	...	...	...	19.05	+.15	21.75	18.75	1,036
Mar	...	...	...	19.23	+.33	21.15	18.80	367
May	...	...	...	19.48	+.39	19.65	18.85	711

Est vol 117; vol Thurs 50; open int 11,134, .

COTTON (NYBOT)-50,000 lbs.; cents per lb.

	Open	High	Low	Settle	Change	Lifetime High	Lifetime Low	Open Interest
Mar	50.45	51.10	50.35	51.07	+.33	67.90	48.60	32,867
May	51.96	52.50	51.85	52.36	+.20	75.40	50.10	13,381
July	53.45	53.75	53.35	53.71	+.10	73.25	51.50	8,016
Oct	55.40	55.40	55.20	55.10	-.40	68.50	53.45	425
Dec	55.90	55.90	55.55	55.63	-.44	64.50	53.75	3,837
Mr01	...	...	...	56.83	-.45	60.45	55.25	338
May	...	...	...	57.43	-.45	59.70	56.50	105

Est vol 9,500; vol Thurs 11,062; open int 59,058, -1,940.

ORANGE JUICE (NYBOT)-15,000 lbs.; cents per lb.

	Open	High	Low	Settle	Change	Lifetime High	Lifetime Low	Open Interest
Jan	85.50	88.60	85.50	88.25	-.70	133.00	76.00	735
Mar	83.75	85.20	83.75	84.05	-1.15	131.00	78.75	13,341
May	83.75	84.60	83.75	83.80	-.70	101.60	81.10	3,479
July	84.50	84.50	83.90	83.55	-1.20	102.50	82.80	1,602
Sept	84.00	84.25	84.00	83.50	-.75	102.95	84.00	653
Nov	84.50	84.50	84.00	84.20	-1.05	103.95	84.00	148

Est vol 2,500; vol Thurs 5,526; open int 20,099, -1,954.

METALS AND PETROLEUM

COPPER-HIGH (Cmx.Div.NYM)-25,000 lbs.; cents per lb.
Market Closed

OTHER FUTURES

Settlement prices of selected contracts. Actual volume (from previous session) and open interest of all contract months.

	Open	High	Low	Settle	Change	Lifetime High	Lifetime Low	Open Interest
CALIFORNIA/OREGON BORDER ELECTRICITY (NYM) 736 mwh; .01 per mwh								
Feb	na	na	na	na	na	30.00	25.25	na
CORN (MCE) 1,000 bu.; cents per bu.								
Mar	0	...	...	200¾	-3¾	289⅞	185⅝	4,225
EURO-STERLING (NYBOT)-100,000 Euros; Euro per Pound								
Mar	0	.6305	.6245	.6305	+.0023	.6361	.6245	6,235
EURO-U.S. DOLLAR (NYBOT)-200,000 Euros; Pounds per Euro								
Mar	3	1.0220	1.0212	1.0322	+.0161	1.0367	1.0098	2,619
EURO-YEN (NYBOT)-100,000 Euros; Yen per Euro								
Mar	1	103.70	101.62	103.60	+1.10	105.29	101.40	6,041
FLAXSEED (WPG) 20 metric tons; Can. $ per ton								
Jan	386	212.00	211.10	211.10	-5.60	276.50	215.00	4,201
LUMBER (CME) 80,000 bd. ft., $ per 1,000 bd. ft.								
Jan	1	354.80	350.20	350.30	+1.30	371.70	305.10	2,789
PALLADIUM (NYM) 100 troy oz.; $ per troy oz.								
Mar	na	na	na	na	na	455.00	335.00	na
PALO VERDE ELECTRICITY (NYM)-736 mwh; .01 per mwh								
Feb	na	na	na	na	na	28.00	23.25	na
PROPANE (NYM) 42,000 gal.; cents per gal.								
Jan	na	na	na	na	na	46.50	23.00	na
RICE-ROUGH (CBT) 2000 cwt; $ per cwt								
Jan	0	5.250	5.180	5.230	+.090	8.650	5.060	6,405
SILVER (CBT)-1,000 troy oz.; cents per troy oz.								
Jan	0	536.0	534.0	536.5	-1.5	535.0	530.0	985
SOYBEAN OIL (MCE) 3,000 bu.; cents per lb.								
Jan	0	...	...	15.44	-0.31	19.22	15.55	147
SOYBEANS (MCE) 1,000 bu.; cents per pu.								
Jan	0	...	...	456½	-5¼	626½	416	6,415
TREASURY BONDS (CANTOR) $100,000; pts. Per 32nds								
Mar	70	90-06	89-28	89-28	-3	93-26	91-03	825
WESTERN NAT. GAS (KC) 10,000 MMBtus; $ per MMBtu								
Feb	0	...	...	na	...	.895	.895	1
WHEAT (MCE) 1,000 bu.; cents per bu.								
Mar	0	...	...	247½	-1	340	219	1,608
EUROTOP 100 INDEX (CMX)-$100 times index								
Mar	na	na	na	na	na	3698.0	2800.0	na
Index: Hi na; Lo na; Close na, na.								
BRIDGE CRB INDEX (NYBOT)-500 times index								
Jan	na	na	na	na	na	211.70	192.00	na
Index: Hi na; Lo na; Close na, na.								
MINI VALUE LINE INDEX (KC)-$100 times Futures price								
Mar	15	1045.0	1009.5	1012.3	-18.1	1087.2	898.0	353
Index: Hi 1029.29; Lo 1005.52; Close 1010.53, -15.27.								
NYSE COMPOSITE INDEX (NYBOT)-500 times index								
Mar	700	659.90	641.25	644.15	-11.05	660.75	476.00	3,364
Idx prl: Hi 650.35; Lo 635.93; Close 639.52, -10.78.								
RUSSELL 1000 (NYBOT)-$500 × index								
Mar	300	782.00	760.50	768.00	-7.50	782.00	654.35	4,929
Index: Hi 772.86; Lo 752.00; Close 761.50, -6.46.								

EXCHANGE ABBREVIATIONS
(for commodity futures and futures options)

CBT-Chicago Board of Trade; CME-Chicago Mercantile Exchange; CSCE-Coffee, Sugar & Cocoa Exchange, New York; CMX-COMEX (Div. of New York Mercantile Exchange); CTN-New York Cotton Exchange; DTB-Deutsche Terminboerse; FINEX-Financial Exchange (Div. of New York Cotton Exchange); IPE-International Petroleum Exchange; KC-Kansas City Board of Trade; LIFFE-London International Financial Futures Exchange; MATIF-Marche a Terme International de France; ME-Montreal Exchange; MCE-MidAmerica Commodity Exchange; MPLS-Minneapolis Grain Exchange; NYFE-New York Futures Exchange (Sub. of New York Cotton Exchange); NYM-New York Mercantile Exchange; SIMEX-Singapore International Monetary Exchange Ltd.; SFE-Sydney Futures Exchange; TFE-Toronto Futures Exchange; WPG-Winnipeg Commodity Exchange.

FIGURE 3–4: Daily Commodity Futures Price Quotations. Source: *The Wall Street Journal*. Reprinted by permission of Dow Jones and Co. All rights reserved .

seller (or buyer) when he or she decides to offset his or her position. The trader merely executes an equal and opposite transaction, usually with an entirely different opposite party, to end up with a net zero position. The Clearing Corporation serves as the intermediary. By rule, all members of the exchange must clear their trades on the day the trade is made and settle disputes before trading resumes the next day.

The Clearing Corporation guarantees financial performance of all contracts traded and cleared. Financial integrity of futures trading is thus insured by the Clearing Corporation. Each trader must deposit original margins with the clearinghouse and maintain a minimum balance in the event of adverse price fluctuations. At the close of the trading day, all settling of accounts is done with the Clearing Corporation; brokers do not make settlement with each other. If wide fluctuations in price occur during a trading day, the Clearing Corporation can call for additional margins throughout the day without waiting for routine settlement. Thus, a trader is not concerned about possible financial embarrassment of other traders with whom trades have been made. Every trader's position is with only one party, the Clearing Corporation, which is financially responsible. The financial position of the Clearing Corporation is assured (1) by the original margin deposits on each contract; (2) by guaranty funds deposited by each clearinghouse member firm; (3) by the surplus reserve of the clearinghouse; and (4) by the joint financial responsibility of all Clearing Corporation members based on their proportion of the total business done over a period of time. "Even in periods of heavy trading and wide fluctuations, original margins alone have almost always been adequate to cover demands made on commodity clearing houses. No semblance of strain has ever been placed on their resources. Not a penny has ever been lost as a result of a commodity clearing house being financially unable to meet its obligations."[10] These provisions guarantee traders will enjoy financial performance.

Opening an Account. Commodity exchanges are generally membership associations. Only members owning seats on the exchange are permitted to participate in the auction on the trading floor. The largest participation in trading, however, comes from individuals outside the exchanges. Futures commission merchants represent these individuals on the exchange floor. They solicit customers from the public and maintain the commodity accounts of these customers.

The first step by an individual toward using the futures market as a marketing tool is to select a brokerage firm, to select an account executive (broker), and to open an account with the firm. Many highly visible financial firms offer commodity brokerage services. A reply to newspaper advertising, public meetings, or trade literature will elicit responses from them. A review of their advertising, market letters, and services available gives a feeling for how they operate. Most emphasize the quality of their research department and their ability to fill orders quickly and efficiently. Since all of them are highly regulated and closely supervised by the exchange and the government regulatory authority (the Commodity Futures Trading Commission), there is little difference between their operating procedures. But farmers may prefer to deal with firms that specialize in agricultural commodities.

Brokerage houses charge commission fees for the service they perform. These fees ($25 to $100) cover both the initial and liquidating trades—a "round turn" trade. On March 6, 1978, commission rates became negotiable rather than fixed. Cut-rate brokerage firms offering fewer services, at lower rates, have sprung up as a result. Some of them advertise rates as low as $15 per "round turn" trade.

Selecting an account executive (broker) is probably more important than selecting the firm. He or she is the individual who will provide day-to-day contact with the market for the customer. Account executives are usually called brokers, but it should be remembered they are distinctively different in function from floor traders who are also referred to as brokers. Account executives (brokers) come in all shapes, sizes, colors, and philosophies. Some are highly aggressive in providing a continuous flow of market tips and information; others are reserved in their recommendations. Personal compatibility and acceptance of customer objectives are important traits between the customer and broker. They should enjoy working together.

After choosing a brokerage firm, the customer fills out information forms and signature cards required by the firm to open an account. The customer must specify whether he or she is opening a hedge or speculative account. These accounts have different margin deposit requirements and tax implications. Hedging account margins are typically half the amount required for speculative accounts.

Tax Treatment of Futures Transactions. Whether a futures market participant is a hedger or a speculator affects his or her federal income tax status. Transactions that reduce price risk of commodities used or produced in a taxpayer's business are considered, for tax purposes, legitimate business transactions. The costs of such transactions, including losses sustained, are deductible in full amount as a business expense. Likewise, profit realized from hedging in futures contracts must be included in the gross income of the taxpayer. Gains or losses in futures positions are thus treated as ordinary income for hedgers. Proof that trades were for hedging rests with the trader. Even though a trader identifies the transaction as a hedge, the trader must be able to prove that the purpose for entering the market was for price protection on commodities that were part of the business operation. Accurate inventory records should be maintained to show the actual cash position at the time of the hedge.

Commodity futures contracts traded for speculative purposes are considered capital assets for income tax purposes.[11] Speculative transactions produce either short-term or long-term capital gains or capital losses. At the end of the year, gains or losses are allocated 40 percent to short-term and 60 percent to long-term categories. Commodity gains and losses are added to gains or losses in other investments for tax calculation purposes. Net gains from commodity speculation must be included in full as gross income, but a speculator is limited to a total capital loss deduction (from all investments) of $3,000 per year. Losses greater than $3,000 may be carried forward to later tax years. At the end of the tax year, open traders are "marked to the market" to eliminate the effect of unrealized losses or gains of switching income between tax years.

Margins. An initial margin deposit must be made with the brokerage firm before a trading order can be placed. Margins on commodity trades are money deposits by the customer to guarantee financial integrity. Inasmuch as the brokerage firm must deposit and maintain cash margins with the Clearing Corporation, the brokerage firm protects itself by securing original margin deposits from the customer. The customer is asked to place from 5 percent to 20 percent of the total value of the commodity contract involved as original good faith money. Minimum margins, as set by the various exchanges, vary in amount between commodities in accordance with prevailing price and price volatility.

Brokerage firms usually require a margin deposit larger than the exchange minimum. The margin money is deposited in a segregated fund at the brokerage firm's bank. A segregated account means that the money will be used only for margin and not for expenses of the brokerage firm. Most firms do not pay interest on the account balance to the customer, but some offer money market or T-Bill accounts that earn interest on customers' balances.

After the original margin is deposited and a trade is made, a minimum margin equity must be maintained in the event of adverse price fluctuation. The broker reserves the right to call for additional margin, called "maintenance" margin, at any time to maintain the customer's minimum equity. The initial margin is usually sufficient to cover small fluctuations, but the trader must be prepared with liquidity to satisfy a call for additional margin should the market move against his or her position. Otherwise, the broker has the right to close out the position on behalf of the customer.

On the other hand, if the customer's transaction shows a profit, the customer can ask for and receive the surplus balance in cash from the broker without closing out the futures position. Usually, the customer should leave only enough cash in the account to maintain the current minimum margin requirements of the broker. The surplus should be withdrawn and placed in an interest-bearing account.

Types of Orders. A market participant can place an order for a futures transaction with the broker in several ways. Usually, a market participant places a "market order" that instructs the broker to execute the trade at the most advantageous price obtainable at the moment the order reaches the pit. The customer will not know the execution price until it is confirmed by the broker.

A "limit order" instructs the broker to buy or sell only if the market price reaches a certain price. The order will not be filled until the price reaches the limit specified by the customer. A trader who gives a limit order to sell 5,000 bushels of May wheat at $4.50, for example, is saying in effect that he or she does not want to sell unless the price is at least $4.50. Such an order is given when the market is trading below $4.50. To protect against the event that the market may have exceeded the $4.50 price by the time the order arrives, the customer may add on "OB" (or better) designation. It insures that the trade will not be filled at a price lower than $4.50 and will be filled above that level if the market is strong. Limit orders are usually considered good only during a specified trading session, but an order may be marked "GTC" (good till cancelled) to indicate that it is a standing order until cancelled by the trader.

"Stop-loss" orders are used by traders who want to limit their risk or stop their losses. For example, a trader who is "long" at $4.20 may enter a stop-loss sell order at $4.15. The trader hopes that the market will rise; but if it slumps instead, he or she does not want to hold the position if the market declines below $4.15. By the same token, a trader who is "short" may enter a stop-loss order to buy if the market advances to the price level specified. Stop-loss orders are entered by traders who want to pre-set how much loss they are willing to incur before they will cancel their position, but they are not close enough to the market action themselves to be ready to place market orders when their price is reached. A stop-loss order does not guarantee, however, that the order will be filled at the price specified. When a market is moving fast, a stop-loss price may be reached and passed before the floor broker can execute the trade. The price may jump over a sell-stop or fall right through a buy-stop. Accordingly, the broker cannot be held responsible for filling stop-loss orders. Since such an order immediately becomes a "market order" once the stop-loss price has been reached, the sale may be made at a price lower—or the purchase higher—than the order. One of the best uses of stop orders is to initiate a position at a particular price the customer wants rather than what the market is offering. There is no assurance the order will be filled at the objective price, but it usually will be within some reasonable range. It gives the trader control over the price at which to enter the market.

By adding "MIT" (market if touched) to a limit order, the customer specifies that the order will have the effect of a market order when the limit is reached or touched even though price may move away quickly. This order can be useful in a thinly traded market that has a particular price objective as defined by some method of analysis. The customer wants an order filled near the objective price, but executing the order is more important than exact price. If trading touches the stated price but quickly moves away, the broker can fill it as a market order as soon as possible.

"Spread" orders initiate simultaneous long and short positions. The trades may be placed in related months of a single commodity, in products of a commodity, or between related commodities. For example, a spread order would buy one contract month (December corn) and sell another contract month (March corn). It might specify buying soybeans and simultaneously selling soybean oil and soybean meal. A related commodity spread might buy feeder cattle and sell live cattle. Traders place spread positions, because they expect the relative prices to change. They feel more confident predicting the difference between prices than predicting their price level. Whether the general price level moves up or down, the spread position will show a profit if the relative prices converge or diverge in the direction the trader expects.

A spread order is useful for farmer-hedgers who are rolling their positions forward out of a maturing contract month into another more distant one. Sometimes they prefer a maximum spread between the two contracts; sometimes they prefer a minimum. Rather than placing one market order to buy and another to sell, they specify a spread they want to achieve. For example, the order would state, "buy Feb. lean hogs and sell July lean hogs at $2.50 premium on the Feb." This order instructs the broker to execute the trades whenever the spread between the two contracts reaches the stated difference.

A customer should never assume that an order has been filled until an oral confirmation or paper from the broker is received. A ticker-tape or a price-reporting board can run several minutes behind and does not determine whether a trade was executed or not. Until the customer receives confirmation, he or she should never re-enter an order against that position. Paper confirmations always follow oral confirmations. The written confirmation of a trade should be checked for accuracy in type of commodity, month, trading price, and quantity of contracts. Any discrepancy should be reported to the brokerage firm immediately, regardless of whether there was a profit or loss associated with the error. If the customer does not receive written confirmation of a trade following an oral confirmation, the customer should contact the account executive immediately to correct the oversight.

All mechanical details and records of a transaction are handled by the brokerage firm for its trading customers. For every trade made, the customer should receive a confirmation, and for every closeout a profit and loss statement showing the financial results of the transaction. In addition, the broker should send the customer a monthly statement showing the ledger balance, the open positions, the net profit or loss in all contracts liquidated since the date of the previous statement, and the net unrealized profit and loss on all open contracts figured at the market settlement price on the last business day of the month.

Risk Transfer

Futures markets transfer price risks from those who wish to avoid them to those who make it their business to assume those risks. Use of futures contracts permits hedging to minimize the effects of price change. Futures markets are thus an important tool for risk management.

Risk and Uncertainty. These are inherent with producing and owning agricultural products. Risks can be classified into two types:

> Physical – quality and quantity variation in commodities as affected by natural conditions

> Price – rising or declining value of production or inventory due to changing economic conditions

Farmers have faced the threat of insects, disease, and weather for centuries, and they continue to expose their production efforts to these risks. But modern farmers have also become production masters. When Columbus came to America, approximately 1 million Native Americans lived in the area that was to become the United Sates. The largest portion of their energies was directed toward providing food for their families. As America's original farmers, they fished, hunted, and raised crops of corn, yams, tomatoes, beans, and squash. These natives barely subsisted, even though most of their time was spent gathering food. Today, there are about two times as many farm workers in the United States as there were Indians in Columbus's time. Today's farmers provide 265 million Americans with an infinite variety of quality food and have enough left over to be the world's largest food exporter. Two centuries after Independence, the United States became the world's breadbasket.[12] Farmers have become experts at overcoming production risk.

Economic conditions of the 1970's, 1980's, and 1990's added a new dimension that farmers had to cope with: price risk. Wide fluctuations in commodity prices mean a potentially wide margin for profit or loss. When a commercial family farm of 600 acres faces gross income fluctuations of $60,000 annually due to price variability, a new form of risk threatens financial security and survival on the farm. Successful production alone is not adequate for business success; profitable marketing is also required. The risk challenge of the new century is not production risk, but price risk.

Production and Product Risk. Participants in the food chain from production through marketing face the possibility of loss from natural hazards that may destroy or damage the production or inventory on hand. Each firm accepts many of these risks as part of the production process and accounts for them as a cost of production. However, to the extent possible, many farmers transfer physical risk to an insurance company that specializes in accepting risk for a fee. Buying insurance is often more economical for an individual than attempting to provide for his or her own protection. Farmers are accustomed to purchasing coverage for stored commodities against fire, wind, and other natural hazards. Crop insurance permits farmers to protect anticipated production.

Another risk every producer and marketer faces is physical deterioration of stored commodities. A rapid change in temperature, a breakdown in equipment or facilities, and moisture or insect damage to stored commodities are risks that cannot be shifted. This threat of loss is one of the factors that increases the cost of owning perishable commodities.

Price Risk. The combination of inventory ownership and uncertain price creates the need for price risk management. Value loss, arising from price change in agricultural products, may originate from many sources. An elevator operator who purchases large amounts of grain for storage in the fall, at what he or she considers low prices, may face a highly unfavorable spring market because of unforeseen declines in general business conditions. An untimely frost early in the fall or late in the spring may greatly increase the value of current orange juice stocks by sharply reducing the growing crop. A series of rains may make or break the wheat crop, greatly affecting the fortunes of wheat farmers producing a crop or holding large inventories of grain in storage.

Various marketing devices either minimize price risk or shift it from one person to another. Vertical integration of production and marketing reduces or transfers risk. Poultry production contracts by farmers typically specify a fixed return and thereby transfer risk of changing broiler prices to the integrator. The integrator in turn may attempt to obtain control over the kind and amount of production coming forth to reduce supply risks. Selling in advance of product availability eliminates price risk. Farmers can forward contract anticipated production of grain with their elevator for fall delivery. Elevators, in turn, may book delivery later in the season from their storage with a terminal elevator or exporter.

Government price-support programs transfer price risk from producers to taxpayers. Non-recourse loans and deficiency payments establish minimum prices for farmers who participate in government programs. Government purchase, sale, or storage of commodities also helps reduce variation in available market supplies and thereby reduces price volatility. Since price fluctuations can also arise from inadequate or inaccurate information, government efforts that improve the gathering and dissemination of market news and statistics help reduce this risk. Equally informed buyers and sellers can more closely negotiate a true price.

Each of these devices helps reduce farmer price risk by either minimizing the risk or transferring it to someone else.

The futures market also provides a mechanism for shifting price risk to individuals outside the commodity production and marketing chain. Buying or selling futures contracts allows producers, owners, and buyers of commodities to hedge against price change. The effects of price change are accepted by speculators.

Hedging. Hedging is defined as taking opposite positions in two different markets with equal quantity commitments so that losses in one market are offset by gains in the other. The two markets are the cash market and the futures market.

Hedging provides protection against price change for both sellers and buyers of commodities. The owner of a physical commodity hedges against price change by "selling" futures contracts equivalent to his or her production or inventory of the same commodity. This is called a "short" hedge. It is used by farmers, elevators, processors, and exporters. They "own" the commodity and they "owe" delivery of it against a futures contract. Commodity buyers "buy" futures contracts to protect against a possible advance in price of the commodity not yet owned but needed to fill feeding, processing, or export commitments. This is a "long" hedge. Long hedgers "own" the right to receive delivery of the commodity, but they "owe" the physical commodity to their production process or buyer.

Who Uses Short Hedges and How? Hedging is not an automatic operation with a stereotyped procedure. Each farm or firm hedges on the futures market in a manner that corresponds to its respective cash position. Following are some of the trade interests who use short hedges when they wish to be protected against possible market price declines:

a. The grower, who sells futures contracts against crops or livestock in production.
b. The farmer-merchandiser, who sells futures contracts against grain in storage.
c. The grain elevator operator, who sells futures against current purchases from farmers, while awaiting an opportunity to sell to consuming, processing, or export interests.
d. The exporter, who sells futures against purchases of commodities.
e. The processor, who sells futures against purchases of actual commodities pending resale in the form of a manufactured product.
f. The manufacturer, who sells futures against unsold manufactured products.

Who Uses Long Hedges and How? Producers, shippers, importers, manufacturers, and others use the futures market to place "long" hedges when they wish to be protected against possible price advances in commodities they intend to buy. Following are some of the trade interests who use the "long" hedges:

a. The livestock producer, who plans to purchase grain and feed supplements.
b. The cattle feeder, who intends to buy feeder animals.
c. The processor, who buys futures against possible price rise in required raw materials.
d. The importer and merchant, who makes forward sales for specific shipment or delivery and buys futures as protection against market changes pending the purchase of the actual product.

Selling Hedge. One way to understand the "equal but opposite" requirement of hedging is to trace the results through examples. Consider a farmer who expects to produce 100,000 bushels of corn for November delivery. About mid-July the price level is high enough to cover his production costs and yield a reasonable profit. The producer contacts the local country elevator to find that the current cash price quoted for new crop corn is $3.30 per bushel.

Cash Market	*Futures Market*
July 19 Price of New Crop Corn $3.30/bu. 100,000 bu. of expected production	No Position

The producer does not yet have the corn harvested. However, since the quoted price meets a pricing target, the opportunity presents itself to eliminate the risk of declining price between now and the time he expects to deliver corn. Setting a hedge will protect the corn crop against price change before harvest. The corn grower contacts a commodity account executive and places a short hedge in the December futures (currently trading at $3.55, for example) per bushel by selling 20 futures contracts. As a hedger, the farmer has an account that now looks like this:

Cash Market	*Futures Market*	*Price Spread (Basis)*
July 20	July 20	
New crop corn @ $3.30/bu.	Sell 20 Dec. corn contracts @ $3.55/bu.	$0.25 under Dec.
100,000 bu. of expected production	100,000 bu.	

Let's move forward in time to November, when the corn is ready for delivery. To lift the hedge, the farmer simultaneously offsets the futures obligation by buying 20 December futures contracts and sells the corn for cash by delivering it to the local elevator. The hedger could have fulfilled the short futures obligation by delivering corn against it, but for a corn grower in central Minnesota, it is usually more convenient to buy back the futures and sell the cash corn locally. In practice, few futures contracts are settled by delivery.

If the price of corn declined, as the farmer expected, the cash price was protected because the same market forces affected both the cash and futures prices of corn. If both declined by an equal amount, meaning that the basis remained the same, the farmer would realize the $3.30 per bushel cash price he was trying to protect.

The accounting follows:

Cash Market	*Futures Market*	*Basis*
July 20	July 20	
New crop corn @ $3.30/bu.	Sell 20 Dec. corn contracts @ $3.55/bu.	$0.25 under Dec.
Nov. 20		
Sell 100,000 bu. cash corn @ $3.14/bu.	Buy 20 Dec. corn contracts @ $3.39/bu.	$0.25 under Dec.
Result:		
$0.16/bu. loss in value	$0.16/bu. gain on futures	change $0.00

Cash sale price	$3.14
Futures gain	+0.16
Realized sale price	$3.30

The hedger realized $0.16 per bushel less in the cash market than the desired price of $3.30 per bushel. However, he recovered that $0.16 from the gain in the futures market position. This represents a perfect hedge, with both prices declining exactly the same amount. They moved parallel to each other over time; i.e., no change in basis.

The price levels might have risen following the placement of the hedge. What would happen if both cash and futures prices rose by equal amounts of $0.30 per bushel? The accounting would look like this:

Cash Market	Futures Market	Basis
July 20 New crop corn @ $3.30/bu.	Sell 20 Dec. corn contracts @ $3.55/bu.	$0.25 under Dec.
Nov. 20 Sell 100,000 bu. cash corn @ $3.60/bu.	Buy 20 Dec. corn contracts @ $3.85/bu.	$0.25 under Dec.
Result: $0.30/bu. gain in value	$0.30/bu. loss on futures	change $0.00

Cash sale price	$3.60
Futures deficit	-0.30
Realized sale price	$3.30

The farmer had a deficit of $0.30 per bushel on the futures market transaction because the December futures were repurchased at a higher price. However, since the cash price of corn also increased by $0.30 per bushel, the cash market gain offsets the futures market loss. The important result is that the corn grower received the desired $3.30 per bushel for the corn.

Obviously, the farmer would have received a higher net price by not hedging. However, when the hedge was initiated on July 20, the farmer was satisfied with the $3.30 per bushel price and was concerned that a drop in price would have serious financial consequences. Therefore, the hedge protected the profit margin that the $3.30 per bushel net price offered. The 30 cents per bushel foregone was the "opportunity cost" of insuring a $3.30 price.

Buying Hedge. The buying hedge fixes input costs by protecting against a possible rise in the commodity's cash price. When a hedger places a "long" or buying hedge, the first transaction in the futures market is a "buy." The futures market "purchase" is a temporary substitute for the actual purchase of the cash commodity the long hedger intends to acquire at a future date.

Soybean meal is an important feed ingredient in broiler production. Therefore, poultry producers face significant risks from rising feed costs. Producers are able to control part of their feed costs by placing "long" hedges in soybean meal futures.

Suppose that in early October a broiler producer projects soybean meal needs for January to be 100 tons. Cash meal is currently selling at $238 per ton. The producer likes the price, unfortunately she does not have any space available to hold the meal until January. Concerned about rising meal prices, she decides to hedge by buying a single contract (100 tons) of January soybean meal. On October 4, the January soybean meal futures are trading at $243 per ton. The accounting follows:

Cash Market	Futures Market	Basis
October 4 Need 100 tons of meal in Jan. Price quotes @ $238/ton	Buy 1 Jan. soybean meal contract @ $243/ton	$5 under January

Assume that during the intervening months there was a greater than anticipated demand or a smaller than expected supply, causing soybean meal prices to rise. On December 30, the poultry feeder buys meal to meet feed needs and lifts the hedge.

What happened if both soybean meal cash and futures prices rose by $8 per ton? The accounting looks like this:

Cash Market	Futures Market	Basis
October 4 Soybean meal quoted @ $238/ton	Buy 1 January meal contract @ $243/ton	$5 under January
December 30 Buy meal @ $246/ton	Sell 1 January meal contract @ $251/ton	$5 under January
Result: $8/ton higher cost (loss)	$8/ton gain	change $0.00

Cash purchase price	$246/ton
Futures gain (lower cost)	−8 /ton
Actual cost	$238/ton

In this example, the futures market gain exactly offsets the cash market loss (increased cost) because the basis was the same $5 under January, when the hedge was placed and lifted. The poultry feeder would have paid $8 more per ton without the hedge.

What would happen if both cash and futures prices fall by an equal amount? The accounting follows:

Cash Market	Futures Market	Basis
October 4 Soybean meal Price quoted @ $238/ton	Buy 1 January soybean meal contract @ $243/ton	$5 under January
December 30 Buy soybean meal @ $233/ton	Sell 1 January soybean meal contract @ $238/ton	$5 under January
Result: $5/ton lower cost (gain)	$5/ton loss	change $0.00

Cash purchase price	$233/ton
Futures loss (higher cost)	+5/ton
Actual cost	$238/ton

Without a change in basis, the futures deficit is exactly offset by the cash gain. In this situation, the unhedged poultry feeder would have paid less. However, since the objective was to fix input cost, hedging fulfilled that objective.

These are examples of perfect hedges—the basis is the same the day the hedge is initiated and lifted. These examples are used to illustrate how hedging reduces price fluctuation risk. The price level may vary considerably but, as long as basis does not change, the price objective will be realized through hedging.

Hedging fixes a price. The opposite positions in the cash and futures markets eliminate the effects of price changes, both up and down. It protects the hedger against adverse price movement, but at the same time, it also eliminates the opportunity to enjoy favorable price changes.

Financing Price Protection. Where does the money for price protection for the hedger come from? Every buyer of a futures contract must have a seller, and every seller must have a buyer. Futures trading is a zero-sum game: If someone profits from a futures trade, someone else must lose.* For a short hedger who judges price to be high and likely to go lower, there may be a long hedger who believes it is low and likely to go higher. One enters the market as a seller and the other, a buyer. If the price level declines, the short hedger enjoys the protected high price, and the long hedger incurs higher input cost than later markets offer. If price level increases, the short hedger pays the piper by having foregone the opportunity of higher price, and the long hedger enjoys the protected low price.

Seldom does a farmer offering 20 corn contracts as a hedger-seller match with a corn processor buying 20 contracts as a hedger-buyer. Most likely, the other party to any hedge trade is a speculator. For the short hedger in a declining market, the speculator finances the price protection by suffering losses in the futures position. In a rising market, the short hedger pays the speculator the extent of the price rise in the futures market. Only the speculator profits or loses from being correct or incorrect in judging direction of price change. The hedger receives a fixed price, neither profiting nor losing from the hedge position. The changing value of the hedger's inventory offsets the gain or deficit in the futures portion of the hedge. For a short hedger, a falling market generates declining inventory values that are offset by gains in futures. In a rising market, a deficit in futures position is offset by gains in inventory value. In the end, when the hedge is canceled by selling cash and simultaneously buying futures, the net price to the short hedger will be what he or she had decided would be satisfactory when the hedge position was initiated. Selling hedgers give up the opportunity to enjoy higher price in return for protection from experiencing low price. Buying hedgers give up the opportunity to buy inputs at lower price in return for protection against having to pay higher price.

Hedging fixes a price (ignoring basis change) that will be realized eventually. Hedging enables the producer to shed the effect of price change he or she would otherwise have to carry; an uncertain outcome is replaced by a certain one for the hedger. The speculator either profits or loses from being right or wrong in his or her position; speculators do not have the offsetting inventory value of the commodity. Speculators finance the hedgers' price protection.

---

*Actually, it is a negative-sum game because of transaction costs and commissions.

Types of Hedging. Hedging can be categorized into three types: routine hedging, selective hedging, and carrying-charge hedging.

Routine hedging is a common practice in the U.S. grain and cotton trade. It "covers" the trader's position in the cash market by executing offsetting transactions in the futures market. Routine short hedging is attractive to an intermediary, elevator, processor, or manufacturer whose main business activities require holding stocks in inventory. They do not wish to risk business profitability in the face of uncertain price movements. Routine hedging assumes no need for forming judgement on price movement. When inventory is on hand, it is automatically hedged, either in futures or in forward contracts. Elevators that buy grain from farmers at a price, add on an operating margin, and contract sale at the higher price essentially practice routine hedging by contracting. Banks encourage routine hedging when they require that borrowers' inventory be hedged so that their solvency is not impaired by a drastic fall in the market price.

Selective hedging differs from routine hedging because the trader exercises judgement about expected price changes. A selective short hedger hedges only when prices are expected to fall. A selective long hedger hedges only when he or she expects a rise in price. According to the confidence in one's market judgement, the extent of one's cash commitment, one's ability to withstand losses, and the likely cost of hedging, one may hedge all or only part of the cash commitment. The selective hedger sometimes take hedging action to reduce exposure to price risk and sometimes does not. The selective hedger's decisions on when and how much to hedge depend heavily on personal price expectations. This type of hedging should be termed "selective hedging/selective speculating." The secret to being successful in a selective hedging program is having the tools and/or insight to make profitable decisions as to when to hedge and when to be unhedged.

A selective hedger enters a futures position for the same reasons as a routine hedger. However, the selective hedger is not committed to that position unalterably until the cash market risks have been satisfied. A $3.00 December futures price may be very attractive for a farmer planting corn in May. However, dry weather that pushes prices to $3.30 by mid-July may encourage the hedger to reevaluate the original position. If the farmer anticipates that a smaller corn crop could push prices to $4.00, he or she would lift the hedge at $3.30 in mid-July. The net effect, if correct, is $3.70 corn in a $4.00 market since the selective hedger incurred a 30 cent loss before lifting the hedge. Neighbors who are cash market speculators enjoy $4.00 corn prices because they were not hedged.

Both routine and selective hedging represent responses to uncertainty about changes in market price. They attempt to fix price level and reduce uncertainty. The third type, carrying-charge hedging, differs from the first two types because it is not a response to market uncertainty. Carrying-charge hedging takes advantage of expected favorable changes in the relative prices of the futures market and the cash market. Short hedging of this type is used by those who expect the spread between two prices (basis) to narrow so that the overall result of the hedged transaction will be profitable. A profitable outcome can be anticipated when the hedger expects the favorable change in basis to exceed the costs of carrying inventory over the period involved. In carrying-charge hedging, the hedger reacts to prospective changes in the basis, rather than the expectation of changes in price level. This hedger does not care whether price level rises or falls; the bet is that basis will narrow. Carrying-charge hedging is used to capture a storage profit. It is a common practice among grain elevator operators. It should be practiced by farmers who own on-farm storage facilities. Its implementation will be developed in Chapter 7.

Hedging Hazards. Hedging with futures contracts can be hazardous. Some hazards that producers face are:

1. Changing basis,
2. Failure to distinguish between hedging and speculating,
3. Hedging too early in rising markets or too late in declining markets,
4. Inadequate liquidity to finance margin calls,
5. Production shortfalls.

In the hedging illustrations outlined above, it was assumed that the price difference, or basis, between cash and futures remained the same at all times. This is not usually the case. The difference may narrow or widen, as one price changes relative to the other. Such disparity prevents price protection through hedging from being 100 percent perfect at all times. Hedging with futures permits fixing price level but does not fix basis. Changing basis is a variable that can either improve or reduce the price protection of a hedge.

Failing to accurately distinguish between hedging and speculating is a hedging hazard. The definition of hedging requires **equal but opposite** positions in two markets to counterbalance the effect of price change. Without fulfilling both requirements, a commodity user or buyer is a speculator. A farmer who places seed in the ground or livestock in pens is speculating in the cash market that he or she can sell the output at a profit. The farmer who sells coffee futures or buys silver is a futures market speculator. A hedger who makes a cash sale but does not cancel the futures position is speculating in futures. A producer who buys futures while owning cash commodities is a "Texas hedger." Texas hedgers have two long positions (one in the cash, one in the futures) and are doubling the effect of price change rather than removing it. A commodity producer or user is not perfectly hedged unless there is a futures market position equal to, but opposite, the cash market position. This requirement is a serious impairment to farmer use of futures in hedging growing crops or anticipated livestock production. The growing production is not yet on hand; actual production may not equal expected (hedged) production.

At times profits are increased by hedging; at other times they are reduced. A seller who hedges before the market top or a buyer who hedges before the market bottom will receive less of a benefit from a hedge than one established at the market extreme. If price level rises after a hedge is initiated, a seller will receive a lower net price than the market later offers. If price level declines after a buying hedge is established, the net cost will be higher than that of not hedging. With routine hedging, the hedger must decide before the fact that he or she will be satisfied with the price offered by the hedge. The hedger gives up the opportunity to realize a more favorable price for the comfort of avoiding an unfavorable one.

Cash liquidity to meet margin calls poses a hazard for hedgers. Since only one of thousands of market participants will place hedges at the market top or bottom, all others must be ready with cash to finance holding their position if the futures move against them. At the time of the cash sale, a more favorable cash price will cover futures account deficits, but the futures market will not wait for settlement until the cash sale occurs. It settles accounts every day and does not finance negative balances. The role of the banker in providing market liquidity for a hedger is important and will be discussed later.

Hedging sometimes fails the farmer because of yield fluctuation. A farmer who hedges corn at $3.00 per bushel may see a widespread drought reduce production and push prices to $4.00. Caught by poor weather, anticipated production of 100,000 bushels might be reduced to 70,000 bushels. The $1.00 inventory value grain on only 70,000 bushels does not offset the futures deficit of $1.00 on 100,000 bushels. Hence, farmers are reluctant to hedge 100 percent of their anticipated production without some recourse for exiting their hedge.

Hedging is not a simple process. It requires understanding, planning, patience, and perseverance. Without them, hedging can increase risk rather than reduce it.

Investment Medium

Three major types of participants use commodity exchanges.

1. Hedgers, who may be short or long, as noted above.

2. Large speculators: businesspersons who make trading in futures their principal business. These professional speculators are ready at all times to buy or sell futures contracts, provided that the price offers the prospects for profits.

3. Small traders: the "public" whose gambling instincts draw them into speculation in search of a "quick kill."

Hedgers and futures market speculators are in two different worlds. One depends upon the other, but each has its respective objectives.

Futures market speculators neither own the commodity nor expect to take delivery of it. They are willing to assume the risk of price change in the hope of profiting by it. Speculators do not care whether price is high or low; they merely want to be right in gauging the direction of price movement. The speculator is, in effect, the risk bearer who assumes the risks that the hedger seeks to transfer. Speculators are like insurance underwriters. They stand between producers who want the highest price for their production and users who seek the lowest price for inputs. Without them, the farmer anticipating 100,000 bushels of corn for sale for November delivery might not have a market at favorable prices. Farmers do not care whether they sell to a processor, cattle feeder, grain merchant, or speculator. Their single-minded objective is to sell for the highest possible price. But in times of high price, user-buyers may prefer to delay their decision in hope that price will decline. The farmers' search for buyers then shifts to the commodity exchange, where they find speculators who are willing to take the other side of their trade. By providing market liquidity, speculators perform an important economic function.

Participants in Futures Trading. An interesting set of characters exists in the drama of the action on the floors of the commodity exchanges. They are described by Hieronymus by the type of trading in which they specialize.[13]

"*Scalpers* busily inject themselves into as many trades as possible. They stand ready to buy a quarter cent below the last trade or sell a quarter cent above. They follow the principle that an incoming buy order will raise the price slightly, or conversely, an incoming sell order will depress price slightly, but shortly, price will reverse and return to the previous equilibrium level. Scalpers are always ready to buy or sell at small price concessions; they bridge the time gap in the flow of orders. They add liquidity to the market and enable orders to be filled without delay.

"*Pit Traders* are grown up scalpers. They come into the pits each day with no positions and leave with no positions. Their objective is to profit from intra-day price changes. They buy when they think price is going up and sell when they think it is going down. They watch the other traders in the pit trying to anticipate what trades others will make before the end of the session. Pit traders want to have a handful of contracts to trade at the close. They collect lots of 5,000 bushel trades and dump them in 100,000 bushel trades. At one time they may trade with the market; at another, against it. They try to sense the mood of the market and buy or sell quickly to get in on the bottom or top of the move and liquidate after the move is made.

"A third group of traders are *Spreaders*. They hope to profit from abnormal price differences between commodities, between delivery months, or between commodity markets. Spreaders may buy December cotton and sell May cotton; they may sell soybeans and buy oil and meal; they may buy Kansas City wheat and sell Chicago wheat; they may sell live cattle and buy lean hogs. Spreaders practice arbitrage in hope of profiting from skewed price relationships that move back to normal relative value. In the process, they add liquidity to the markets and stability to price relationships.

"*Floor Traders* are professional speculators who trade mostly for their own account. Their positions are flexible. They may trade with the market or against it, hold positions for long periods or short periods. They often take large positions and add to them as they sense they are right about direction of the move.

"*Floor Brokers* are the largest group on the exchange floor. They execute orders that flow through brokerage houses from outside the exchange or are placed with them by floor traders. Each broker holds an individual seat, but it may be financed by a firm, so in essence, they work for the firm. They act as agents and are paid a fee rather than a salary for each contract they execute. They can and do trade for their own account, but they must execute consumer orders first. The broker might wish to buy 25,000 bushels for his or her own account and then get an order to sell 500,000 bushels for the firm. The broker must execute for the firm first, then execute his or her own, no doubt after the price has fallen. They have decks of orders tucked in their fingers. Some are sell orders; and some, buy orders, all at different prices. If they make a mistake in executing a trade, the loss is covered out of their pocket. If an order was for 10,000 bushels and the floor broker sold 20,000, the broker makes up the difference if losses result."

For the observer in the visitors' gallery, the activity on the floor resembles a beehive. With hundreds of brokers shouting and gesturing wildly in the pit, each attempting to effect a quick sale or purchase before the price changes, the scene looks more like a yelling convention than an organized market. But they are normal, sane people participating in an orderly arrangement for trading that makes possible the completion of an enormous number of individual transactions within a remarkably short time and in a most effective manner.

In addition to the participants on the floor of the exchange, small trader investors from all walks of life are attracted to the profit potential of commodity speculation.

In order to better understand the nature of market participants, a special study of the corn futures market was made on January 27, 1967. Virtually every occupation was represented among the nonmember traders that day—farmers and ranchers, architects, physicians, secretaries, students, plumbers, and barbers. There were 16 homemakers with positions in excess of 100,000 bushels; a large European trader was short 200,000 bushels; and a South Vietnamese trader was long 10,000 bushels. The number of speculators outnumbered hedgers by 8 to 1. Farmers and food intermediaries were among the most active speculators; more than one-third of the long speculative positions were held by farmers.[14]

A later study took a closer look at farmer participation in the futures market.[15] This survey indicated that almost 6 percent of U.S. farmers with gross sales over $10,000 traded futures contracts in 1976. Large farmers and grain producers were more active in the market than smaller farmers or livestock producers. Over two-thirds of the participating farmers traded commodities they neither used nor produced; this group was speculating rather than hedging. It was estimated that only 1 percent of the farmers were actually using the futures market for hedging in 1976. Their most frequently stated reasons for not trading futures contracts were (1) not acquainted with how the futures market operates, (2) farm too small, (3) futures market too risky, (4) lack of capital for trading, and (5) limited time to devote to trading. A larger group of farmers used the market in other ways. An estimated one-third of all farmers watched market prices as a barometer of commodity values, even though they did not trade futures contracts.

It is important to note that no commodity exchange or reputable commission house encourages uninformed public speculation. The modern commodity trader is, for the most part, very well informed and a keen student of markets. He or she may well be considered akin to any experienced, resourceful businessperson. The traders' business is to analyze all pricing-making factors, buying when he or she thinks prices are too low and selling when prices are too high. In addition to foresight, the trader must have courage and resources and know how to limit losses when his or her judgement is wrong.

Two views have been held on the role of the small speculative traders in the futures market. On one hand, they are seen as a disturbing influence in the market, accentuating price swings, and on occasion contributing to wild and disastrous price fluctuations. On the other hand, such traders are seen as a necessary element in the market, since their presence makes it possible for the expert trader and hedger to find traders to take the opposite side of a transaction. This view holds that small traders, through their losses, "supply the income which is necessary to support the continued trading activity of the professional."[16]

Stewart conducted a pioneering research project on the trading behavior of small speculators in grain futures over a nine-year period.[17] Information on trading activities of nearly 9,000 traders and more than 400,000 individual transactions were analyzed. The study confirmed a number of commonly held opinions on results of speculative trading, but it tended to disprove others that had been widely accepted.

The first conclusion was that the great majority of small speculators lost money in the grain futures market. There were 6,598 speculators with net losses, compared with 2,184 with net profits. About 75 percent of small traders lost money speculating. The total net losses of almost $12 million were six times as great as profits of about $2 million.

The data showed that the great majority had relatively small profits and losses. The average loss for unprofitable traders was $1,812, and the average profit for successful traders was only $945 per trader. The profits of 84 percent of the profitable traders were less than $1,000 each; the losses of 68 percent of the losing traders were less than $1,000 each. Obviously, a large percentage of the traders in the sample operated on a small scale, and many of them discontinued trading before realizing large profits or suffering large losses. At the extreme, the most successful trader had profits of $395,718; the most unsuccessful trader had losses of $413,212 even though four of the first five trades showed profits. The major reason for the high ratio of losses was the small speculator's characteristic hesitation to close out losing positions. An often-quoted maxim for speculative trading is "Cut your losses short and let your profits run." Contrary to this philosophy, speculators in the study exhibited a clear tendency to cut their profits and let their losses run. Losing positions were usually held for longer durations than profitable trades.

The study also confirmed the commonly held impression that the amateur speculator is more likely to be long than short in the futures market. About half of the speculators in wheat and corn held positions only on one side of the market, mostly long. The other half of the traders who held both long and short positions accounted for a larger proportion of the trading, however. From the standpoint of market activity, 63 percent of the positions were long and 37 percent short in wheat futures, but in corn, 58 percent were long and 42 percent were short. Short positions tended to show profits more frequently than long positions.

In the study, people from all walks of life participated in trading, as shown in Table 3–1. Businesspersons were the largest occupational group, and farmers were next. There were 316 persons engaged in the grain business who also carried personal, speculative accounts with the brokerage firm from which the sample was obtained. Among managers of business concerns, 840 were profitable traders, compared to 2,563 losing traders. Persons with occupations "unknown" had the greatest proportion of profitable trades, 32.3 percent. Farmers had the lowest proportion of profitable trades, 21.2 percent. Special knowledge of the commodity being traded seemed to have little effect on the outcome of speculative trading in the group studied.

The study showed the tendency of long speculators to buy on days of price declines and for shorts to sell on price rises. Trading against the current movement of

TABLE 3–1: Number of Speculative Traders with Profits and with Losses, and the Percent with Profits, by Grain and Major Occupational Group

Commodity and Occupational Group	Traders with—		Total	Percentage with Profits
	Profits	Losses		
	(number)	(number)	(number)	(%)
WHEAT				
Business managers:				
Grain business	83	189	272	30.5
Other	702	1,993	2,695	26.0
Professional	179	498	677	26.4
Semiprofessional	27	64	91	29.7
Clerical	130	349	479	27.1
Farmers	200	697	897	22.3
Manual workers	214	621	835	25.6
Retired	240	576	816	29.4
Unknown	270	509	779	34.7
Total	2,045	5,496	7,541	27.1

(Continued)

TABLE 3–1 (Continued)

Commodity and Occupational Group	Traders with—		Total	Percentage with Profits
	Profits	Losses		
	(number)	*(number)*	*(number)*	*(%)*
CORN				
Business managers:				
Grain business	76	96	172	44.2
Other	577	806	1,383	41.7
Professional	133	221	354	37.6
Semiprofessional	16	20	36	44.4
Clerical	76	159	235	32.3
Farmers	151	261	412	36.7
Manual workers	163	289	452	36.1
Retired	189	280	469	40.3
Unknown	144	271	415	34.7
Total	1,525	2,403	3,928	38.8
OATS				
Business managers:				
Grain business	23	54	77	29.9
Other	223	374	597	37.4
Professional	45	74	119	37.8
Semiprofessional	7	7	14	50.0
Clerical	37	50	87	42.5
Farmers	57	112	169	33.7
Manual workers	53	108	161	32.9
Retired	69	111	180	38.3
Unknown	75	107	182	41.2
Total	589	997	1,586	37.1
RYE				
Business managers:				
Grain business	22	37	59	37.3
Other	201	303	504	39.9
Professional	36	78	114	31.6
Semiprofessional	2	12	14	14.3
Clerical	33	49	82	40.2
Farmers	43	94	137	31.4
Manual workers	52	88	140	37.1
Retired	62	79	141	44.0
Unknown	46	76	122	37.7
Total	497	816	1,313	37.9
ALL GRAINS				
Business managers:				
Grain business	92	224	316	29.1
Other	748	2,339	3,087	24.2
Professional	185	583	768	24.1
Semiprofessional	23	79	102	22.5
Clerical	121	433	554	21.8
Farmers	218	810	1,028	21.2
Manual workers	212	736	948	22.4
Retired	254	700	954	26.6
Unknown	331	694	1,205	32.3
Total	2,184	6,598	8,782	24.9

Source: Blair Stewart 65.

prices was the dominant pattern. This indicates that traders in the sample were predominantly price-level traders. Longs tended to buy when prices fell below levels that they considered proper, and shorts sold when prices advanced above levels that they believed justified.

The results of the study confirm the saying that small traders are usually wrong in their analysis of market direction. But someone has to supply grist for the commodity mill.

Attractions of Speculating. Huge potential profits earned quickly are the greatest lure to speculation. Books have been written on *How I Made a Million Dollars Trading Commodities*.[18] Speculators must keep in mind that losses occur as quickly as profits. Financial leverage works both ways. In the zero-sum game of futures trading, someone's gain is always someone else's loss. Either the hedger is gaining price protection at the expense of speculators or hedgers are sending speculators reward for their efforts.

In addition to possible high profits, other positive factors attract speculative participation in the market:

1. Financial leverage. A small amount of money controls a larger value. Typically, speculative margin requirements are 5 to 10 percent of the contract value. At $1,200 per contract on corn, it is possible for $24,000 to control the price effects on 100,000 bushels of grain worth maybe $300,000.

2. No long-term dollar commitments. It is possible to be in and out in one day.

3. Choice in the degree of risk assumed. Investments can be made in commodities with relatively large or small fluctuation in price. The speculator can trade 5,000 bushels or 500,000 bushels in a single trade.

4. Profits come as easily from price decline as from price increase. Either short or long positions may be taken, depending on which way the speculator thinks price will move.

5. Commissions are low relative to asset value. Trading fees range from $20 to $100 on a soybean transaction worth perhaps $35,000.

6. Price information is public, and expert advice is available to all traders. Most knowledge of factors affecting prices (politics, weather, etc.) is public, so everyone has an equal opportunity to analyze such factors. Anyone trading the market has access to reams of newsletters, books, and pamphlets providing information and offering advice.

Trading Units. Each commodity has its specific unit of quantity traded. For example, the cotton futures contract on the New York Cotton Exchange is 50,000 pounds (100 bales). The live cattle contract on the Chicago Mercantile Exchange calls for 40,000 pounds (about 40 head @ 1000 pounds each). A list of agricultural commodity contract trading facts is available at the end of this chapter.

In most cases when placing an order with a broker, the customer does not specify the number of pounds or bushels. A person who places an order to buy "one cotton contract" on the New York Cotton Exchange bears the understanding that is for a regular contract unit of 50,000 pounds. Placing an order to sell "one February cattle contract on the Merc." Means that the order is for the Mercantile Exchange unit of 40,000 pounds of live cattle.

Trading Price Fluctuations. Minimum price fluctuations in many commodities are in one-hundredths of a cent; others are in quarters of a cent. For cotton, the market fluctuates in "points," each equal to one-hundredth of a cent per pound. Therefore, each point equals $5 in the value of the contract, and a fluctuation of a full cent, or 100 points, equals $500 per contract. Livestock contracts also fluctuate in one-hundredths of a cent. Grains are quoted in quarters of a cent per bushel, the minimum fluctuation in most markets. With a full contract representing 5,000 bushels, a fluctuation of a full cent in the price of wheat equals $50 per contract.

In addition to minimum fluctuations, daily trading limits specify the maximum price fluctuation that can occur from the previous day's settlement price. Limits restrict how much price can change from one day to the next. Trading limits on corn are 12 cents per bushel or $600 per contract; wheat, 20 cents or $1,000; soybeans, 30 cents or $1,500 per contract; and cotton, 2 cents (300 points) or $1,000. Live cattle can fluctuate by a maximum of 1.5 cents (150 points) or $600 per contract; lean hogs, 2 cents (150 points) or $800 each day. It is possible for a commodity to trade limit up and limit down during the same day. The change in value of a contract under those circumstances would be double the limits stated above.

It is also possible for limit price moves to occur on several consecutive days. After three limit days, limits are modified or removed as the exchange governing board deems it necessary in order to allow the severe condition to work itself out in the market. Three days of 30-cent–limit moves on 5 soybean contracts (25,000 bushels) produces a $22,500 change in value of the contracts; three limit moves of 200 points on 10 hog contracts means a $24,000 change in value—plus or minus. The fact that markets can, and sometimes do, change drastically emphasizes the need for market participants to have cash liquidity to meet their financial obligations. Limit day moves often prevent any trading from taking place. The trader on the wrong side of the move cannot cancel the position because no one will take the other side of the offer. As a consequence, margin calls can require lots of money in a short period of time.

Regulation of Futures Trading[19] Public concern with futures markets, especially the price effects of alleged "excessive" speculation, has resulted in state and federal regulation of these markets. The regulations complement the very extensive self-regulating efforts of the commodity exchanges. Regulation is intended to protect participants and eliminate abuse from unfair practices. But it cannot prevent traders from incorrectly assessing price movement and losing money as a consequence of misjudgment.

The history of futures trading is marked by stories of price fixing, corners, squeezes, and market manipulation. The allegations continue to this day. The markets attract high rollers, because the profits to be won by successful manipulation by "gunslingers" are enormous. The market is fueled by greed, which can be controlled only by competition and by trading rules.

The Grain Futures Act of 1922 was the initial federal regulation of futures markets. In 1936, this law was amended to create the Commodity Exchange Act. These laws granted authority to the federal government to prevent market abuses, price manipulation, cheating, dissemination of false information, and fraud in futures markets. From 1936 until 1975, futures trading was regulated by the Commodity Exchange Authority (CEA), a division of the U.S. Department of Agriculture. Regulatory authority over these markets was shifted to the Commodity Futures Trading

Commission (CFTC), an independent governmental regulatory agency, by the Commodity Futures Act of 1974.

The Commodity Exchange Act provides that no organized futures trading may be conducted in any commodities, except on an exchange that is officially designated by the CFTC. Exchanges are required to demonstrate that futures contracts for which they seek trading approval will not be contrary to the public interest. To maintain trading approval, the exchanges must comply with certain stipulations of the Act and the rules and regulations thereunder for the protection of customers, who trade through member brokers.

The Act provides that commission merchants trading in regulated commodities must be registered. The registrations must be accompanied by financial statements to demonstrate the financial integrity of the registrant. Registrations expire as of December 31 of each year. Thus, a new registration and new report are required each year. A new registration is also required if any important changes of partnership or ownership occur during the year, as set forth in the rules. Moreover, a new section of the Act requires registration of persons associated with a futures commission merchant who are involved in soliciting or accepting customers' orders. It also requires registration of commodity trading advisors and commodity pool operators. Registration provides a means for detecting and eliminating previously fraudulent operators.

Regulations provide for complete segregation of customers' funds from brokers' funds. When depositing customers' funds, brokerage firms must notify the bank of the nature of the funds by appropriately labeling the account and obtaining an agreement with the bank, waiving any claim or right of offset that the bank might otherwise have against such funds.

The Act gives the Commodity Futures Trading Commission the authority to establish position limits on the number of contracts held by individual speculators (spec limits). For example, the spec limit on the net long or short positions for individual speculative accounts has been set at 3 million bushels for wheat, corn, and soybeans in the spot month and 20 to 45 million bushels in all months combined. The limit on the maximum net long or net short position that any person may hold or control for lean hogs is 650 contracts in the expiring, spot month and 2,000 contracts in any contract month. These limits upon position and daily trading do not apply to bona fide hedging transactions.[20]

Despite the efforts of regulators, several famous market "corners" and "squeezes" have occurred. A corner results when an individual (or group acting in concert) holds a large portion of the outstanding long positions and also controls most of the cash commodity that could be delivered by short sellers at contract maturity. In liquidating their position as contract maturity approaches, the shorts can either buy back their speculative positions (from a long) or purchase the commodity to deliver against the contract. But in a squeeze, they face the same seller in both cases and must pay the price of the seller. This is reason enough that farmer-hedgers are advised to lift their positions or roll forward their contract obligations before the beginning of a delivery month. It is not wise to get caught in a squeeze.

Extensive regulation has made market manipulation difficult and hazardous. Limits on the quantity of contracts, increases in the number of delivery points, and careful scrutiny of the markets for monopoly behavior diminish the opportunities for manipulation of future markets. On balance, they are probably no more subject to manipulation than other markets.

Farmer Trading. Farmers who become intrigued with commodity marketing face the choice of whether they should become futures market hedgers or speculators. They often fail to make that important distinction and lose the farm before they understand the difference. For example, they accept double jeopardy if they institute a "Texas hedge." Here they own the commodity and buy futures contracts, feeling that the market is going up. If they are wrong and the market declines, they lose on both positions; they suffer from decline in value of their inventory and losses on their futures contracts.

There is one good reason for farmers or ranchers to open a speculative account: It may encourage them to become better students of the market. What they learn from "playing" with $5,000 in a speculative account may cause them to make advantageous sales of their products that could earn $50,000 additional income.

When farmers make a conscientious decision to become speculative traders, they have a new set of rules to learn and apply. "Paper trading" is a good place to start. It provides practice without incurring costs. Writing down the position taken and reasons for it forces the beginner to keep score on gains and losses. In six months' time, a paper trader will determine whether the plan is working and whether he or she has the emotional make-up to be successful. The next step into the real world is to open an account on the MidAmerica Exchange to trade mini-contracts. The initial investment is low, and extent of loss exposure is relatively small. A producer should trade only the commodities he or she actually produced.

Why speculate on futures? For the thrill of it. The thrill encourages farmers to follow the market more closely as they read about the subject and gain understanding of fundamental and technical market forces. The farmer who limits a speculation account to $5,000 is not going to lose much money, but it will provide an outlet to his or her speculative urges. Hedging and speculating are two different worlds. One depends upon the other, but each has its own objectives. Futures trading as a marketing tool should reduce risk, not increase it. As the studies cited show, most small traders lose in their speculative ventures. The objective of farmer speculating is to learn to be a better hedger.

Rules for Speculating[21] There are a number of basic rules for commodity speculating. An analysis of fundamental and technical conditions is the first step toward successful speculation. Price forecasts must then be combined with a trading plan or strategy that the individual is capable of adhering to and that will generate profits. As a "position" trader, a person must have reasons for his or her trades. By limiting oneself to one or two commodities, learning all one can about them, and studying the technical factors, one can develop a strategy. Emotions should not change the plan; common sense and logic should prevail.

Traders generally believe that profit potential should be large in relation to risk. Traders should specify their profit objective as well as the maximum loss they are willing to sustain on each trade they initiate. They must learn to practice the old adage: "Limit your losses and let profits your run." *The challenge is knowing when the profits have stopped running.* Their own personal preferences will determine if the objective for profit is the maximum potential profit or some lesser figure.

Once the trading guidelines have been determined, the trader should specify the amount of capital to be risked. Although it is implicitly assumed that speculators should be using funds that are totally risk capital, it is still essential to maximize the return from this risk capital. A farmer who chooses to be a futures market trader should open two accounts, one for hedging activities and one for speculating. The two accounts should be kept distinctly separate. Farm production money should not be used for speculation. Trading money should be limited to the amount that can be lost without jeopardizing the business.

Experienced speculators recommend that a great proportion of an individual's capital should not be risked on a single trade. Also, the trader should limit the open positions as to many commodities as he or she can adequately follow, keeping some capital in reserve for additional opportunities. Successful speculators often recommend (1) that additions to an initial position should be made only in the event that profits have already been made and (2) that subsequent commitments should not be in amounts greater than the initial position. Liquidating a position should be based on the original plan. Market conditions may change, however, and it is essential that the speculator maintain some degree of flexibility.

Only the speculator's own experience and preferences can be relied on to assess the desirability of a trade based on profit potential relative to risk, to determine profit objectives and loss limits, and to decide on additions to the original position and final liquidation. Successful price forecasting and trading ultimately depend on the individual's own temperament and objectivity, as well as the economic models and trading plans that have been developed.

In addition to the numerous rules implicit in a systematic approach to speculation, other guidelines, representing the trading knowledge of successful speculators, are helpful. The following list of such guidelines is intended for reference and not necessarily as recommendation.[22]

1. "Markets in which trading is contemplated should be carefully analyzed. Quick actions based on rumors, tips, etc., should be avoided.

2. "Speculation should not occur in the presence of doubt about the price forecast or without a trading plan.

3. "Speculation should usually follow the market trend.

4. "Speculators should generally avoid attempts forecasting reversal points of trends.

5. "Forecasts of price declines, and consequently, short positions should have the same priority as forecasts of price increases, which call for long positions.

6. "Speculation should be initiated when the potential profit is great relative to risk.

7. "Successful speculation requires limiting losses and letting profits run. Traders should be prepared to accept numerous small losses. A limited number of highly successful trades will offset those losses.

8. "In addition to a careful analysis of commodities, successful speculation requires a well-developed trading plan. This includes limiting the amount of capital on any single trade and maintaining capital in reserve."

A careful approach to the market, coupled with the appropriate temperament and psychology, will increase the probability of successful trading.

Is Futures Trading Gambling? Much of the historical controversy about commodity speculation has related to the confusion of speculation and gambling. Gambling is almost universally condemned as wasteful and as working great evil on the gambler himself or herself. Commodity speculation has often been branded as the biggest gamble of them all.

Commodity trading is a zero-sum game; for every winner there is a loser. Credits in the brokerage account of one trader are covered by debits in the account of another.* However, brokerage account credits and debits may or may not be profits or losses. And it is inaccurate to characterize the zero-sum nature of trading as an unnecessary exercise for shifting money between winners and losers. As illustrated in Figure 2-9, short hedgers accumulate brokerage account credit balances as price level declines. Speculators ante up the money transferred to hedgers. On the other hand, as price level rises, speculators accumulate credit balances as hedgers pay up. Depending on price direction, the speculator loses or profits. But the hedger neither profits nor loses; hedgers merely fix a price on their commodity inventory. Because the hedger owns inventory in addition to futures position, the value of the inventory fluctuates in the opposite direction of the brokerage account balance. Losses in one account (inventory) are offset by gains in the other (brokerage). Thus the futures market fulfills its role in shifting the financial effects of price change from owners of commodities (hedgers) to others (speculators) who are willing to risk loss in hope of gain due to price change.

Speculation is like gambling in several ways:

1. Both depend upon uncertainties,

2. Both involve risk of loss for the sake of possible gain,

3. Occurrence of certain events results in losses to one player while other events have the opposite effect,

4. Both may involve careful calculation of probabilities and developing forecasts, or both may be based on pure chance.

Nonetheless, certain essential distinctions can be made in defense of speculation:

1. Gambling involves creating risks that would not otherwise exist; speculation involves assuming necessary and unavoidable risks of commerce.

2. In every futures transaction, the speculator incurs the duties and acquires the rights of a holder of property and thus is an integral part of commerce. Whether the impact of the speculator's activities are good or bad is neither here nor there—they are inevitable and necessary conditions of business practice.

3. Every trade on futures markets has a buyer and seller; the number of contracts bought equal the number of contracts sold. Without speculation, hedging is impossible.

Gambling is the creation of games of chance; futures trading is a mechanism to transfer unavoidable business risk from one who seeks to avoid it to another who spe-

---

*Actually, commodity trading is a negative-sum game. Trader A's debits offset Trader B's credits, but brokers collect commissions from both.

cializes in it. Futures trading has a social and economic justification that gambling lacks. Therefore, futures trading is not the same as gambling.

Summary

Futures markets serve three important roles: price discovery, risk transfer, and investment medium. In a capitalistic economy, value is determined by the interaction between buyers and sellers. Governmentally dictated prices do not work. Neither do monopoly pricing practices. A free and open market available to all buyers and sellers more accurately reflects true supply/demand conditions. Futures markets bring buyers and sellers face to face in their quest for price discovery. They also increase the level of pricing information as the results of trading flash worldwide almost instantaneously.

Farmers are the ultimate speculators because of the nature of their business. They commit thousands of dollars to purchase facilities and equipment, seed, fertilizer, herbicides, pesticides, feeder livestock, feed, and supplements. They and their families work all year tilling the soil, feeding the livestock, while wearing out their machinery and facilities. They risk disease, insects, weather, fire, and theft to produce a crop or pen of livestock. When their product is ready for market, there is no guarantee that their efforts will be adequately rewarded. Market price may offer significant profits, or it may fall below cost of production. Futures markets allow farmers (and

Commodity Futures Contract Specifications

Commodity	Exchange	Hours (CST)	Size	Months	Min. Price Change	Daily Price Limit*
Corn	CBOT	9:30-1:15	5,000 bu.	H,K,N,U,Z	¼ cent = $12.50	12cents/bu.
Soybeans	CBOT	9:30-1:15	5,000 bu.	F,H,K,N,Q,U,X	¼ cent = $12.50	30cents/bu.
Wheat	CBOT	9:30-1:15	5,000 bu.	H,K,N,U,Z	¼ cent = $12.50	20cents/bu.
Cotton	NYCE	9:30-2:00	50,000 lbs	H,K,N,V,Z	1 cent = $5.00	Variable
Soybean Meal	CBOT	9:30-1:15	100 tons	F,H,K,N,U,V,Z	10 cents = $10.00	$10/ton
Oats	CBOT	9:30-1:15	5,000 bu.	H,K,N,U,Z	¼ cent = $12.50	10cents/bu.
Orange Juice	NYCE	9:15-1:45	15,000 lbs	F,H,K,N,U,X	5 cents = $7.50	None
Live Cattle	CME	9:05-1:00	40,000 lbs	G,J,M,Q,V,Z	2.5 cents = $10.00	$1.50/cwt.
Lean Hogs	CME	9:05-1:15	40,000 lbs	G,J,M,Q,V,Z	2.5 cents = $10.00	$1.50/cwt.
Feeder Cattle	CME	9:05-1:00	50,000 lbs	F,H,J,K,Q,U,V,X	2.5 cents = $12.50	$1.50/cwt.

Months: F= Jan., G= Feb., H= Mar., J= Apr., K= May, M= June, N= July, Q= Aug., U= Sept., V= Oct., X= Nov., Z= Dec.

* Daily price limits subject to change when several specific contracts are limit up (or down) for several days.

others) to transfer the effects of price change and guarantee themselves profit before delivering their products to market.

Speculation in commodity markets does not depend upon the existence of futures markets. Speculation is a part of any economy in which economic decisions have to be made in the face of uncertainty of future events. But futures markets do contribute to speculative activities by making it possible for individuals to engage in speculation without having to become engaged in producing, handling, storing, or processing commodities. Specialization in practicing speculation is facilitated by futures markets. Transaction costs and capital commitments are minimized. Standardization of futures contracts enables the speculator to establish either long or short positions in the market with equal ease. Organized exchanges allow speculators to practice their trade efficiently.

The commodity exchanges thus provide a market that is not matched by markets without futures trading: markets for houses, works of art, or antique automobiles. Agricultural producers should learn to use these exchanges efficiently to minimize the producers' speculation rather than lobby for their elimination.

Questions for Study and Discussion:

1. Trace the evolution of grain trading practices that became part of the features of futures trading.

2. What characteristics qualify a commodity for futures trading? Why are some products unsuitable for trading?

3. Discuss the three major roles of futures markets and the significance of each one.

4. What is price discovery?

5. What functions does the Clearing Corporation perform?

6. How does the Internal Revenue Service treat profits and losses from futures trading for both the hedger and the speculator?

7. List the various types of market orders placed by a futures market trader.

8. What types of risk are associated with ownership of agricultural commodities?

9. Give examples of buying and selling hedgers and how they fix a price.

10. Do future market gains and deficits produce profits or losses for hedgers?

11. Discuss the three types of hedging.

12. Who are the participants in futures trading?

13. What are the features that make commodity trading attractive to investors?

14. What is the regulating agency of futures trading, and what does it do?

15. What generalities serve as rules of successful speculation? Why are they important?

16. Is futures trading gambling?

Futures Markets Exercise: Paper Trading

On any day, you should have a reason to believe the present futures price of a commodity is too high or too low, relative to what its true equilibrium price will be at a future date. "Paper trading" is a technique to test your analytical ability and to gain experience in the mechanics of futures trading without having to put money on the line. It also requires you to develop a trading strategy and test it in real time.

You will be supplied with trading forms and a record sheet. You should make decisions to sell or buy a futures position and record them on the forms. You should record your position, commodity, number of contracts, and REASON for your position after the market close on one day and turn it in before the market opens the next day. Your reason should be based on a fundamental or technical fact rather than "I think the market is too low (or high)." Why is it high or low? Maintain an additional record of profits, losses, and account balance over the period of time of the exercise. In the end, you should write a summary of your results and what you learned about both market mechanics and your speculative abilities. Grading should be upon student justification of trades rather than amount of trading profit.

Futures Trading Exercise: Record of Trading

The following form can be used to record daily activity in your account. Use a photocopy of this form for recording your transactions. To illustrate the process, open an account with $20,000 of paper money. On the next day, suppose you decide the price of corn is too low and is likely to rise because dry weather threatens production. To initiate a position, you buy one December corn (5,000 bushels) by completing the order form and explaining the reason for the decision. On the accounting form, record the position as **Buy 1 Dec. Corn** under the "Transaction" column. Assume it is Tuesday afternoon and the closing price is $3.05. The transaction must be recorded before the opening at 9:30 a.m. CST on Wednesday. An initial speculative margin requirement of $1,200 per contract calls for a "Margin Deposit" of that amount, dropping the cash balance to $18,800.

Each day following initiation of the position, the account is settled by adding gains or deducting losses due to price change. The effect is recorded in the "Change in Value of Position" column. "Equity in Excess of Maintenance" refers to the balance in excess of minimum maintenance margin requirements. Minimum maintenance on a corn position is $600. Should "Equity" fall below minimum maintenance, additional deposits are required to satisfy "margin calls."

Suppose that on Wednesday the market closes at $3.08. This figure entered in the "Price" column translates into a $150 gain as recorded in "Change in Value of Position." The "Equity" column notes the $150 of "profit." On Thursday price falls to $3.07. The "Change in Value of Position" drops $50 and "Equity" changes to $650. A limit down move on Friday is recorded in the "Change in Value of Position" as $-500 and drops "Equity" to $150. On Monday, another limit move to $2.87 loses $500 and drops "Equity" below maintenance to $-350. A margin call for $400 is issued to bring the equity balance back above minimum; the cash account balance drops to $18,400. You panic and close out the trade with a "sell" order. You deduct the commission paid to the broker (assume $50) and absorb the 18 cents per bushel ($900) loss. You are left

FUTURES TRADING EXERCISE

Name _____ Date _____

New Position_____ Cancelling Out Old Pos._____

Transaction	Kind of Grain	Quantity	Month	Price
_____	_____	_____	_____	_____

Explain why you made this decision and give a complete and detailed description including *dates* of the *sources* of information you used in making this decision.

ACCOUNTING FOR TRADING EXERCISE

Transaction	Date	Price	Margins: Initial (I), Deposits (D), or Withdrawals (W)	Change in Value of Position	Equity in Excess of Maintenance	Trade Profit or Loss	Commission	Balance in Cash Account
Initial deposit	6/1							20,000
B1 Dec corn	2	3.05	1200 (I)		600			18,800
	3	3.08		150	750			
	4	3.07		−50	650			
	5	2.97		−500	150			
	8	2.87	Margin call 400 (D)	−500	−350 50			18,400
S1 Dec	9	2.87			650	(900)	50	18,350

(Students can record trades of their own in the form on the following page.)

ACCOUNTING FOR TRADING EXERCISE

Transaction	Date	Price	Margins: Initial (I), Deposits (D), or Withdrawals (W)	Change in Value of Position	Equity in Excess of Maintenance	Trade Profit or Loss	Commission	Balance in Cash Account
			(Students can use this form to record trades of their own.)					

with $18,350 in the cash account, since $600 of initial maintenance remains with the broker.

Now it is your turn to enter a trade and continue the accounting throughout the term. You will be assigned three or four commodities to trade, accounting for each one on a separate page.

Endnotes for Chapter 3:

1. _____, *An Architectural History, The Chicago Board of Trade* (Chicago, IL: The Chicago Board of Trade, n.d.).

2. Lloyd Besant, *Action in the Marketplace: Commodity Futures Trading* (Chicago: Chicago Board of Trade, 1983).

3. Thomas A. Hieronymus, *Economics of Futures Trading* (New York: Commodity Research Bureau, 1971).

4. Hieronymus, *Economics of Futures Trading* 14.

5. Julius B. Baer and Olin G. Saxon, *Commodities Exchanges and Futures Trading* (New York: Harper & Row, 1948) Chap. VI.

6. Frederick L. Thompson, *Agricultural Marketing* (New York: McGraw-Hill, 1951) 204.

7. Besant, *Action in the Marketplace* 24.

8. This discussion is based on copyrighted material from Chicago Board of Trade publications and is used with permission. Besant, *Action in the Marketplace.*

9. Material for this section was gleaned from William J. Jiler, *Understanding the Futures Markets* (Jersey City, N.J.: Commodity Research Bureau, 1983) 23. Used by permission.

10. Jiler, *Understanding the Futures Markets* 24.

11. Commerce Clearing House, *Standard Federal Tax Reporter* (Chicago: Commerce Clearing House).

12. USDA, *The Secret of Affluence* (Washington, D.C.: USDA, Office of Communication, June, 1976) ii.

13. Hieronymus, *Economics of Futures Trading* 47.

14. Commodity Exchange Authority, *Trading in Corn Futures* (Washington, D.C.: USDA, Aug. 1967).

15. John W. Helmuth, *Grain Pricing,* Economic Bulletin No. 1 (Washington, D.C.: Commodity Futures Trading Commission, Sept., 1977).

16. Blair Stewart, *An Analysis of Speculative Trading in Grain Futures,* Tech. Bulletin No. 1001 (Washington, D.C.: Commodity Futures Trading Commission, Oct., 1949) 4.

17. Stewart, *An Analysis of Speculative Trading in Grain Futures.*

18. Larry Williams, *How I Made $1,000,000 Last Year Trading Commodities,* 3rd ed. (Brightwaters, N.Y.: Windsor Books, 1979).

19. Information for this section was gleaned from two sources: William M. Phelan, *How Commodity Futures are Regulated* (Chicago: Chicago Mercantile Exchange, Aug., 1973), and Jiler, *Understanding the Futures Markets.* Used by permission.

20. _____, Chicago Board of Trade and Chicago Mercantile Exchange Rulebooks, CBOT and CME, Chicago, IL, 1999.

21. For a discussion about speculating from a producer's perspective, refer to Merrill Oster, *How to Multiply Your Money—A Beginner's Guide to Commodity Speculation* (Cedar Falls, Iowa: Investor Publications). Materials used by permission.

20. Richard I. Sandor, *Speculating in Futures* (Chicago: Chicago Board of Trade, 1977). Information comes from copyrighted material taken from Chicago Board of Trade publications and is used with permission.

Chapter 4

Price Determination: Market Fundamentals

Price Determination: Market Fundamentals

Farmer and rancher marketing decisions and price forecasting are rooted in analysis of market fundamentals of demand, supply, and prices. The dual questions of "What determines value?" and "How are prices established?" stimulate interest in market analysis. In capitalistic economic conditions, answers are supplied by economic principles of demand and supply. But probably no more overused term is less understood than the "law of supply and demand." To form a foundation for understanding and evaluating commodity prices from the fundamental perspective of demand/supply, we need to sort out the meanings and implications of the terms.

The Meaning of Demand[1]

The economic definition of demand is **a schedule of the quantity of a commodity which buyers will purchase at different prices in a given market, at a given time.** The numbers in Table 4–1 present a hypothetical demand schedule for corn to illustrate that relationship. In any production period with the price at $4.00 per bushel, buyers would be willing to take 6.1 billion bushels; if price were $3.00 per bushel, they would take 7.0 billion bushels; if $2.00, 8.5 billion bushels, and so on for all possible quantities and price combinations.

It is important to emphasize that the demand schedule does not indicate what the market price and quantity taken actually are; it indicates only the effect different prices would have on quantity purchased. A demand schedule expresses a price-

TABLE 4–1: Hypothetical Demand Schedule
for Corn

Price	Amount Purchased
($/bu.)	*(billions of bushels)*
4.00	6.10
3.75	6.30
3.50	6.50
3.25	6.75
3.00	7.00
2.75	7.30
2.50	7.60
2.25	8.00
2.00	8.50

quantity relationship from a buyer's point of view. The price that will exist in a market situation is not known; a demand relationship alone cannot establish price. What it does show is how quantity demanded will rise or fall as price decreases or increases.

The Demand Principle

Demand expresses the relationship between quantities purchased by consumers and the price of the commodity: The lower the price, the more that will be purchased; the higher the price, the less that will be purchased. The demand concept is formalized into the Demand Principle. It says that the quantity of a commodity that potential buyers are willing to take varies inversely with price. The principle may be stated another way: The greater the quantity of a commodity offered for sale in a given market at a given time, the lower must be the price per unit at which the entire amount can be sold.

The demand relationship can be illustrated as a demand curve as in Figure 4–1. Three reasons cause the characteristic downward, negative slope. The first derives from the diminishing marginal utility of a consumption good. The more consumers have of any given item during a given period of time, the less enjoyment they get from each successive unit of it. The first ice cream cone on a hot day is delightful, but the second and third produce successively less satisfaction. Because the enjoyment is less and less, a customer would be willing to purchase additional cones only if they were reduced in price. As stated by economists, diminishing marginal utility requires lower and lower price to induce people to consume more and more of a product during any one time period.

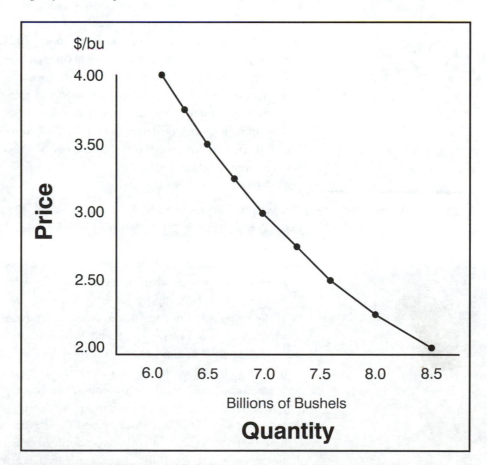

FIGURE 4–1: Hypothetical Demand Curve for Corn. Source: Kohls and Uhl 149.

Second, people have different purchasing power. Wealthy people can afford to buy steaks at $9.00 per pound; those with lower income, having just as great a liking for steak, perhaps, could not afford to pay more than $4.00 per pound. Thus, the lower the price, the greater the number of potential customers who can afford to buy steak, and therefore, the greater will be the quantity purchased at lower prices.

Finally, people differ in their liking for a commodity and thereby in the price they are willing to pay for it. One person thoroughly enjoys lamb chops, possibly because they were a prestige dish during childhood. A second person eats lamb upon occasion and a third may not like it at all. Lamb would have to be sold at a moderately low price, relative to other meats, to get the second person to buy it, and at a very low price to induce the third. The lower the price, the greater the number of potential customers who will pay that price, and hence, the greater the sales.

Aggregate Demand

Demand principles explain how consumers pick and choose the products they buy as they allocate their income among the various items available for purchase. Consumers can often substitute one commodity for another, according to relative prices and the respective want-satisfying power. If pork is plentiful and cheap, consumers will buy more of it, relative to beef, than if it were scarce and expensive. Each individual buys more and more of a particular product until the satisfaction he or she gets from the last unit of each one is proportional to its price. If pork is $2.50 per pound and beef is $5.00 per pound, the consumer will buy enough of each one until the additional satisfaction he or she gets from the last pound of beef is twice the satisfaction he or she gets from the last pound of pork. The consumer allocates income among all the different goods and services bought until he or she gets the same additional satisfaction from the last nickel spent on each one. In so doing, each consumer establishes an individual demand equilibrium for the goods and services purchased. An individual consumer's willingness or reluctance to part with spendable income establishes demand for a product or service.

By adding together the demands of individual consumers, the aggregate demand for the market is established. It is the quantity of a product that all consumers in a market purchase at different prices.

Demand Elasticity

Elasticity of demand is the extent to which the quantity bought changes in response to a change in price. Demand elasticity is the relative, or percentage, change in the quantity bought resulting from a given percentage change in price. With some products, a change in price will bring about a proportionately large increase or decrease in the quantity that can be sold, while with others the quantity is only slightly responsive to change in price.

If the response in quantity taken is relatively greater than the change in price, demand is said to be **elastic**. In general, luxury or semi-luxury items have the most elastic demands. Consumers can do without the item or easily substitute another item for it. Leather coats provide an example of a luxury good with elastic demand.

Commodities for which consumption is relatively insensitive to price change have **inelastic** demand. The quantity of consumption of these products will hardly change as a result of higher or lower price. The classic example of a commodity with inelastic demand is salt. If salt price were doubled or tripled, people would consume about the same amount. If the price were halved, consumption would not increase materially. Salt has few substitutes and is a minor cost item. It fits the classification of necessities, all of which have inelastic demands.

The transition point between elastic and inelastic demand is called **unitary** elasticity. Here the proportional change in price is matched by the same proportional change in quantity.

Price elasticity of demand can be calculated mathematically by dividing the percentage change in quantity by the percentage change in price.

$$E = - \frac{\text{Percentage Change in Quantity}}{\text{Percentage Change in Price}}$$

The resulting value is always negative because demand curves slope downward and to the right. The size of the number calculated defines the degree of elasticity. Unitary elasticity has a value equal to –1; elastic demand has negative value from –1 to –infinity; and inelastic demand has values from –1 to 0. The importance of demand elasticity for farm commodities will be emphasized later in discussion.

Changes in Demand

A **change in demand** is a shift of the **entire** demand curve. An increase in demand shifts the curve to the right, a decrease to the left.

If the quantities that can be sold at each price are larger after a shift in demand, demand has increased. If the quantities that can be sold at each price are smaller, demand has decreased. Both changes are illustrated in Figure 4–2. The line D_ID_I illustrates a demand increase; D_DD_D illustrates a demand decrease relative to the original demand curve D_OD_O. With price fixed at P_O, quantity purchased is Q_I and Q_D respectively.

Note that a change in demand is distinctively different from movements along a single demand curve. The changes in quantity demanded are price responses to **only**

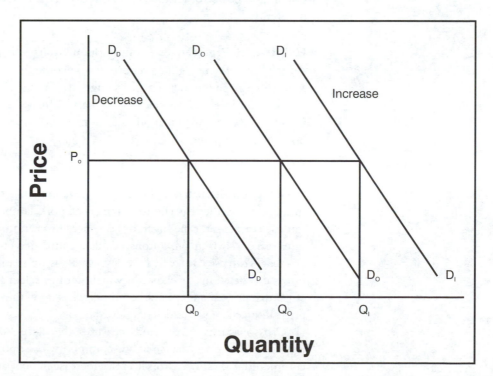

FIGURE 4–2: Changes in Demand.

different supplies available on the market, while change in demand are shifts of the entire curve.

Changes in demand for a commodity that moves the entire curve to the right (increase) or to the left (decrease) are responses to change in:

1. Population, or the number of consumers in the market;
2. Income, or purchasing power of consumers;
3. Habits, tastes, and preferences of consumers; and
4. Price of substitute and complementary products.

Population. The effect of number of people on demand is obvious—more mouths to feed require more baskets of food. Over time, as population in the world increases, there is increasing demand for food. The opening of trade between two countries increases the demand in the exporting country as a larger population enters its market equation.

Income. The quantity of a commodity that is purchased is affected by level of income. Income elasticity is the term used to express relative change in quantity taken in response to relative change in income. In one case, consumer purchases rise with increased income; and in the other, purchases decrease. The first is called a "superior" good and the second, "inferior." Meals away from home, highly processed items, and red meats qualify as superior goods; bread, rice, and potatoes are examples of the inferior goods. Income elasticity is high or positive for superior goods, indicating increased purchases in response to higher incomes. Income elasticity is low for inferior goods, indicating reduced purchase with higher income. Thus, the changing purchasing power of customers influences demand levels of agricultural products differently.

Habits and Preferences. Consumption habits and personal preferences affect demand by influencing the proportion of income that consumers will be willing to spend for a given product. Thanksgiving and Christmas holiday periods create different demand levels for turkey and cranberries than exist at other times of the year. Fads in consumption, nutrition education, or food scares can suddenly shift demand one way or the other.

Prices of Substitute Products. Virtually every commodity has substitute items as consumption alternatives. Grocery shoppers may select protein from vegetables or meat. In meats, they can choose among beef, pork, lamb, or poultry. Food processors can use corn sugar, cane sugar, or beet sugar in a canning process. They may substitute corn oil, soybean oil, cottonseed oil, or palm oil in preparing salad dressings. Cattle feeders can use corn, wheat, barley, or oats as alternative feed ingredients. Relative prices dictate a buyer's choice of which item to select among the alternatives. The effect that price change in one commodity has upon the demand for another is called price-induced cross-elasticity of demand.

Cross-elasticity of demand measures the nature of the effect and degree of response that a price change in one commodity has upon demand for another. Two commodities are **substitutes** if a price increase in one causes increased consumption of the other. Beef and pork are examples of substitute meats. Rising beef prices encourage consumers to eat more pork. Some foods are **complementary** in that they are eaten together, such as bacon and eggs. An increase in the price of one causes a decrease in consumption of both.

Understanding Demand

There are three important distinctions that must be made in understanding demand. These distinctions are:

(1) demand versus consumption,

(2) need versus effective demand, and

(3) farm level versus retail level demand.

Demand and consumption are often confused. Consumption is the amount of a commodity that is actually used; demand is a schedule of the various quantities that will be consumed at various price levels. Demand is a barometer of consumer buying pressure; consumption is actual use of a commodity. Consumers will consume what farmers produce–at some price level. For example, pork consumption may increase significantly from one period to another. Does that mean pork demand has increased? Not necessarily! Consumption may have increased because hog slaughter increased. Because hogs must be marketed and moved into consumption at the time they reach proper weight, an increase in pork supply will cause prices to drop to encourage higher consumption. Thus, increased consumption implies large supplies, not necessarily strong demand.

On the other hand, a low rate of consumption may not mean slack demand; it may reflect reduced quantities available for consumption. Pork producers may reduce production in response to low prices. In time, consumption figures fall as a result. In this case, reduced consumption implies short supplies, not weak demand. Price will rise to limit consumption and prevent pork from disappearing from grocery cases as production decreases.

Demand indicates the differing quantities that will be purchased at differing prices, not the amounts needed by consumers. Distinctions must be made between need, want, and effective demand. **Need** is what the other person thinks is good for you. It is what we mean when we talk about millions of people being hungry in the world; they need an adequate diet. **Want** is what I would like to have if I had my choice. I may not be able to pay for it, but I strive to get it. **Effective demand** is what I choose to buy when I can afford it. For demand to be *effective,* consumers must have the income (purchasing power) to pay for it. In the 1960's and 1970's, farmers were encouraged to increase production to feed a hungry world because people needed it. The result was surplus production, depressed prices, and continued need for food for hungry people. It was a mistake to call that need "demand" or encourage farmers to produce to fill the need. There was inadequate purchasing power to back up the need to make it effective demand.

Farm level demand for agricultural commodities is different from retail level demand of grocery products. Consumers do not buy significant amounts of corn; rather, they purchase corn products in the form of corn flakes or corn-fed beef. The demand for corn is therefore derived demand, derived from the multitude of products that require corn as a basic resource. Derived demand has important implications upon elasticity designations and determination of important demand factors, as we shall see later.

The Meaning of Supply

Supply is a schedule of the varying quantities that will be offered for sale by producers at varying prices at a given time and place. Supply represents the price-quantity relationship from a producer-seller perspective.

The Supply Principle

The principle of supply states that more will be offered for sale at higher prices and less at lower prices. That is, the quantity of the commodity that will be produced and offered for sale varies directly with price.

The relationship can be illustrated in a hypothetical supply schedule of quantities and prices as in Table 4–2. It shows the amount of corn that would be produced and offered for sale at different prices in a production period. At $4.00 per bushel producers would supply 8.45 billion bushels; at $3.00, 7.65 billion bushels; at $2.50, 6.95 billion bushels, and various combinations in between. The figures can be plotted on price and quantity axes to show a supply curve, as in Figure 4–3. The supply curve slopes upward and to the right, since larger quantities are induced as a result of higher prices and smaller quantities result from lower prices.

**TABLE 4–2: Hypothetical Supply Schedule
for Corn**

Price	Amount Offered
($/bu.)	*(billions of bushels)*
4.00	8.45
3.75	8.35
3.50	8.20
3.25	7.95
3.00	7.65
2.75	7.30
2.50	6.95
2.25	6.55
2.00	6.00

There are three factors that affect the extent of producers' increased supply in response to higher prices. They are (1) willingness to sell an existing supply from storage, (2) increased use of variable inputs to increase production from a fixed resource base, and (3) the shifting of resources between alternative uses.

With storable commodities, the total amount available is fixed by the production cycle. The supply cannot be changed until the next harvest. However, the day-to-day availability of a portion of the total supply, or the rate at which it filters into the market place, creates an important sub-market for commodity producers. Higher prices entice grain out of storage and movement into marketing channels; lower prices encourage continued withholding of grain from the market. A farmer's assessment of prospects for higher price in the future and the cash flow requirements of the business affect the decision of whether to sell now or continue to hold for a higher price

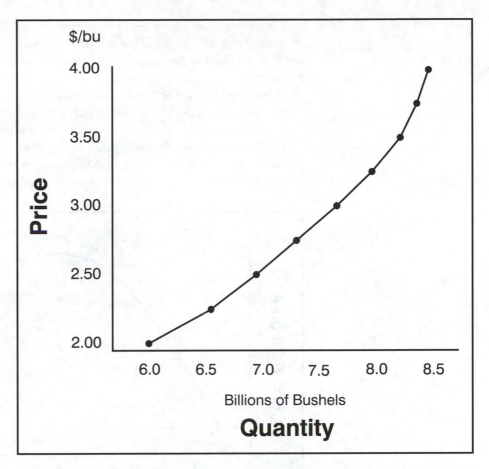

FIGURE 4–3: Hypothetical Supply Curve for Corn. Source: Kohls and Uhl 152.

in the future. A loan that is due and must be paid now often forces poor decisions about when to sell.

Withholding storable commodities affects only short-run supplies of a commodity. Production was initiated by the farmer to generate income. It is simply a matter of time, usually within the production cycle, before the crop must be marketed. Few farmers can finance the accumulation of several years of production in hopes of higher and higher prices being offered by the market. Furthermore, holding commodities in inventory incurs significant costs for facilities and for interest on money tied up in inventory. Prospective price increases can be quickly offset by these holding costs.

For non-storable commodities, day-to-day market supplies are more rigidly fixed by the production process. Producers do not have the opportunity to significantly influence the time span when non-storable commodities flow through the marketing system. Hams, peaches, and tomatoes can be canned and stored, but the bulk of fresh commodities flood onto the market when they are market-ready. The system accepts all the beef cattle, hogs, eggs, fluid milk, peaches, tomatoes, lettuce, and strawberries offered and rapidly moves these quantities to consumers, regardless of how low price must fall to clear the market or how high price must rise to ration a limited supply. Non-storable commodities necessitate a different approach to market decisions than storable commodities. The differences will be examined later.

The second supply response to higher price arises from a producer's ability to increase output by increasing the amount of variable inputs committed to the production process. The classic example is nitrogen application on corn. Corn yield can be increased with nitrogen application, all other things remaining the same. The physical transformation of nitrogen into corn defines how much additional production arises from additional nitrogen application. As depicted in Figure 4–4, each additional unit of nitrogen produces a smaller and smaller increase in additional corn. This is referred to as the diminishing return from additional units of a variable resource. The ability to increase production is limited by physical relationships inherent in nature. Excessive quantities of nitrogen will actually decrease yield.

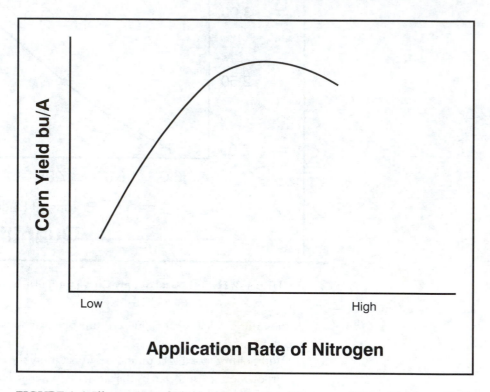

FIGURE 4–4: Illustration of Physical Production Response of Corn Yield to a Variable Input, Nitrogen.

If the price of corn increases and nitrogen price remains the same, the additional production is worth more. Thus, a farmer can afford to apply more nitrogen to increase production and increase the supply available. On the other hand, a lower corn price makes the additional bushel of corn produced worth less. Consequently, the producer is encouraged to reduce the nitrogen application and be satisfied with a lower yield. The market consequently receives lower production. With changes in the profit-maximizing yield level, the market, therefore, receives increasing or decreasing supply in response to changing price of a commodity.

The production relationship described for nitrogen exists for all variable production inputs (e.g., feed, fertilizer, seed, chemicals, fuel, and extra labor). If the crop is worth more, there is more effort and expense applied to increase yield. Thus, total aggregate production increases as more bushels are produced on each acre in response to higher price. If crop value is decreased by lower price, the variable inputs are reduced and a smaller aggregate supply is available for sale.

The third supply response involves the relative market prices of competing enterprises. An acre of cropland can be utilized to produce cotton, soybeans, or rice in the South; corn, wheat, or soybeans in the Corn Belt; and wheat, barley, or grain sorghum in the Great Plains. Given time for adjustment, land use will be dictated by relative prices of the competing crops. To maximize profit, each producer will calculate the net return to land from the various alternative uses and plant the most profitable crop.

Relatively higher soybean price will encourage acreage shifts out of corn and cotton and into soybean production. Table 4–3 shows a simplified budget of comparative returns to corn and soybeans. The important parts of a budget are prices, yields, and production costs. Within any single production period, relative production costs of each crop do not change materially. Relative yields are also fairly constant. The only decision variable in determining what to plant is, therefore, relative price. If a soybean price 2.5 times as great as corn generates the same $100 per acre net return to land, a price ratio higher than 2.5 would encourage more soybean planting at the expense of corn acreage; less than 2.5 would encourage a shift to more corn. Most crop acres have several alternative uses and therefore have critical price ratios that will encourage shifting use from one crop to the other. Relative price dictates land use and the resulting supply of competing crops.

TABLE 4–3: Illustration of Net Return to Corn Compared to Soybean Production

Item	Corn	Soybeans
Price per bushel	$2.50	$6.25
Yield (bu. per acre)	150	50
Gross Receipts	$375.00	$312.50
Production Costs	$275.00	$212.50
Net Return to Land	$100.00	$100.00

Relative Prices: $6.25/$2.50 = 2.5:1.0

Supply Elasticity

Supply elasticity is a measure of the extent of increase or decrease in quantity supplied in response to higher or lower market prices. Supply responses have the same elasticity designations as demand except that they are positive since increased price encourages increased production. Commodities that are highly responsive to price change have **elastic** supply curves. With them, a small change in price will encourage relatively larger increases in supply. The supply responses that react relatively little to price changes are **inelastic.** A large price change produces only a small supply change. In between elastic and inelastic is **unitary** response, where price change and supply change are proportionately the same.

Time designations are very important in supply elasticity considerations. Time is classified as short run, intermediate run, and long run. Corn, wheat, and cotton have annual production cycles and relatively short harvest periods. Once the planting period has passed, supply is fixed for the most part until after harvest of the next planting period. Weather conditions during the growing season and imports can modify the available supply within a country, but changes in supply during a production period are generally relatively small and are largely beyond control of producers.

Livestock supplies are also fixed by production cycles from breeding to market weight. Regardless of price change, little can be done to change the number of animals coming to the market during that period. It is only after a production cycle, when females can be diverted from the slaughter market into the breeding herd, that livestock supplies can be increased. In the short run, supplies are largely fixed by production cycles. Changes in supply are confined to changes in market weight and numbers of animals kept for breeding. Short-run supply is almost perfectly inelastic; it is a vertical line in Figure 4–5.

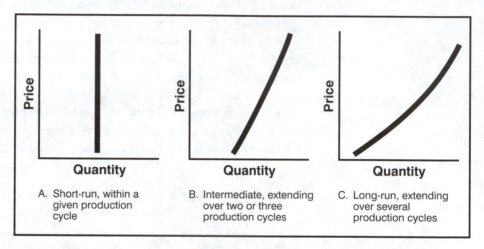

FIGURE 4–5: Hypothetical Supply Curves Illustrating the Effect of Time on Elasticity. Source: Kohls and Uhl 161.

In the intermediate period, production can be increased or decreased within the limits of existing facilities. Crop producers can apply more production inputs, which increase yields; livestock producers can add females to the extent they have surplus grazing or building space. Supply increase, due to changes in the application of variable input, occurs over two or three production cycles. It is more responsive to price change than short-run adjustments but is still relatively inelastic.

Once existing production facilities are used to capacity, it takes longer-run adjustments, over several production periods, to further increase supply in response to higher price. This consideration represents long-run supply adjustments in Figure 4–5. In this case, supply increases depend on building more facilities, bringing more land into production, adding equipment, and/or applying technology.

Within the intermediate and long-run adjustment periods, supply elasticity depends upon the ease with which supply adjustments can be made and the costs associated with changing production levels. If changes can be made easily and inexpensively, supply curves will be more elastic. If changes are difficult and expensive, supply responses will be more inelastic.

Supply reductions are fairly inelastic in response to price decline. It takes a relatively long time and continued low prices to force farmers to reduce production. This occurs because of the mix of fixed and variable costs making up total production cost. Land, labor, and machinery charges represent a large proportion of the total cost of producing agricultural products. They constitute the fixed costs of production and are incurred whether or not there is any production. The variable out-of-pocket production costs (e.g., seed, fertilizer, fuel, chemicals, and other inputs) are incurred only if production is initiated. Cutting production, by leaving resources idle, eliminates these variable costs. However, fixed costs go on. Thus, farmers will continue to produce if they can get more for the product than they spend for variable out-of-pocket costs. Even though they are not covering total cost of production, they minimize their losses by producing when they cover variable costs and have something left over to apply to a part of these fixed costs.

The time period of adjustment is important in supply response to lower price. In the short run, farmers may actually increase production in response to lower price in hopes of maintaining income. Since price per unit is lower, they reason that more units need to be produced in order to maintain a cash flow. In the intermediate run, production will decline as machinery wears out and facilities deteriorate. In the long run, producers quit agriculture, and land and facilities are taken out of production or revert to some other use. Supply eventually adjusts to equate total production cost with price, but the time period may be long and the human cost of adjustment may be extremely painful.

The almost totally inelastic supply of agricultural commodities within a production cycle and the highly inelastic responses in both intermediate and long-run periods help explain much of the volatility of agricultural prices. Price changes are much more severe because of inelastic supply responses.

Shift in Supply

Up to now we have looked at the change in quantity supplied as it relates to price level. We looked at one supply curve and the relationship between changes in quantity and price. Figure 4–6 illustrates a shift in the entire supply curve. The original supply curve is S_O; the curve S_I shows an increase in supply; and the other curve, S_D,

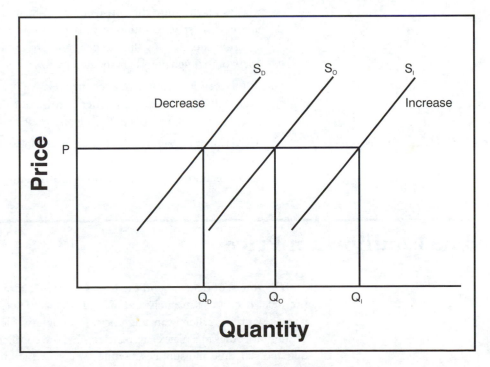

FIGURE 4–6: Changes in Supply.

shows a decrease in supply. Notice that price, P, does not change. An increase in supply is defined by producer's willingness to provide a larger quantity to the market at the same price levels. A decrease in supply is defined as a reduction in quantity supplied at the same price levels.

In contrast to a change in quantity supplied, a shift in supply is a different economic concept. A shift in supply means that the entire supply curve moves to the right (increased supply) or to the left (decreased supply). To identify shifts in supply, the price level stays constant, but the quantity of supply available in the market changes.

Several factors lead to shifts in supply:

1. Technology: By definition, technology means more production from the same or similar resources. Thus, cost per unit of output is lowered as higher yield is produced. Producers are willing to sell more at each price level. Hybrid seed is often cited as the example of technological improvement that increased corn yield per acre and reduced per-unit production cost. Its adoption as a production practice led to increased supply. Farmers were willing and able to produce more at the same price because technology increased output and lowered average cost.

2. Innovation: This refers to the improvement in production practices that increase yield or reduce cost. Yield increases due to fall tillage of heavy soils are but one example. Plowing costs do not change by adopting the practice, but weather effects enhance the soil's productive capacity. Adoption of the innovation lowers unit cost of production and, thereby, shifts the supply curve to the right.

3. Rules and regulations: Compliance with governmental regulations affects costs and, in turn, supply. Regulations that specify the handling of livestock wastes may require the construction of facilities for handling effluent. An increased cost of operation shifts the supply curve to the left (reduce supply) as producers are unwilling to maintain production at the same price level in view of the higher costs mandated by new regulations.

4. Input cost: As resource costs change relative to each other, economic advantage shifts between crops. Higher nitrogen fertilizer price increases corn and cotton production costs to the advantage of soybean production.

5. Government programs: Incentives to improve drainage, develop irrigation, or grow conservation crops and idle cropland shift supply curves right or left and affect how much production is forthcoming at a given price level.

The Equilibrium Price

The forces of demand and the forces of supply acting alone affect consumer or producer decisions separately and individually as they react to market price. Buyers want low prices; sellers want high prices. Somehow the two sets of forces must be brought together in a market arena to resolve the conflict. Table 4–4 shows both the demand schedule and supply schedule for corn which might exist in a market. At a price of $2.75 per bushel, buyers will take 7.3 billion bushels and sellers will produce and offer for sale 7.3 bushels. The amount supplied **equals** the amount demanded; this is called the equilibrium quantity. The equilibrium price is $2.75. It is the market clearing price level that will be established by demand and supply forces in a competitive free market economy. Figure 4–7 presents a picture of the relationship. Where the demand curve, D, crosses the supply curve, S, the two are equal. P_E is the equilibrium price, and Q_E is the equilibrium quantity.

TABLE 4–4: Hypothetical Demand and Supply Schedules for Corn

Demand	Price	Supply
(billions of bushels)	($/bu.)	(billions of bushels)
6.10	4.00	8.45
6.30	3.75	8.35
6.50	3.50	8.20
6.75	3.25	7.95
7.00	3.00	7.65
7.30	**2.75**	**7.30**
7.60	2.50	6.95
8.00	2.25	6.55
8.50	2.00	6.00

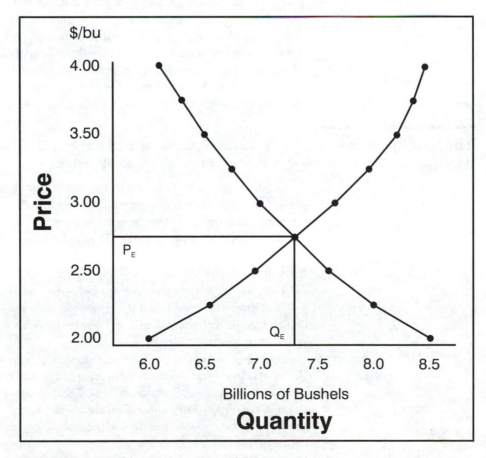

FIGURE 4–7: Equilibrium Quantity and Price.

Sellers would prefer a higher price, such as $3.50; buyers would prefer $2.25. Not all of the participants are satisfied with $2.75, but there is no other acceptable compromise. At a higher price, say $3.50, sellers would increase production to offer 8.2 billion bushels; but buyers would be willing to take only 6.5 billion bushels. A surplus of 1.7 billion bushels would hang over the market and begin to drive price down. In a market with price above the equilibrium price, it is a buyer's market in which more is offered than is demanded. Price cutting occurs, and the general price level declines to move the available supply.

Should the price be below equilibrium at a level of $2.25, sellers offer only 6.55 billion bushels. Buyers want 8.0 billion bushels at that price. A shortage of 1.45 billion bushels exists and pulls up the price level. It is a seller's market, and the price level rises to ration the limited supply.

At $2.75, buyers and sellers who enter the market are satisfied; 7.3 billion bushels change hands. An equilibrium price and quantity are established.

Changing Price from Shifting Demand or Supply

Various opposing factors may be present in a market at any particular time to shift demand or supply curves, possibly both at the same time. Fortunately, the forces causing shifts in demand are distinctly different from those that change supply. It is, therefore, possible to trace the expected results of changes on equilibrium price and quantity by shifting one of the curves and moving along the other. Figure 4–8 illustrates whether rising or falling price can be expected due to demand or supply shifts. The solid lines represent the original situation, and the dotted lines represent the shift in either demand or supply, or both at the same time.

Either increased demand or reduced supply can be expected to increase equilibrium price. Likewise, decreased demand or increased supply will reduce price. If both curves are shifting at the same time, the price effects of one curve may be offset by the other, or they may magnify price effects of each other. Further on, we will discuss the important factors that shift demand curves or supply curves of farm commodities and try to assess the extent of price change to expect from those changes.

The Supply-Demand Balance

In the real world, a stable demand-supply equilibrium is seldom achieved. Supply and demand conditions are constantly shifting.

On the supply side, variations in production decisions, government policy, technology, weather, disease, and unpredictable events affect acreage, yield, and total production. Furthermore, the time span required to adjust to changing price levels greatly affects the supply response. Demand changes occur from shifts in exports, population, consumer incomes, tastes and preferences, employment, business conditions, and political events.

The forces at work in achieving a supply-demand balance are illustrated in Figure 4–9. Equilibrium is established in relative terms by the balance between supply and demand. As the balance is tipped one way or the other by supply or demand factors, price simultaneously changes.

The price signals elicit opposing reactions from consumers and producers. Reduced supply increases the relative price and signals producers to increase production. At the same time, the price increase signals consumers to reduce consumption. In time, as a response to the increased supply and lower consumption, the price indicator swings back toward the right of the scale. The resulting lower price signals the consumer to increase consumption while at the same time signaling producers to reduce production.

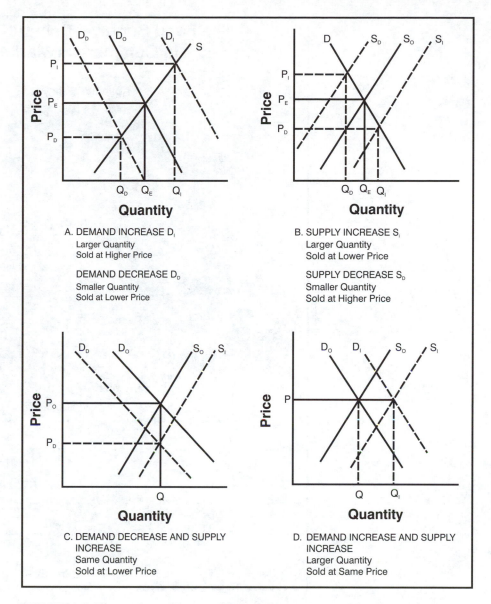

FIGURE 4–8: Illustration of Expected Price Change Due to Changes in Demand or Supply.

The supply/demand balance illustrates both the communication role of price in marketing and the built-in stabilizing feature of price-coordinated markets. Higher price simultaneously tells producers to increase production and tells consumers to use less. Lower price tells producers to reduce production and consumers to buy more. Adjustments to price changes maintain a supply/demand balance. But in the meantime, prices rise and fall in order to induce the adjustments.

The location of the fulcrum on the indicator needle of the supply/demand balance emphasizes the inelastic demand for agricultural commodities. Small changes in supply or demand produce large changes in commodity prices.

In theory, equilibrium price and equilibrium quantity are specific and clearly defined. In reality, prices undershoot or overshoot the mark in search of the price that will clear the market. Prices are dynamic and change day to day, hour to hour, minute to minute. No on really knows exactly how much supply is available or required.

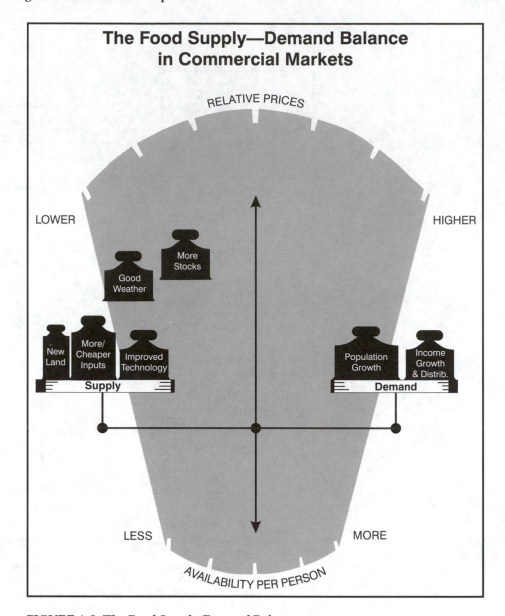

FIGURE 4–9: The Food Supply-Demand Balance.

Furthermore, the time element greatly complicates the process. Large receipts on today's market may or may not depress price level. A large crop might be followed by gradually increasing price throughout the year, and a small crop may experience a falling price level following harvest. Because of imperfect knowledge, day to day observations may be the opposite of what theory suggests should be true. It is more nearly correct to say that equilibrium price is that price toward which price level tends to move but somehow never reaches. Understanding the principle is nonetheless important in trying to sort out price-making factors.

Price Discovery

At each point in the food chain from farm to fork, whenever title to a commodity changes hands, a transaction price must be determined. This price determination process needs to be related to the economic principles that explain equilibrium price. To begin, we must distinguish between price equilibrium and price discovery. **Price**

equilibrium is the process by which the broad forces of demand and supply establish a general, market-clearing price for a commodity. **Price discovery** is the process by which buyers and sellers arrive at a specific price for a given lot of a commodity at a given location on a particular day. The overall demand-supply equilibrium price must be "discovered" and applied to each transaction in the market place.

The process of price discovery has two distinct phases:

1. Evaluating the conditions of demand and supply to determine the general level of price for the commodity. This is the price level around which prices for particular lots of the commodity, at different locations, of differing qualities, and in different time frames, will fluctuate.

2. Determining the value of a specific lot of the commodity being exchanged, relative to the general market level. This phase takes into consideration grade differences, location, buyer and seller services, and relative bargaining power of seller and buyer.

"Price discovery is a human process, beset by errors of fact and judgement."[2] There is no guarantee that buyers and sellers will always discover the equilibrium price, but the automatic stabilizing features tend to move in that direction. In real life, each buyer and seller operates in his or her own best interest according to the profit motive. The variations and imperfections in the price-discovery process make it profitable for both buyers and sellers to shop around among alternatives and to bargain on price and other terms of trade. This bargaining process leads to price discovery and price equilibrium.

Pricing errors occur at all stages of the price-discovery process. But the errors can be minimized and pricing improved by high-quality market information and commodity grading programs so that all parties to the trade are acting on comparable information.

Two people in the food chain are most concerned about price level: the farmer on one end and the consumer on the other. Intermediaries usually operate on spreads, or price margins, between what they receive for the goods as they pass them along the food chain and what they must pay their suppliers. Intermediaries are often considered the "bad guys" who pay producers too little and charge consumers too much. Actually, intermediaries do not "determine" prices, since they do not determine consumer demand or market supplies. They merely "discover" prices based on their evaluation of the quantities of supply available and what customers are likely to pay for the amounts offered to them. Since intermediary profit margins are relatively stable, the **fluctuations** in price received by producers reflect, for the most part, changes in consumer demand and changes in producer supplies available on the market. The farmer must decide if the current price level makes it worth the effort to allocate resources to produce commodities for the market. The consumer decides if the products offered in the supermarket satisfy the family's needs and desires at an acceptable price. If both concur, the system will continue to function as it is; if either does not concur, there will be a change. This is how individual decisions are translated into consumer demand, market supply, and equilibrium price.

Law of One Price[3]

Despite the tendency for market-making forces to establish an equilibrium price, numerous different prices exist in various markets. The *law of one price* provides an explanation of these differences to confirm the existence of one equilibrium price for the market as a whole.

The Law of One Price (LOOP) states that "all prices in a market are uniform, after taking into account the costs of adding place, time, and form utility to products within the market."[4] The law defines the relationship among prices in terms of space, time, and form utility created by the marketing process. Understanding the law of one price is important in understanding differences in price at different points in the country or world, at different times of the year, and in different product forms available in the market.

Figure 4–10 illustrates the law of one price in terms of location differences in the international corn market. The map shows local prices from three shipping points in the Corn Belt (Columbus, Champaign, Farnhamville), transportation costs to different markets, three export markets (Chicago, New Orleans, Norfolk), and a major port of entry in European trade (Rotterdam). Grain tends to flow from surplus production areas through the marketing channels to meet the needs of deficit production areas. Prices at the different locations are aligned according to the law of one price. They differ only in the cost of transportation, or the cost associated with providing location utility.

Under competitive market conditions, prices tend to maintain the same relative values over space, time, and form. If European customers are willing to pay an additional 50 cents per bushel, grain traders see the extra profits that could be made by increasing the grain trade from the United States to Europe. Assuming that the shipping charges are constant, traders willingly increase their bid price to entice a larger flow through the system. Competing traders make successively higher bids to sellers trying to increase their volume of supply going to the European market. After the adjustments are made, the U.S.-Rotterdam price difference will return to the cost of transportation, but both the Rotterdam and U.S. price levels will be higher, U.S. price pulled up by a higher Rotterdam price. Traders bidding against each other to capture the additional volume serve to reduce the extra profits promised to them. Price differences within the market return to levels consistent with the law of one price. In the meantime, the traders perform a useful function in shifting the corn from the lower-value market to a higher-value market.

LOOP is useful in explaining differences in price over time. Crop harvest occurs in relatively short periods of time, but consumption is fairly stable each day of the year. The harvest surplus must be carried forward to satisfy daily consumption requirements before the next crop is harvested. There are costs of building storage facilities to hold the crop, and there are interest costs incurred by owning the commodity in storage. The law of one price leads to the obvious conclusion that the difference between market price late in the crop year and the earlier harvest-time price will have to be large enough to pay the costs of holding the commodity in storage. If the difference is narrow, grain merchants will sell immediately rather than hold until later because there would be no profit to be realized from storage. The increased volume coming to the market at harvest tends to drive harvest-time prices lower. Processors and other users, seeing a potential shortage of supply later in the season, raise offers for later delivery. Thus, the price difference will tend to widen to cover the costs of storage and reestablish relative prices between harvest and later consumption, consistent with LOOP.

Finally, the law of one price explains the addition of form utility in the creation of different products from a common source commodity. A corn producer can sell cash

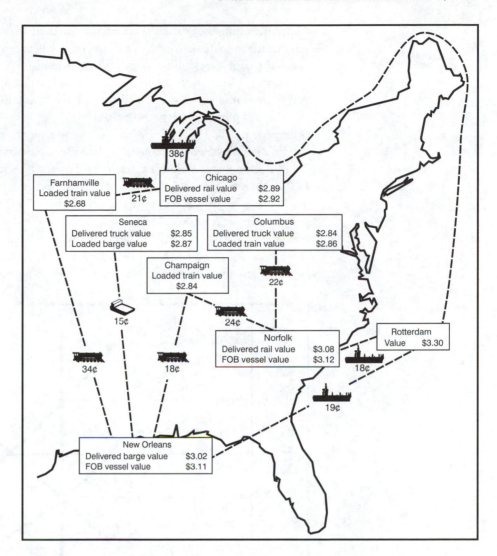

FIGURE 4–10: Relative Prices in the International Corn Market.

grain or change the form of corn grain into milk, pork, beef, or poultry products. Producers will allocate their supply of the raw commodity to alternative products in such a way that the price they receive for the products, less their respective production costs, will equal the cash corn price. Whichever product will generate a larger return above cash costs and additional production costs will be increased in volume. When cash corn is worth $2 per bushel and hogs are worth $60 per cwt., farmers will convert corn into hogs and "walk" $5 per bushel corn to market. Again, LOOP explains the different forms of a commodity.

Market Forces in Action

An interesting analysis for understanding basic price-making forces of demand and supply was presented by Robert L. Heilbroner in tracing market interactions of the Soviet Grain Sale.[5]

In midsummer of 1972, the Nixon administration startled the world by announcing that it had sold one-fourth of the American wheat crop to the Soviet government. It was the largest agricultural commodity sale in history, worth more than $1 billion. As a result, wheat prices began to soar from a level of $1.70 per bushel in mid-July before the announcement to an all-time high of $5.00 in late summer of 1973.*

The first point to understand is that the 1972 crop had been planted and, to a large degree, completely harvested when the announcement of the sale was made. Thus, the 1972 supply of wheat was fixed; the supply curve was totally inelastic. Production could not be increased until the next crop year. The additional demand from the USSR, therefore, made its effect wholly as an increase in price, as shown in Figure 4–11. Had the Soviets contracted to buy grain for 1973 delivery rather than 1972 delivery, the situation would have been quite different. More grain could have been planted in the next crop year to satisfy the additional demand. As a result, the equilibrium price, P_2, with sufficient forewarning to increase supply, would have been considerably less than the equilibrium price, P_1, given no advance warning.

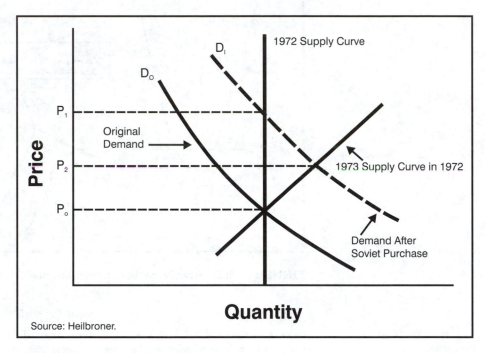

FIGURE 4–11: Illustration of Price Effect of Increased Demand from Russian Wheat Deal. Source: Heilbroner.

The rise in wheat price was only the initial shock. The ripples spread to other commodities through the substitution effect. Since corn, oats, and rye are close substitutes for wheat in many uses, the cross-elasticity of demand is high. That is, an increase in the price of one feed crop will cause substantial increases in the quantities demanded of other feed crops. As a result, the rise in wheat prices caused a shift in the demand curves for other grains, as Figure 4–12 shows for soybeans. Although the

---

*Farmers called the sales "The Great Russian Grain Robbery" because price was so low. On the other hand, the Soviets paid for the grain with $35-an-ounce gold. Subsequently, gold price rose to about $850 per ounce. Farmers thought they were underpaid; probably, the Soviets thought they overpaid.

increased demand was confined to wheat, the purchase quickly induced price increases among all the grains that could be substituted for wheat. Combined with other factors that changed demand, the effect was to boost the price of oats from $0.80 to $2.06 only 13 months later, rye from $1.01 to $3.86; and soybeans from $3.50 to a peak of $12.00.

The Russian Wheat Deal, though important in the commodity price explosion, was not the only cause. Because of El Niño, the Humboldt Ocean current off the West Coast of South America shifted position in 1972. The change caused a significant drop in the anchovy crop—so significant, in fact, that the Peruvian government ordered a halt to all anchovy fishing lest anchovy breeding capacity be severely damaged. Anchovies are an important source of world protein supplies and a direct substitute for soybean meal. Once again, the substitution effect on prices was important. As fishmeal prices soared in response to reduced supply, there was another induced demand for soybeans and feed grains as substitutes for anchovies.

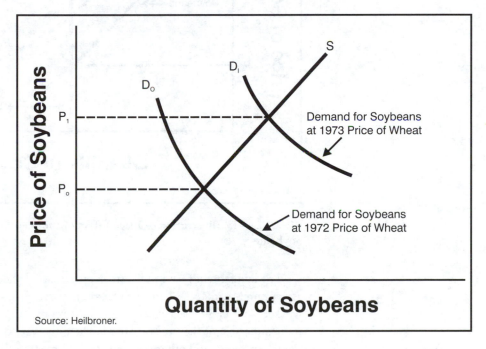

Source: Heilbroner.

FIGURE 4–12: Illustration of Increased Demand for Soybeans Due to Cross-Elasticity of Wheat Demand. Source: Heilbroner.

A final influence on commodity prices was devaluation of the U.S. dollar in 1972 and 1973. Devaluation made exported items less expensive to U.S. trading partners. Devaluation induced a shift in the dollar demand curve for all American crops. Japanese buyers may have been purchasing U.S. grain at an exchange rate of 375 yen to the dollar. For them, $4.00 soybeans cost 1,500 yen per bushel. Devaluation meant that the dollar was cheaper; it took fewer yen to exchange for $1.00. At 300 yen to the dollar, $4.00 soybeans cost only 1,200 yen. In response to the effectively lower price, the Japanese increased their soybean purchases. Because the supply of soybeans was fixed by the crop year, large increases in price occurred due to the change in demand. The response is illustrated in Figure 4–13. The graph shows the rise in soybean prices from P_O to P_1, as induced demands shifted the demand schedule to the right.

The effects on commodity prices did not stop with a grain sector of the economy. Soybean products are used for animal production as well as for human consumption. Higher soybean prices increased the cost of producing meat. Other things being equal, higher costs reduced profitability of beef production and caused feeders to decrease production. The supply of meat output was reduced, or in economic terms, the meat supply shifted to the left.

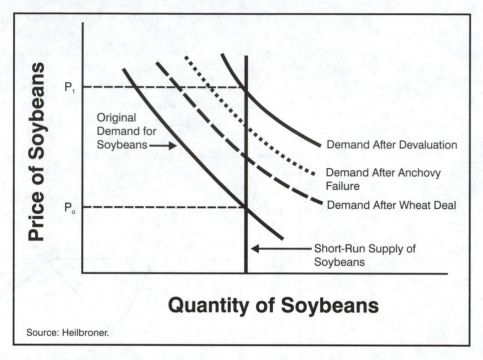

FIGURE 4–13: Illustration of Price Effect of Induced Shift of Demand Schedule for Soybeans.

The adjustment phase in livestock production created one of the peculiarities of agricultural markets. When manufacturers want to reduce production, they lay off workers and close the factory. To reduce production, the livestock breeder must reduce the size of the breeding herd. The only way to do that is to sell females for slaughter that would otherwise be used for breeding. In other words, to get a long-term decrease in the supply of meat brought to the market, livestock producers cannot avoid a short-term increase in the supply of meat. Thus, the price of beef or pork first goes down and rises only after breeding herds have been reduced. These effects are shown in Figure 4–14. They created a lagged response between the occurrence of events and the price effects of these events.

These interactions help us to understand what happened to beef prices. In mid-summer 1972, Omaha steers were selling for $37.50 per hundredweight. A few months later, price had fallen to $33.00. Thereafter, prices rose to well over $50.00 as the supply curve for beef shifted to the left in response to reduced supplies.

In this review of effects of the Russian Wheat Deal, we have applied the economic terms of demand and demand schedules, supply and supply schedules, prices, substitution, elasticity, and lagged responses. Economic principles explain price-making events in the real world as the free markets act and react to changing conditions.

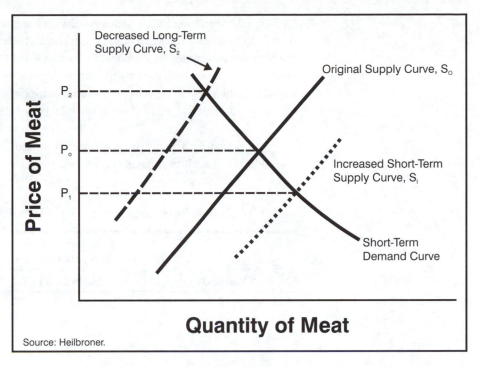

FIGURE 4–14: Illustrations of Price and Quantity Responses in Meat Production Due to Changing Profitability. Source: Heilbroner.

Applications of Demand and Supply Analysis

A study of economic principles forms the foundation for understanding demand and supply influences upon price determination. However, definitions of economic terms and concepts are rather sterile until they are applied to real-world situations. We will develop some of those applications to improve our understanding of the principles and to provide a framework for analyzing effects of changing conditions on prices.

Effects of Demand Elasticity on Price and Gross Receipts

Commodity producers need to understand the price elasticities of agricultural commodities if they wish to project the effects that changes in supply will have upon price and income.

Table 4–5 reports price elasticities of agricultural commodities at both the farm level and the retail level. There are several important implications that can be gleaned from the elasticity coefficients. First, all food commodities have inelastic demands at the farm level and almost all at the retail level. The elasticity coefficients show that a 1 percent price change is associated with less than a 1 percent change in quantity sold at the new price. Inelastic demand means that consumers are relatively insensitive to food price changes. It also means that increased sale volume can be attained only with relatively large price concessions. On the other side of the coin, inelastic food demand means that relatively small decreases in market supplies will produce pro-portionately larger price increases.

Second, farm-level price elasticity is always lower than the retail-level price elasticity. On graphs of the same scale, the more inelastic demand curves can be illustrated with steeper slopes, as shown in Figure 4–15. More elastic demand curves at comparable prices are depicted with flatter slopes. Food products generally increase in price elasticity as they move closer to consumption. This reflects the greater substitution possibilities for branded items at a local store compared to raw commodities at the farm gate. Raw commodities at the farm gate have more inelastic demands. As a consequence, the price volatility for farm commodities will be much more severe than the price volatility at the grocery store.

TABLE 4–5: Retail and Farm Price Elasticities for Selected Foods

Commodity	Effect of a 1 Percent Change in Price on the Percentage Change in Quantity Demanded	
	Farm Level	Retail Level
Turkey	—	−1.56
Margarine	−.69	− .84
Chicken	−.60	− .78
Apples	−.68	− .72
Butter	−.46	− .65
Beef	−.42	− .64
Ice cream	—	− .52
Cheese	—	− .46
Pork	−.24	− .41
Fresh milk	−.32	− .34
Eggs	−.23	− .31
Potatoes	−.15	− .30
Bread, cereals	—	− .15

Source: P. S. George and G. A. King. *Consumer Demand for Food Commodities in the U.S. with Projections for 1980,* University of California, Giannini Foundation Monograph No. 26 (Berkeley: University of California Agricultural Experiment Station, March, 1971). Reprinted by permission.

The final note is the wide range in elasticities associated with different products. Consumption of basic foods such as bread and potatoes is almost totally insensitive to price change. Fruits, vegetables, and meats are somewhat more price-responsive. Thus, the kind of commodity that a farmer produces determines the price volatility he or she might experience as a consequence of demand elasticity.

From the producer's perspective, demand elasticities provide a measure of the price change that can be expected from supply changes. Inelastic demand coefficients indicate that relative price changes will be much greater than relative supply change. The price responses in Figure 4–16 bear out this point. Commodities with very low demand elasticities, such as bread and potatoes, are subjected to large price changes as the result of a small change in supply. A lesser inelastic demand for chicken produces less severe price changes.

Are farmers better off selling a large quantity at a lower price or a smaller quantity at a higher price? Demand elasticity explains how receipts will be affected by price and quantity changes. Total receipts from a sale are computed by multiplying market

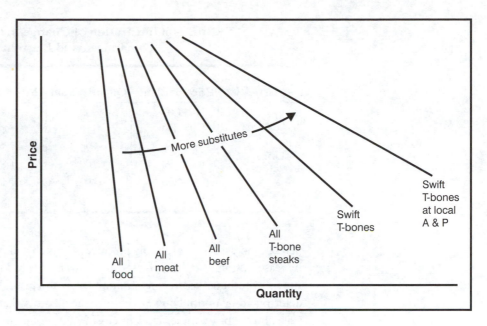

FIGURE 4–15: Demand Elasticity at Various Market Levels.

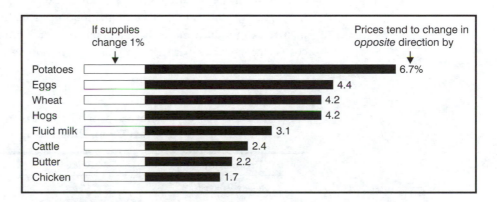

FIGURE 4–16: How Change in Supply Affects Price Received by Farmers.

price by quantity sold. For a commodity with unitary elasticity, a 10 percent decrease in price increases the amount purchased by 10 percent, and the gross receipts are exactly the **same** before and after the price cut. The increase in volume of sales exactly offsets the price decline and thus gross receipts are constant. For a commodity with elastic demand, the amount purchased increases by more than 10 percent and thus gross receipts are **greater** than before. A commodity with inelastic demand experiences less than a 10 percent increase in quantity purchased, so that gross receipts **decline.** These calculations are illustrated in Table 4–6.

The impact that inelastic demand for farm commodities has on farmers' gross receipts is especially important in commodity marketing. Demand elasticity is important to farmers in that it tells them what to expect as price and income change as supply increases or decreases. A farmer thinks that producing more product will increase his or her income. But the world does not work that way. More total aggregate production clears the market at a lower price. Inelastic demand means that a 10

TABLE 4–6: Illustration of Change in Gross Receipts
Due to Demand Elasticity

Elasticity	Price per Unit	Quantity Sold	Gross Receipts (Price × Quantity)	Change in Gross Receipts
Elastic	$1.00	90	$90.00	Increase
	.90	110	$99.00	
Unitary	1.00	90	$90.00	Same
	.90	100	$90.00	
Inelastic	1.00	90	$90.00	Decrease
	.90	95	$85.50	

percent increase in total supply produces as much as a 20 percent price decrease. An increase in supply and proportionately lower price reduces total revenue. Table 4–7 shows hypothetical corn consumption levels and price. The total value of the crop is determined by multiplying price times the quantity that would be consumed at that price. From the example, a low price of $2.00 per bushel on a large crop of 8.5 billion bushels has a gross value of $17.0 billion. A high price of $4.00 per bushel on a small crop of 6.1 billion bushels has a gross value of $24.4 billion. Clearly, there is more income from the smaller crop. Furthermore, the larger crop probably came from more acres in production and/or more inputs per acre. Either way, greater costs were incurred to produce more. Farmers' incomes suffer from lower price and higher costs. Farmers work harder to produce more and end up earning less. Why would they do so?

TABLE 4–7: Hypothetical Corn Production, Market Price, and
Gross Value of Corn Crop, Illustrating Inelastic Demand

Production	Price	Gross Value
(billions of bushels)	($/bu.)	($ billions)
6.10	4.00	24.4
6.30	3.75	23.6
6.50	3.50	21.1
6.75	3.25	21.9
7.00	3.00	21.0
7.30	**2.75**	**20.1**
7.60	2.50	19.0
8.00	2.25	18.0
8.50	2.00	17.0

Conflict Between the Individual and All Farmers

The aggregate demand for corn is highly inelastic. Relatively small increases in total supply must be sold at large price concessions to encourage additional consumption. An individual farmer, however, faces a totally elastic demand curve, as illustrated in Figure 4–17. One farmer can sell the entire production at the market price without the necessity of cutting price to encourage buyers to take a few more bushels. By the same token, an individual producer cannot raise the price level by withholding corn from the market. There are so many corn producers and each one produces such a small portion of the total crop, no one of them acting alone appreciably affects the supply. While 200,000 bushels of corn represent a fairly large farming operation, 200,000 bushels are minuscule in a 10 billion-bushel crop. When corn production is profitable, it behooves the individual farmer to produce as much as possible. Total profit is increased by producing more bushels. Furthermore, since many costs, such as labor and machinery, are fixed, the farmer can lower total costs per unit by increasing production. Thus, a farmer has double incentive to produce more. But when all producers increase production, the larger total supply lowers farm price. A farmer, as an individual, makes rational, profitable decisions to produce more. However, more total output is not necessarily in the best interest of farmers as a whole. The collective impact of all those individual decisions to produce more serves to drive down the market price, and consequently total income and profit.

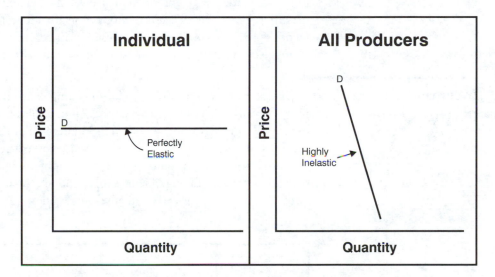

FIGURE 4–17: Demand Curves for Corn for the Individual Producer and for all Producers.

The conflict in objectives arises because agricultural producers come the closest of any industry to operating under economic conditions of pure and perfect competitions. Those conditions are:

1. A great number of buyers and sellers so that no one is sufficiently large to influence price through individual actions, and therefore, farmers are price takers;

2. Both buyers and sellers have complete and perfect knowledge of all factors that affect market conditions. Furthermore, each one will utilize the information in a rational manner to maximize individual profit;

3. There is no collusion or operating restriction among either buyers or sellers to limit the supply of product coming to the market (except when the government attempts to intervene with production controls);

4. The product is homogeneous in that one lot is exactly like another lot within quality specifications. There is no product differentiation;

5. The factors of production are perfectly mobile and free to enter or leave an industry in order to achieve greater returns.

These competitive conditions significantly impact the way the farmers operate. Because no single individual can influence the market, farmers are price takers; they take the price dictated by aggregate demand and supply conditions. Furthermore, individuals feel that they can produce and sell as much as they like without influencing the price. Thus, a farmer's attention is focused on production efficiency to improve profit. There is a strong incentive to lower operating costs in hope of improving profitability. That explains why agriculture is continually subject to the price-cost squeeze. Anytime that profits increase, additional production is stimulated. More production, in turn, reduces market price back to "normal" profit levels.

The conditions of pure and perfect competition are modified to a degree by real-world situations. Some farmers and marketing organizations are large enough to influence market price; information is never 100 percent perfect; there can be product differentiation; and factors of production are not perfectly mobile. These imperfections provide farmers with opportunity to increase profits through improved marketing decisions.

Incentives to Restrict Output

Inelastic demand provides producers, acting collectively, an opportunity to raise price by restricting output through production quotas or acreage controls, and shifts the supply curve leftward, as shown in Figure 4–18. Net income is boosted by the combination of higher price and lower production costs.

Some farm groups promote the idea of output restriction by their members. But they are not successful in reducing supply enough to raise total return. It is against a farmer's nature to leave land or other resources idle. As an individual, a farmer has the

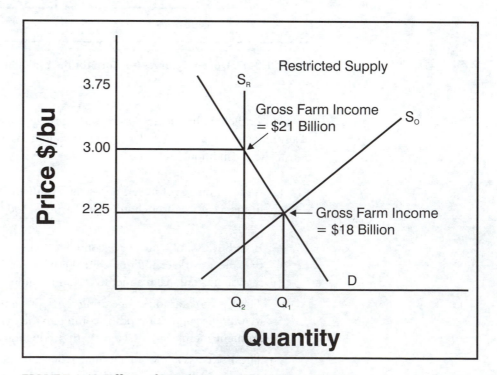

FIGURE 4–18: Effects of Supply Restriction on Gross Receipts.

economic incentive to produce more in response to higher prices. Thus, there are strong incentives not to cooperate in voluntary group action to reduce production or to cheat against those who do. A less-than-altruistic farmer might hope that neighbors or some other region might cut acreage or suffer poor yield to reduce the overall crop, but the individual producer prefers bumper crop yields. A large crop sold at a high price is the best of all possible worlds for the individual.

Successful agricultural supply control programs usually require governmental authority to determine output restrictions and to enforce compliance. Any kind of market order or production quota, in essence, restricts supply. Both "carrot and stick" approaches are used. Land retirement programs, in the name of conservation, are encouraged by payment schemes to farmers to hold land out of production. Commodity groups with specialized products and limited production areas form producer marketing cooperatives and subscribe to production quotas. Consumers, of course, pay higher food costs as a result of successful supply control programs. Furthermore, many of the benefits are subsequently capitalized in the value of the scarce resource. Tobacco and peanut allotments, for example, have taken on values all their own, separate from the land to grow the crop.

Effects of Price Ceilings and Floors

Price ceilings or floors are government-imposed interference in the market place to limit price rise or prevent price decline. Price ceilings to hold prices low are set below equilibrium price, and price floors are set above equilibrium price. Price ceilings are imposed to control inflation and hold down consumer costs, although it is debatable how effective they are in the long run. Price floors are used to support farm prices and bolster farm income.

Price floors and ceilings prevent buyers and sellers from reaching equilibrium price and quantity. Surpluses or shortages are created by market interference. As shown in Figure 4–19, a price floor set above equilibrium at P_f signals buyers to consume Q_m, but it signals producers to produce Q_g. The difference is surplus produc-

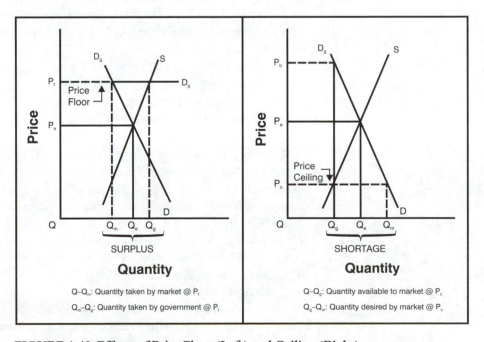

FIGURE 4–19: Effects of Price Floor (Left) and Ceiling (Right).

tion, which must be controlled by marketing quotas, stored, dumped, plowed under, or sold in noncompeting markets.

A price ceiling, P_c, set below equilibrium price creates a shortage. The low price encourages buyers to consume Q_m but encourages farmers to produce only Q_g. A rationing procedure has to be devised to allocate the restricted supply. Black markets tend to develop to circumvent the rules and regulations and raise the price to P_b.

Relationship of Prices to Costs of Production in Agriculture

The question is often asked, "Is there a relationship between commodity price and cost of producing the commodity?" Economic principles state that equilibrium price is established where demand and supply are equal for short-run periods as well as long run. In a free market economy, with strong demand, price rises high enough to ration the commodity to highest-value users. Price will rise as long as buyers continue bidding for the existing short supply. But as price rises, more and more potential buyers are priced out of the market. High price will thus ration a limited supply to those who valued it the most or who have the means (income) to acquire it. At the other end of the scale, a low level of consumption relative to available supply means that the commodity must be sold at a low price to dispose of it. Low price encourages consumption and discourages future production.

The time frame is important in its effect upon the level of the equilibrium price. Once a crop such as corn is harvested, the operator can either sell it at the prevailing market price or store it to hold it off the market temporarily. The farmer grew it to sell, and eventually grain will move into the market. Because there are so many sellers, one producer cannot influence price by either selling his or her product or withholding it from the market. At harvest, the demand situation may allow the producer to enjoy large profits from high price or substantial loss from low price, but he or she cannot alter the previously made decision to produce. Therefore, in the short-run period of a production cycle, no relationship exists between cost of production and market price. Perishable production and competitive economic environment force producers to be price takers. **The market does not care what it costs to produce the product.** Its only concern is the price necessary to balance available supply with effective demand. With a short supply, price rises to ration it, and with a surplus supply, price falls to move it.

In the intermediate time period before planting occurs, the producer can make certain adjustments in response to price level. He or she may plant only the best land and reduce fertilizer rates and chemical applications if prices are low. A producer will restrict inputs and consequently reduce production somewhat. But as long as he or she covers the variable, or direct out-of-pocket, costs of production, the producer will continue to produce even though he or she is not covering the fixed portion of costs. Thus, in the intermediate time frame, production will continue as long as a producer covers variable production costs, that is, seed, fertilizer, chemicals, and fuel, even though fixed costs of capital, labor, land, and management are not paid. At the other extreme, if price is high, farmers exert every effort to produce as much as possible by applying more inputs. They perform those production practices that return more than cost in the manager's effort to balance added returns against added costs. Thus, in the intermediate period, high or low price encourages some modest supply adjustment within the fixed farm plant.

In the long run, larger changes in production are possible. In response to low prices, poor quality land will be taken out of production and either lie idle or be put to some less intensive use, e.g., forestry. Some farmers will not replace facilities or equipment. If product price is persistently low, high-cost producers will be forced to quit. The supply of the commodity will decline and prices will rise until remaining producers share the demand at a price where there no longer are losses. On the other hand, if price is high, operators will maintain land in high productivity and expand the operating base. Producers will clear forests, drain swamps, and irrigate new land, expand facilities, and adopt new technologies to share in the profits. The supply curve will be shifted to the right, causing price to be under pressure from the increased production. Therefore, in the long run, production is significantly affected by price level. Supply will adjust so that average price will tend to equal average total cost of production. All production costs will be covered, and returns to capital, land, labor, and management will be no more, or no less than they could receive if they were used in other activities.

Again, the market does not care what it costs a producer to supply commodity. The "invisible hand" of the market concerns itself with moving the available supply at whatever price is determined by the interaction of supply and demand.

Delay in Producer Response: The Cobweb Theorem

The principles of supply and demand explain how price reaches an equilibrium point to allocate supply between buyers and sellers. The **cobweb** theorem explains why price may not settle at an equilibrium point but fluctuate about it and constantly rise and fall to create commodity price cycles. If the quantity of a commodity could be adjusted instantaneously to price change, price cycles would not occur. That is, however, not the case. The biological nature of agricultural production gives rise to production cycles, and production cycles lead to price cycles.

Price cycles run counter to production cycles. When supplies increase, prices fall; when supplies decrease, prices rise. Price changes are much greater in magnitude than production changes because of the inelastic nature of demand and supply of farm products.

Periodic expansion and contraction of supply of the product causes price cycles. Price cycles are, therefore, supply-based; they reflect producer output decisions. Suppose, for example, hog production was low at Q_1, and price was high at P_1 in period I, as shown in Figure 4–20. Hog producers look at the favorable price and base their future production plans in Period II upon it. High price encourages the retention of breeding stock to expand production. New producers are encouraged to enter the business. Breeding, gestation, farrowing, and finishing the market animals require about nine months before production decisions result in increased output. As a result, producers bring Q_2 to the market in period II. Since they must move animals when they are ready for market, the increased output is literally dumped on the market. In response to the larger supply, price falls to P_2. Producers now reevaluate the situation in light of depressed prices and decide to raise fewer hogs. Many quit production altogether. Consequently, production falls to Q_3 in period III. This is the same production level as period I, where the cycle began. Price rises as high as P_1, and the stage is set for the cycle to repeat itself.

Agricultural price cycles are caused by the tendency of farmers to base future production plans on current price and profit rather than on future price expectations. They act as though present price will continue indefinitely into the future. Production response is often off target as a result.

Four conditions are necessary for price cycles to occur. First, there must be a time lapse between a change in price and producers' production response to the change. Second, producers must base future production plans on current prices rather than

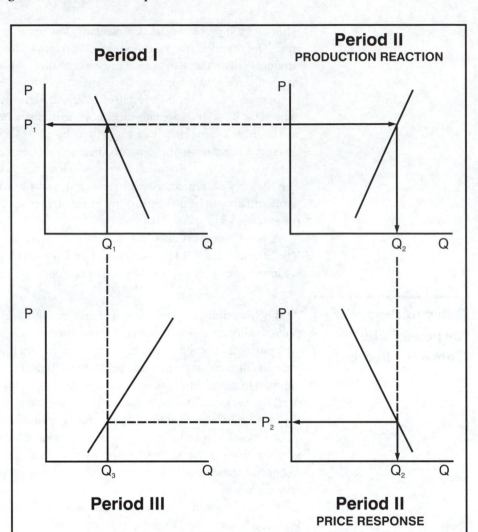

FIGURE 4–20: Illustration of Supply Response to Market Price in Explaining Price Cycle.

future price expectations. Third, producers must have reasonably good control over output as they do in livestock. There are no well-defined price cycles in crops because producer decisions are only partially responsible for crop size (with nature being the major force). Fourth, the product is not storable and is therefore dumped on the market. Finally, price cycles develop in competitive industries because each producer believes that his or her own production decision will not affect the overall supply; and he or she will not, therefore, affect price.

Length of agricultural price cycles depends upon both biological and psychological time lags. The biological lag in livestock is the time period from expansion of the breeding herd to market maturity of the resulting offspring. The cycle is longer for cattle than chickens because of biological differences. Agricultural price cycles are usually longer, however, than the biological time lag. Though the biological production cycle in hogs is 9 to 10 months, the hog price cycle usually runs 3 or 4 years. Cattle production cycles are about 4 years, but the cattle price cycle runs 8 to 12 years. Psychological factors account for the remaining time lag. They explain the time required for current market conditions to convince producers that current prices are truly here to stay. The combination of biological and psychological lags explains the length of price cycles. However, the variability in psychological factors makes the length of the price cycle somewhat unpredictable.

Fundamental Approach to Price Level Determination

Economic principles provide essential understanding of demand and supply relationships in establishing price. But they fall short of providing workable information for decision making.

Thousands of supply and demand forces tug and push at equilibrium price; some buoying it up, others pulling it down. Analysis of market price from a fundamental approach attempts to sort out the important price-making forces, while discarding unimportant ones. Price analysis further measures the degree of influence exerted by such forces.

Econometricians combine economic theory and mathematics into analysis of price-determining forces. Agricultural price analysts investigate changes in supply and demand and the resulting changes in commodity price. It is not enough to simply show that agricultural prices decline when the supply increases or demand decreases. Price analysis provides a quantitative estimate of **how much** the price changes in response. Price analysis begins with determining which supply and demand forces are most important in setting price, then it attempts to measure the degree by which changes in demand and supply affect prices.

Once relationships are defined and respective influences measured, price forecasting becomes possible. Statistics showing increased crop acreage, increased yield, and increased carry-over of supplies can then be translated into projected price change as the result of increased supply.

The underlying principles of supply-demand relationships are common to all commodities. However, application of these principles to individual commodities varies greatly because each commodity has its particular set of price-making forces. Furthermore, all commodities interrelate to each other and together fit into the much larger macroeconomic picture.

Darrel Good, Thomas Hieronymus, and Royce Hinton developed mathematical models to measure the impact of individual forces on price.[6] Their master model of Midwest agriculture is presented in Figure 4–21. It is a systematic description of the factors that affect prices and price relationships. Their model is an interrelated system in that a change in one part of the system (such as a new corn production estimate) changes every other value in the system in due course. A goal of price analysis is to quantify the relationships and measure the values so that each new piece of information can be fed into the calculations and its impact on the other factors of the system instantly estimated. Despite the fact that "the current state of the art of price analysis and forecasting falls far short of providing all the data needed to measure all of the relationships in the system, we must be aware of the total system and know where each segment fits in the system if we expect to understand price determination in the market process."[7]

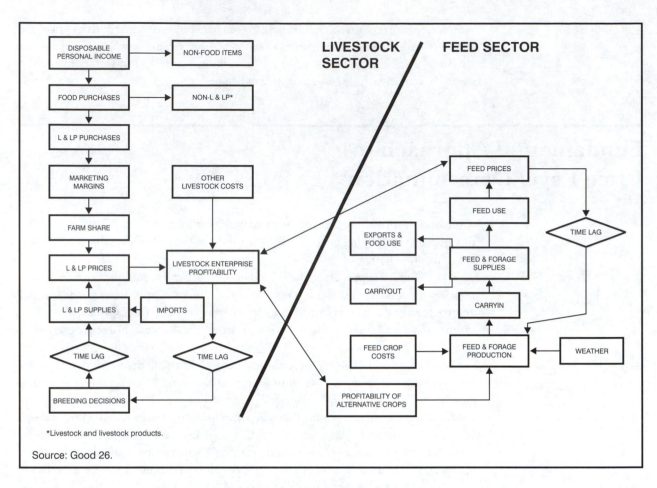

FIGURE 4–21: Master Model of Midwest Agriculture. Source: Good 26.

The two major segments of the Illinois model are the livestock sector and the crop sector. Corn Belt farmers are primarily concerned with crop production, but most of their grain is marketed as livestock or livestock products. The relationship of livestock enterprise profitability and feed costs (crop prices) links the two major sectors together.

The dominating factor that in large measure determines values in the entire system is disposable personal income (DPI). As a measure of consumer purchasing power, it is income remaining after taxes are paid. The amount of disposable personal income that consumers spend for the end products of agricultural production ultimately determines the level of farm income. Out of the total livestock production portion of food expenditures, marketing margins paid to intermediaries are subtracted to determine farm share of consumer dollars. These dollars constitute the demand side of the livestock and livestock product price equation. On the supply side of the equation is the quantity of livestock and livestock products available for consumption. Domestic production plus imports less exports make up the total supply available to consumers. At this point in the model, net consumer expenditure, expressing demand, interacts with livestock product supplies from farms to determine prices.

The Illinois regression analysis explained 79 percent of the variation in consumer meat expenditures by the two variables of consumer income and meat supplies.★ The thousands of other variables accounted for only the remaining 21 percent of expenditures.

Relative to the first variable, meat consumption changes directly with income change. As incomes increase, meat consumption also increases. When incomes decrease, meat consumption decreases.

In the supply relationship, a larger supply sells for less total money and smaller supply sells for more money due to price inelasticities. Consumers are reluctant to change consumption patterns and are encouraged to do so by relatively large price changes. As an example, a 1 percent increase in supply may result in a 2.4 percent decrease in price. As the quantity of meat supply increases by 1 percent, total expenditures fall by 2.4 percent. A larger quantity multiplied by a much lower price produces a decrease in total revenue from meat sales.

In the next stage of the model, livestock production decisions are based upon enterprise profitability. Each producer's assessment of livestock prices, compared to feed costs and other production costs, determines his or her expected profitability. Livestock profitability thus establishes the demand side of the crop price equation. Losses discourage production; high profits encourage livestock expansion. Livestock numbers and feeding rates therefore translate into demand for feed grains.

The crop sector represents the supply side of the crop price equation. Acreage planted multiplied by yield produces annual production figures. Supply carried over from the previous year is added to production to make up total supply available. Total supply then moves in four alternative demand directions: food, industrial use, export, and feed. Carry-over is the amount left over that was not sold at an acceptable price. In the later stages of a crop year, next year's expected production is a factor in determining how much of old crop is used immediately and how much is saved. If prospects point toward a large crop, price declines to encourage immediate feeding. If prospects suggest a small crop, price level rises and more of the old crop is saved for consumption the next year.

It is out of these aggregate market interactions that prices of individual commodities are formed. We shall follow the analysis further to identify important price-making forces in particular commodities. By understanding those forces and assessing their importance, farmers are in a better position to evaluate market information. They can better sort out real influences from market noise.

Understanding Grain Prices

Grain markets must establish a price level at the beginning of the season that will encourage consumption up to an appropriate level of carry-over at the end of the year. If price is too high, usage will be curtailed and carry-over will be too large. If the price is too low, the rate of use will be too rapid and pose a threat of running out before the next harvest. The solution is to establish a market clearing price at the beginning of the year, but the problem is to anticipate the "correct" market clearing price. The market as a whole looks at all factors and forms a judgement about the appropriate price. "It is a collective speculative judgement; it is almost always in error"[8], but it continually redresses the errors as prices fluctuate up and down day to

---

★Meat supplies of beef, pork, veal, broilers, and turkeys.

day and week to week. The collective judgement changes continually as market forces and their interpretation change. Daily price changes are a series of adjustments to changing supply-demand factors.

Corn Price. A balance sheet method was used by Good to estimate availability of corn for feed use, as in Table 4–8. Estimates for food, industrial use, and exports were subtracted from total supply, leaving the remainder available for feed and carry-over. The estimates for the first three uses were made without regard to price. In the case of food and industrial use, the products were high in value and had no close substitutes. The amounts of corn used in these production processes were consumed largely without regard to price.

TABLE 4–8: Corn Balance Sheet: Production, Use, and Carry-over, 1989-2000 (Mill. of Bu.)

	1989-90	1990-91	1991-92	1992-93	1993-94	1994-95	1995-96	1996-97	1997-98	[a]1998-99	[a]1999-00
Carry-in	1930	1342	1515	1074	2080	796	1495	347	791	1208	1644
Production	7532	7934	7475	9477	6338	10051	7400	9233	9207	9761	9386
TOTAL	9462	9276	8990	10551	8418	10847	8895	9580	9998	10969	11030
Seed, Food & Industrial	1370	1425	1534	1556	1609	1704	1612	1692	1782	1860	1950
Export	2368	1725	1584	1663	1328	2177	2228	1795	1504	1900	2000
Feed and Residual	4382	4611	4798	5252	4685	5471	4708	5302	5504	5565	5500
TOTAL	8120	7761	7916	8471	7622	9352	8548	8789	8790	9325	9450
Carry-out	1342	1515	1074	2080	796	1495	347	791	1208	1644	1580
Change in carry-out	xxx	173	–441	1006	–1284	699	–1148	444	417	436	–64
U.S. Ave. Price $/bu.	2.36	2.28	2.37	2.07	2.50	2.26	3.24	2.71	2.45	2.00	2.15

[a]Projected.
Source: USDA ERS.

Foreign demand was highly inelastic. In the past, an important destination of U.S. corn exports was the European Economic Community. Prices there were higher than world price and were maintained at high levels by a variable levy that could be changed daily. When the price outside the EEC Common Market went down, the levy was increased and vice versa. Changes in U.S. prices had little effect on the amount used in EEC. Japan, as the largest importer, did allow prices to fluctuate somewhat but not enough to affect the quantity imported. The USSR was another important destination of U.S. corn. Soviets appeared to buy on the basis of need and political goals rather than price. Other countries that imported relatively small amounts of corn also responded primarily to need rather than price. Only during years of short U.S. crops, with prospects for abundant supplies after the following harvest, did it appear that importers modified their demand.

Since the other demands were relatively fixed, the focal point in forecasting corn price was livestock feed demand. Corn was priced low enough to make its use attractive to livestock producers but high enough to prevent feeders from attempting to use the entire supply. Corn price was consequently demand-determined, given the available supply. The key to corn price forecasting was the calculation of the supply available for feeding. When corn was scarce relative to livestock numbers and level of feed use, its price was high in relation to livestock prices. When corn was abundant, the price was low in relation to livestock prices.

The Illinois study, therefore, tied value of corn to three factors: its feeding value, increase or decrease in carry-over, and speculative error.

The comparison of the feed value of corn and its market price for 10 years is given in Table 4–9. Though feed value and crop price were not the same year to year, corn price tended to return to feeding value as its central value.

TABLE 4–9: Feeding Value and Average Price of Corn, and Change in Carry-over, 1969 to 1978 Crop Years

Years	Feed Value	Average Price	Corn Price/Feed Value Difference	Change in Carryover
	($/bu.)	*($/bu.)*		*(mil.bu.)*
1969-70	1.36	1.24	– .12	– 113
1970-71	1.27	1.41	+.14	– 338
1971-72	1.36	1.15	– .21	+459
1972-73	1.58	1.81	+.23	– 417
1973-74	1.99	2.82	+.83	– 226
1974-75	2.18	3.01	+.83	– 124
1975-76	2.20	2.64	+.44	+ 40
1976-77	1.84	2.19	+.35	+485
1977-78	2.18	2.17	– .01	+220
1978-79	2.49	2.46	– .03	+ .68
Mean	1.85	2.09		

Source: Good 80.

Establishing a central value based on livestock prices is therefore a useful point of departure in forecasting average price of corn for a given year. Feed value is determined by the price livestock feeders can afford to pay for corn as a production input.

Another factor in explaining corn price is projected change in carry-over. When the carry-over supply is expected to increase, we should expect corn price to be below feed value to encourage consumption. When carry-over is expected to decrease, corn price should be higher than feed value to discourage usage. Comparison of the difference in corn price versus feed value and change in inventory for 10 years is also presented in Table 4–9. Again, the relationship was not perfect, but there was a tendency for corn price to be lower than feed value when carry-over increased and to be higher than feed value when carry-over decreased.

The third factor in the analysis is speculative error. It refers to the market's inability to accurately assess the forces at work, and the occurrence of unanticipated events, such as drought, corn blight, early frosts, changes in exports, and political events. Speculative error represents the unexplained portion of price variability.

In summary, the Illinois authors noted that government loans and inventory programs dominated the general price level of corn for many years prior to 1972. During the period 1972 to 1982, the market overestimated and underestimated the price impacts of both shortage and surplus. The market's biggest problem was identifying value in a climate of rapid inflation. An equilibrium value of $2.50 in one year was the same as $2.75 the next if inflation was running at a rate of 10 percent per year. However, when other forces were accounted for, the price foundation of corn depended on its value as livestock feed in the United States, plus or minus changes in carry-over stocks.

In another work, Scott Irwin related carry-over corn stocks to the price/loan ratio.[9] This work has been updated when Allan Lines calculated carry-over stocks as a percent of use ratio.[10] Their work illustrates the effects of both the carry-over supply and the price-supporting effect of a commodity loan on price, as in Figure 4–22. When stocks were high, the loan rate became the market price. At the other extreme, with the small carry-over, corn price was considerably higher than loan. Farmers can use Irwin's analysis to project from expected year-end corn supplies to price level direction. Shrinking carry-over supplies point toward higher prices; increasing carry-over supplies point to lower prices.

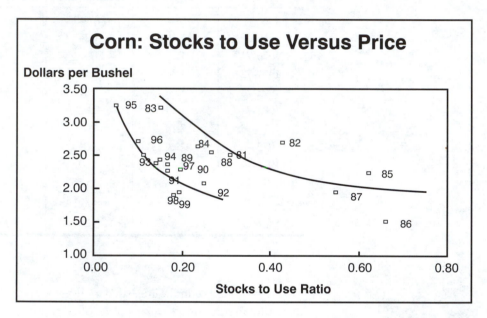

FIGURE 4–22: Corn Prices Relative to Carry-over Stocks. Source: Irwin.

Soybean Price. Good, Hieronymus, and Hinton observed that soybeans are the most complex and difficult of all commodities to forecast. The soybean is not one commodity but two, soybean meal and soybean oil. One cannot be produced without the other. Yet these products have totally different markets and uses. Soybean oil is a high-grade vegetable oil used primarily in shortening, margarine, and salad and cooking oils. Meal is a protein supplement for livestock feeding. Both commodities move extensively into world trade; both have numerous close substitutes in use. This makes supplies of competing products important in price determination. Price elasticities of demand for both oil and meal are low, making prices highly volatile. Market demand for both products expanded rapidly since 1940, but the rate of demand expansion is uncertain in the future.

Since soybean products have different price-making forces, they must be considered separately and their values added together to obtain a value of soybeans. Each 60-

pound bushel of soybeans contains about 11 pounds of oil and 47.5 pounds of 44 percent protein meal, with the remainder being waste or loss. The value of a bushel of soybeans can be calculated from the value of the products. The formula is:

Soybean oil Price _____ x 11.0 = _____

plus SBOM Price/lb. _____ x 47.5 = _____

equals Product Value, (per bushel) _____

The difference between the product value and soybean price is the crusher's (processor's) margin.

The "normal" processing margin averaged about 45 cents per bushel. Deviations from normal were related to farmers holding soybean supplies from the market. In years when prices were high and farmers were ready sellers, margins tended to be high. In years of low price levels and tight farmers holding, margins tended to be small.

Soybean price is, therefore, determined by the world market for its products. Soybean oil prices are a function of world price of edible fats and oils, since exports form a major part of the market. Furthermore, the domestic demand for oil is highly inelastic. Price of oil is relatively cheap and U.S. consumption is constant regardless of price change. Price in the United States must remain close to world price. On the other hand, soybean meal prices are made in the United States. Demand outside the United States is highly inelastic, so the amount exported is not materially affected by price. Quantities imported by other countries are usually tied to perceived needs for the commodity and are not price-sensitive. When the supply situation dictates an adjustment in the use of soymeal, the adjustment is made in the United States. Thus, factors outside the United States tend to set oil price, and factors inside the United States set meal price.

The soybean balance sheet becomes a basic reference point in forecasting soybean price. The projected increase or decrease in carry-over of supply suggests price direction. Table 4–10 is a 10-year record of the soybean balance sheet.

TABLE 4–10: Soybean Balance Sheet: Production, Use, and Carry-over 1989-2000 (millions of bushels)

	1989-90	1990-91	1991-92	1992-93	1993-94	1994-95	1995-96	1996-97	1997-98	a1998-99	a1999-00
Carry-in	182	236	324	270	282	193	314	157	96	159	421
Production	1924	1926	1987	2190	1870	2515	2174	2380	2689	2757	2900
TOTAL	2106	2162	2311	2460	2152	2708	2488	2537	2785	2916	3321
Crush	1146	1187	1254	1279	1276	1405	1369	1436	1597	1550	1625
Export	623	557	684	770	589	838	851	882	870	765	850
Feed, Seed, and Residual	101	94	103	129	94	151	111	123	159	180	160
TOTAL	1870	1838	2041	2178	1959	2394	2331	2441	2626	2495	2635
Carry-out	236	324	270	282	193	314	157	96	159	421	686
Change in carry-out	xxx	88	–54	12	–89	121	–157	–61	63	262	265
U.S. Ave. Price $/bu.	5.70	5.75	5.58	5.60	6.40	5.48	6.77	7.35	6.47	4.90	4.65

aProjected.

Source: USDA ERS.

The first step in calculating a balance sheet is to determine acreage, yield, and production. The first planting intentions report for the year is released in January; the second in April; and a report on planted acres at the end of June. Typically, 98 percent of planted acres are harvested, so planted acreage is multiplied by 0.98 before being entered on the balance sheet. The first yield report is released in August. Until then, yield estimate is based on trend in historical yield and modified by weather conditions to date. Production estimates are based on conditions on the first of the month and are released about the tenth of each month, August through November. The final estimate of production is made in late January.

Current estimates of carry-in stocks are added to production estimate to obtain total supply for the coming year. After harvest, when the consumption year starts, the balance sheet gradually becomes an estimate of carry-over, based upon disappearance to date. Disappearance occurs as exports or domestic use. Exports are reported weekly and fairly quickly set a trend for the year. They must be adjusted, however, by estimated production in South America. In the last three decades, Brazil and Argentina have become important competitors in the world market. Their crop is planted in October-December; January is a critical weather month, and harvest is in the March-May period. Crop estimates for South America are made by the Foreign Agricultural Service of the USDA and announced periodically throughout the growing season. Calculation of domestic processing in the United States is available from the National Soybean Processors Association on a weekly schedule and the Bureau of the Census on a monthly schedule. The rate of processing or crushing to date becomes increasingly reliable and important as the crop year progresses. Carry-out is what is left of supply at the year end.

The carry-out figure gauges whether soybeans are relatively scarce or abundant. For budgeting purposes, 180 million bushels was selected as a reasonably comfortable carry-over. Below 150 million bushels, supplies are light and prices stronger. Above 200 million bushels, large supplies weigh heavily on price.

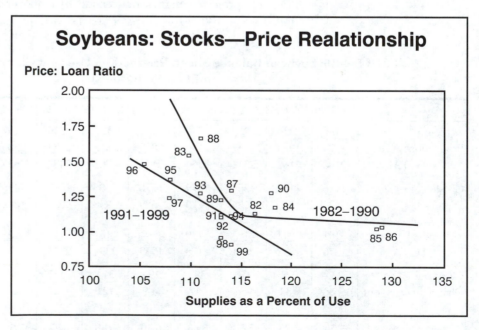

FIGURE 4–23: Soybean Prices Relative to Carry-over Stocks. Source: Irwin and Lines.

Changes in carry-over supplies on the soybean balance sheet, thus, become a basic reference point in forecasting soybean price. The soybean balance sheet is an indication of the demand situation in the product markets of soybean oil and meal as demand interacts with available supply to set price direction.

Irwin and Lines calculated the price level of soybeans relative to loan as compared to stocks on hand at the end of the crop year.[11] The carry-over supplies were calculated as a percent of annual usage. When stocks exceeded 15 percent of annual use, the soybean loan rate became the effective market price (Figure 4–23). With stocks about 10 percent or less of use, price level was 1.25 to 1.50 times higher than the loan rate.

Similar data for wheat price and annual carry-over stocks are presented in Figure 4–24. In the 1970's and 1980's stocks above 1.0 billion bushels caused the loan rate to set the floor on price level. Carry-over stocks below 800 million bushels produced prices 1.5 to 3 times higher than loan rate. In the 1990's, stocks over 600 million bushels forced the loan rate to set the price level.

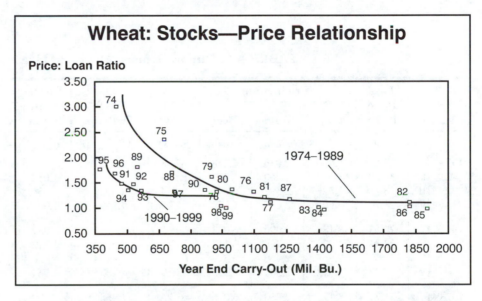

FIGURE 4–24: Wheat Prices Relative to Carry-over Stocks. Source : Irwin and Lines.

Understanding Livestock Prices

Hog Prices. The nature of pork production makes hog price prediction relatively straightforward. Hogs come to market at fairly uniform weights on a biologically consistent schedule. Since there is very little inventory buildup or liquidation, there is hardly any inventory speculation on the part of producers. Pork is consumed shortly after it is produced and is not held in storage. Since no major trend elements in either production or consumption exist, price of hogs "lends itself to systematic quantification and to derivation of functional relationships."[12]

The factors that were found to explain retail pork price include consumer demand for meat in general, supplies of pork, supplies of competing meats, and consumer preferences. A rather consistent marketing margin was subtracted from the retail price to get back to farm level hog price. The Illinois demand formula accounted for about three-fourths of the variability in quarterly average price of live hogs. It showed that the price of hogs depended upon the amount of pork production, beef production, broiler production, and real disposable income. A 1.0 percent change in pork production was associated with a 1.64 percent change in hog prices in the opposite direction; for each 1.0 percent change in beef production, there was a 0.72 percent change in hog prices in the opposite direction; for each change of 1.0 percent in broiler production, there was a 0.77 percent change in hog prices in the opposite direction; and for each 1.0 percent increase in disposable personal income, there was a 1.75 percent increase in price of hogs.

A mathematical forecast, based upon historical price relationships, can be treated as an indicator of value.[13] The Illinois procedure for forecasting hog price is to measure anticipated pork, beef, and poultry production for the current and next three quarters of the year, adjust for seasonally price changes, account for income effects, and adjust for inflation. Results of their work are shown in Table 4–11. Though the projected price seldom matched the actual market price, the forecasts did project periods to expect significant price change due to changes in the factors that affect price.

TABLE 4–11: Illinois Demand Model for Hogs

Period[1]		Hog Price (1972 $)	Production			DPI (1972 $)	GNP Deflator	Adjustment Factor	Hog Price, Current $		
			Pork	Broilers	Beef				Computed	Actual	Error
		($/cwt.)	- - - - - - - (mil. lb.) - - - - - - -			(bil. $)	(1972 = 1.0)		- - - - - - - - ($/cwt.) - - - - - - -		
1976	I	37.17	2,860	2,035	6,228	875.8	1.3055	.939	46.31	48.53	–2.22
	II	36.29	3,044	2,260	6,300	883.3	1.3180	1.012	43.86	47.83	–3.97
	III	35.78	2,827	2,397	6,504	889.4	1.3329	.956	43.35	47.69	–4.34
	IV	25.70	3,605	2,268	6,572	895.5	1.3498	1.063	35.57	34.70	.87
1977	I	28.65	3,188	2,090	6,253	905.7	1.3701	.939	44.10	39.25	4.85
	II	27.84	3,420	2,338	6,169	915.3	1.3920	1.012	41.06	38.76	2.30
	III	31.57	2,965	2,460	6,379	926.5	1.4148	.956	46.14	44.67	1.47
	IV	28.27	3,522	2,304	6,295	940.9	1.4333	1.063	44.46	40.52	3.94
1978	I	31.79	3,172	2,250	6,076	950.6	1.4556	.939	49.38	46.27	3.11
	II	32.07	3,397	2,502	6,049	959.6	1.4831	1.012	46.24	47.57	–1.33
	III	31.61	3,109	2,608	5,909	964.7	1.5170	.956	50.08	47.95	2.13
	IV	32.49	3,507	2,509	6,113	976.5	1.5450	1.063	49.22	50.20	– .98
1979	I	32.97	3,276	2,413	5,669	985.9	1.5786	.939	54.15	52.04	2.11
	II	28.54	3,797	2,809	5,129	986.2	1.6142	1.012	43.85	46.07	–2.22
	III	23.71	3,786	2,932	5,325	986.6	1.6484	.956	39.27	39.08	.19
	IV	21.65	4,230	2,803	5,336	990.3	1.6833	1.063	36.27	36.44	– .17

Source: Good 47.

[1]Quarter I = Dec-Feb, II = Mar-May, III = Jun-Aug, IV = Sept-Nov.

Regression Equation:

Price of Hogs (1972 $) = 70.6357 – .0141 Pork Prod. – .0111 Broiler Prod. – .00375 Beef Prod. + .0610 DPI
(1972 $) "t" values   (6.04)   (–9.86)   (–3.16)   (–2.68)   (3.91)
price flexibilities   Pork: –1.64   Broiler: – .77   Beef: – .72   DPI: +1.75
plus 100 mil. lb.   –$1.41   –$1.11   –$ .38
r^2 = .768 r^2 after adjustment = .937

Live Cattle Prices. Accounting for factors to explain live cattle prices is more difficult than live hog prices. The production cycle is longer and there is a wider range of
sources of supply from dairy, breeding stock, and fed and non-fed steers and heifers.
The mixture of kinds and quality continually changes as feeding profitability
changes.

Several factors were identified as important in affecting cattle prices. Disposable
personal income and consumer willingness to spend money for meat made up the
most important demand variable in the equation. A second factor was beef supply,
which was a function of production. The third factor was the supply of competing
meats: lamb, broilers, veal, and turkey. Other substitutes such as eggs, dairy products,
or vegetables were not found to be important competitors.

The mathematical model explained about 78 percent of the variation in cattle
prices by using the four variables of beef, pork, and broiler production, and disposable personal income. The results showed that a 1 percent change in beef production

TABLE 4–12: Illinois Demand Model for Cattle and Wholesale Beef

Period	Cattle Price (1972 $)	Whlsl. Price (1972 $)	Beef	Pork	Broil.	DPI Annual Rate	GNP Defl.	Com-puted	Actual	Error	Com-puted	Actual	Error
	($/cwt.)	(¢/lb.)	(mil. lb.)			(bil. $)	(1972 = 1.0)	($/cwt.)		(%)	(¢/lb.)		%
75 I	28.76	48.93	5,842	3,146	1,833	825.5	1.2421	41.98	35.72	+14.9	68.64	60.77	+11.5
II	38.22	63.10	5,593	2,991	2,061	867.1	1.2596	48.34	48.14	+ .4	77.00	79.48	− 3.2
III	37.94	63.41	5,942	2,557	2,080	858.3	1.2820	45.57	48.64	− 6.7	73.27	81.29	− 10.9
IV	35.38	58.19	6,295	2,901	1,992	864.9	1.3017	40.02	46.05	− 15.1	65.77	75.74	− 15.2
76 I	29.48	48.68	6,490	2,957	2,116	877.5	1.3130	37.29	38.71	− 3.8	61.55	63.92	− 3.9
II	31.19	49.77	6,144	2,848	2,314	881.2	1.3279	42.87	41.42	+ 3.4	68.89	66.09	+ 4.1
III	27.77	44.41	6,617	3,014	2,371	888.4	1.3430	35.30	37.30	− 5.7	58.03	59.64	− 2.8
IV	28.61	46.03	6,410	3,668	2,185	896.7	1.3634	37.07	39.00	− 5.2	61.18	62.75	− 2.6
77 I	27.38	44.04	6,329	3,276	2,156	903.6	1.3834	41.62	37.88	+ 9.9	67.89	60.92	+10.3
II	28.93	45.90	6,162	3,186	2,399	912.5	1.4093	44.95	40.77	+ 9.3	71.91	64.68	+10.1
III	28.38	45.65	6,320	3,074	2,424	928.0	1.4259	44.38	40.47	+ 8.8	71.11	65.09	+ 8.5
IV	29.29	47.41	6,218	3,499	2,249	939.9	1.4482	46.11	42.42	+ 8.0	74.21	68.66	+ 7.5
78 I	31.13	49.98	6,104	3,243	2,327	948.7	1.4705	50.41	45.77	+ 9.2	80.17	73.50	+ 8.3
II	36.51	57.98	5,936	3,264	2,547	953.0	1.5082	53.60	55.06	− 2.7	84.07	87.45	− 4.0
III	35.03	54.88	5,922	3,158	2,567	962.2	1.5345	55.90	53.75	+ 4.0	87.48	84.12	+ 3.7
IV	34.95	53.86	6,040	3,540	2,442	973.2	1.5668	54.18	54.71	− 1.1	85.52	84.39	+ 1.3
79 I	40.83	62.98	5,546	3,399	2,551	981.3	1.6022	65.40	65.42	0.0	101.18	100.90	+ .3
II	44.26	66.85	5,076	3,760	2,845	977.8	1.6381	71.45	72.51	− 1.5	108.66	109.51	− .8
III	39.40	60.41	5,219	3,779	2,854	980.9	1.6720	70.22	65.88	+ 6.2	107.02	101.00	+ 5.6
IV	39.16	61.39	5,416	4,346	2,665	986.7	1.7058	65.61	66.86	− 1.9	101.73	102.36	− .6

Source: Good 63.

Regression Equations:

Cattle Price (72 $) = 82.1626 − .0116 (Beef Prod) − .0039 (Pork Prod) − .0037 (Broil. Prod) + .0463 (DPI-72 $)

| "t" values | −11.5 | −3.8 | −1.3 | +3.6 |
| Price flexibilities | 1.895 | .368 | .216 | 1.125 |

r^2 = .779

Wholesale Beef Price (72$) = 129.887 − .0164 (Beef Prod) −.0056 (Pork Prod) − .0077 (Broil. Prod) + .0641 (DPI-72 $)

| "t" values | −10.4 | −3.5 | −1.7 | +3.2 |
| Price flexibilities | 1.666 | .347 | .347 | .974 |

r^2 = .736

was associated with a 1.9 percent change in price of cattle in the opposite direction; each 1 percent change in pork production brought a change of 0.39 percent in cattle prices in the opposite direction; each 1 percent change in broiler production brought a change of 0.2 percent in cattle prices in the opposite direction; and each change of 1 percent in real disposable income brought a change of 1.13 percent in the price of cattle in the same direction.

On balance, the formula provided satisfactory ballpark estimates of cattle prices. Results are shown in Table 4–12. In particular years, the forecast differed significantly, but the method established a central value. When prices departed substantially from central value, they subsequently returned and usually over corrected.

Summary

Fundamental supply and demand principles explain equilibrium price and value of an agricultural commodity and commodities in general. However, discovering the market-clearing equilibrium price is not an easy matter in a dynamic, changing real world. Shifts in supply and demand conditions result in new prices being established; inelastic demands and supply responses cause exaggerated price changes. As a consequence, farm prices and incomes tend to be highly variable. But if farmers and ranchers understand price-making principles and their application to various situations, they can improve their skill in making price decisions for the commodities they produce.

Our objective in this chapter was to develop the notion of a general PRICE LEVEL established by FUNDAMENTAL demand and supply considerations. We further sought to identify important price-making forces and evaluate their impact in causing price change.

By sorting out important price-making variables from the unimportant ones, we identify the fundamentals to watch in making price projections. By using the calculated elasticities of the important variables, we can project expected price change based upon change in the fundamental factors.

By knowing what is important and how important individual variables are, farmers are in a better position to project prices from fundamental information.

Questions for Study and Discussion:

1. Define and illustrate the demand principle.
2. What does elasticity of demand show?
3. What is meant by a change in quantity demanded? What does this change depend upon?
4. What is meant by a shift in demand? What factors shift the demand curve?
5. What is effective demand?
6. Define and illustrate the supply principle.
7. What is supply elasticity, and what factors affect it?

8. What is meant by a change in quantity supplied? What does this change depend upon?

9. What is meant by a shift in supply? What factors shift the supply curve?

10. What is the meaning of equilibrium price and quantity?

11. Illustrate the effects on price that arise from (a) shifts in demand and (b) shifts in supply.

12. How does the supply-demand balance operate to achieve equilibrium price?

13. Explain price discovery as distinct from, but a part of, equilibrium price.

14. How is the Law of One Price used to explain price differences?

15. What is the effect on total receipts from increased supply of a commodity with inelastic demand?

16. Farmers work harder to produce more but end up earning less. Why?

17. What are created by price ceilings and price floors? Why?

18. How does the cobweb theorem explain price cycles?

19. What are the objectives of price analysis in a fundamental approach to price determination?

20. What fundamental factors are important determining grain prices?

21. What fundamental factors are important in determining hog and cattle prices?

Exercise: Maintaining a History of Price-Making Events

Each day the commodities page of *The Wall Street Journal* reports on traders' comments about market forces in play for the previous day. Make a copy of those comments and maintain them over a period of time. A periodic review of a month or two of comments is enlightening in evaluating published reports of what were thought to be important market-making forces. Time helps to sort out whether the events were important or not. After a while, you may determine that newspaper reports of current events are really records of history and comments on events were really "noise," not true price-making factors.

One handy way for maintaining this record is to glue the newspaper clipping in a spiral notebook. Headline news stories of demand and supply events from other sources can be attached to the record also.

Students can hold a group discussion or write a paper as a review of their records of market actions after three months. These activities will help students understand the forces at work and their relative importance in price determination.

Endnotes for Chapter 4:

1. Demand, supply, and price sections of this chapter are adapted with permission of Macmillan Publishing Co. from Richard L. Kohls and Joseph N. Uhl, *Marketing of Agricultural Products,* 6th ed. (New York: Copyright 1985 by Macmillan Publishing Co.).

2. Kohls and Uhl, *Marketing of Agricultural Products* 170.

3. Kohls and Uhl, *Marketing of Agricultural Products* 168.

4. Kohls and Uhl, *Marketing of Agricultural Products* 169.

5. Robert L. Heilbroner, *The Economic Problem Newsletter*, Vol. V, No. 2 (Englewood Cliffs, N.J.: Prentice-Hall, Inc., Winter, 1973).

6. Darrel L. Good, Thomas A. Hieronymus, and Royce Hinton, *Price Forecasting and Sales Management* (Champaign: Cooperative Extension Service, College of Agriculture, University of Illinois, 1980). Copyright 1980 by the Board of Trustees of the University of Illinois. Used by permission.

7. Good, Hieronymus, and Hinton, *Price Forecasting and Sales Management* 27.

8. Good, Hieronymus, and Hinton, *Price Forecasting and Sales Management* 67.

9. Scott Irwin, unpublished data, The Ohio State University, Columbus.

10. Alan Lines, *2000 Agricultural Policy Outlook*, AEDA Publication ESO 2555, The Ohio State University, Columbus, OH.

11. Irwin and Lines.

12. Good, Hieronymus, and Hinton, *Price Forecasting and Sales Management* 35.

13. The Illinois analysis of projected price is available on a subscription basis from Darrel Good, "Outlook Update," Department of Agricultural Economics, University of Illinois, Urbana, 61801.

Chapter 5

Technical Price Analysis

Technical Price Analysis[1]

Farmers, grain elevator operators, speculators, and other businesspersons involved in commodity markets are keenly aware that commodity prices are influenced by thousands of variables. These may include number of animals on feed, size of crop, domestic use, exports, or range conditions. Governmental policies to support prices, restrict imports, or establish embargoes also influence prices. International tensions and economic progress in other countries impact on the value of commodities. To complicate matters further, prices are affected by rainfall or drought anywhere in the world, transportation costs, and currency exchange rates.

Forecasting prices based upon these supply and demand factors is called fundamental analysis. It is based upon the economic principle that the value of any commodity derives from a balancing of supply and demand forces registered as market price. But fundamental analysis has inherent problems, which limit its usefulness in forecasting day-to-day or month-to-month price action. Nobody can know all the facts about what a price should be, despite being given all the fundamental information about real value of a commodity. First, factual information is subject to error. Data samples used as estimates of demand and supply statistics are based upon an acceptable statistical probability of being wrong. Second, facts change because fundamental conditions change. Today's valid price projection of expectations six months hence will change in the interim because of drought or changes in breeding intentions, for example. Third, the facts may, by design, not be presented correctly. What poker player will allow the face to show what cards he or she really holds? Reliability of information supplied on Brazilian production or intended Soviet purchases is often subject to question. Finally, the numerous fundamental facts available and the extent of influence each one exerts simply cannot be sorted out in one brain.

Even if all facts were known, true price discovery remains a problem because the constant flow of fundamental information into the market does not form a price until people interpret it. The extent of knowledge and interpretation of what it means differ markedly. Since people are involved in pricing decisions of a commodity it is impossible to rule out human emotions from the decision making process. When a market is going up, people expect it to continue going higher; when it is going down, they expect it to continue moving lower. Before supply and demand facts become price, they are subjected to interpretation and human judgment. It is not necessarily the facts, but what people believe the facts to be, that is important in establishing price.

Because of the problems of fundamental analysis, many market participants pay little heed to fundamentals. They are technicians who rely upon technical analysis to judge price direction and objective. Technical analysis can be defined as **a study of market action with the purpose of determining probable price change.** The working tool of technicians is a graph or chart, which shows price movement over a period of time. Hence, technicians have become known as chartists. Technicians concede that bearish technical information, which drives market price down, or bullish information, which drives it up, sets the tone of the market. However, it is the prevailing consensus of all market participants that determines market direction and extent of movement.

One purpose of chart reading is to measure the relative strength of buying and selling pressure. If chart patterns demonstrate that buying pressure at the prevailing price level is more powerful that selling pressure, it is logical to expect that prices will rise. On the other hand, if selling pressures overpower buying pressures, the expectation is that prices will fall. Technical analysis attempts to measure the collective judgment of buyers, sellers, and speculators. Charts graphically illustrate the collective opinion of the market participants.

Chartists arrive at their conclusions about price by diagnosing formations or patterns that appear on the charts. Formations suggest where other technicians are likely to place buy orders or sell orders—they define price objectives. Furthermore, charts can actually influence the market because thousands of traders make their buying and selling decisions from them. So many technicians are watching the same patterns that their collective analysis causes self-fulfilling prophecies. Charts work because chartists make them work.

Farmers are sometimes suspicious of "crystal ball gazing," seemingly divorced from reality. They are by nature fundamentalists. As practicing economists, they understand that drought-reduced crop supplies will command a higher price, or over-expansion of the beef cow herd will eventually produce too much beef and drive beef prices lower. They fret and complain when the "gunslinger speculators" create distorted low prices on the various commodity exchanges. Their agitation is further compounded when they miss short-lived high prices. But technicians can work to the farmers' advantage. Because so many traders watch chart patterns, there is a tendency to over-buy a market in a bullish trend and over-sell it in a bearish trend. Instead of lobbying to close the exchanges when excessive selling pressure creates "over-sold" low markets, or enthusiastic buying creates "over-bought" high markets, producers should learn to use chart patterns to help them reap the fruits of the speculators' efforts. Producers can be assured that if price levels move excessively high or excessively low, fundamental factors will ultimately correct the distortion by setting a price where demand and supply are in equilibrium. Charts present a road map of how the market price arrives at its final destination, offering high and low market-pricing opportunities along the way.

Charts are not a utopia in price forecasting, but they can be an important tool in a farmer's marketing toolbox. First, they can be helpful in developing good market judgement as a "check" on the accuracy of conclusions that are derived from studying fundamentals. For example, farmers who sold soybeans in the bull markets of 1980, 1983, 1988, and 1997 at prices of $9.00 to $10.00 per bushel had several chart confirmations that market objectives were in the $9.50 to $10.00 range. Bullish fundamentalists expected prices to reach $12.00 or $15.00 per bushel. They clearly misinterpreted supply and demand information. Within a few weeks, the market slumped to the $7.50 price area.

Charts force the user to follow market activity on a daily schedule. This attention makes one a better student of the market. By providing a visual picture of past price activity, a historical record in charts gives one a broader perspective of market activity. A chart reader is not so elated by daily limit up moves, nor as depressed by limit down moves. Finally, following charts and learning their rules can help the producer predict price trends of crops and livestock to establish price objectives where profitable sales and low-cost purchases can be made. With charting, farmers should be better able to know what is a high price or low price and make improved marketing decisions. A cold, hard picture of price action tempers emotions when marketing decisions are being made.

Charting

Technicians base their analysis on three factors: (1) price movement, (2) volume, and (3) open interest. By far, the most important is price movement.

Various charting systems show price movements. They may consist of lines, bars, step diagrams, or different symbols plotted on arithmetic, logarithmic, or square root scales. Bar charts illustrate one method used to define price objectives.

Bar Chart Basics

Each day's futures trading establishes three benchmark prices on bar charts: high, low, and settlement. A line connecting the high-low points corresponds to the price range of the day. It is a gauge of intra-day activity. The top of the line marks the highest transaction price of the day and the bottom, the lowest price. A horizontal "tic" mark indicates the closing (settlement) price as shown in Figure 5–1.

The settlement price is particularly important. Millions of dollars change hands between holders of respective long positions or short positions as accounts are balanced at the close each day. A higher close than the previous day means that holders

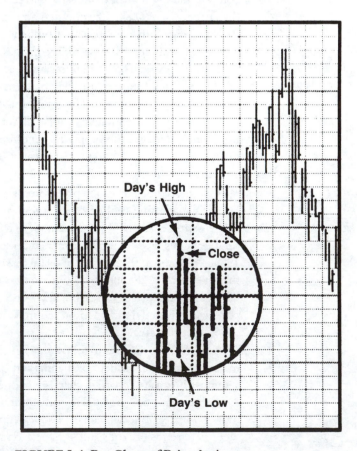

FIGURE 5–1: Bar Chart of Price Action.

of short positions transfer money to longs and vice versa if the close is lower. Real money changes hands based upon the settlement price.

In a broader sense, the close defines the winner in the struggle between bulls and bears. During the trading day, the shifting forces of demand and supply float the market upward or downward, but account balances change only with the trade at the bell. The close represents the sum total of everyone's opinion of value that day and an expectation of what will happen in the interim between the close and the opening the next trading day.

Bar chart lines are plotted on arithmetically ruled paper with the vertical scale referring to price and the horizontal scale referring to time. Commercially available paper has a heavy line for Monday and lighter lines for the remaining days of the week. Most paper has 52 weeks of space. Gradations on the vertical scale should be sufficiently wide to allow easy and accurate plotting, but sufficiently close so that price fluctuations during the year can be contained on one page. Uniformity and accuracy in plotting are extremely important. Inaccurate charting can give false impressions. Because charting can be a time-consuming and tedious process, many traders subscribe to one of several chart services available. These provide current and historic charts and chart analysis on a weekly or monthly schedule. There is no substitute for maintaining one's own charts, however. Without them, the advantage of daily attention is lost. Some publications provide a convenient method for individuals to keep daily records on the nearby contract, and monthly reports from the publication company provide a permanent record of all trading months.

Weekly and monthly charts are drawn in the same way as daily charts, with each bar corresponding to price action covering time of either the week or the month, as the case may be. They are used for the study of long-term trends. Examples of daily, weekly, and monthly charts are shown in Figure 5–2.

Reading Charts

The art of forecasting from bar charts depends upon recognizing and interpreting formations, which have historically been associated with price movement in a particular direction. However, there are sufficient examples of unexpected developments and unusual formations that rule out charting as an infallible forecasting device. Some regard chart analysis as "reading tea leaves."

The prevailing tendency, illustrated on charts, is for prices to move in a particular direction, or trend. A market is in an uptrend when it develops a series of higher highs and higher lows over successive days. A downtrend is just the opposite, lower highs and lower lows. An uptrend line is drawn by connecting the lower points of the overall movement. A downtrend line is drawn by connecting the higher points. High and low points on a sideways trend conform to bounds set by horizontal lines.

A key charting rule is that commodity prices tend to trade in channels. These are the boundaries through which prices pass as they generate a trend line. As long as price activity continues within the trend channels, the channels are useful for suggesting buying and selling levels.

The trend lines are also critical for defining important trend changes. The top line of channels is called "resistance," and the bottom is called a "support" line. As shown in Figure 5–3, each time a price move approached support or resistance, it turned the other direction. The reason for this is that thousands of chart traders are using basically the same information. They buy on "dips" and sell "rallies." Thus, there is buying activity when the market dips to support and selling pressure when it rises toward resistance – until, inevitably, there is a breakout.

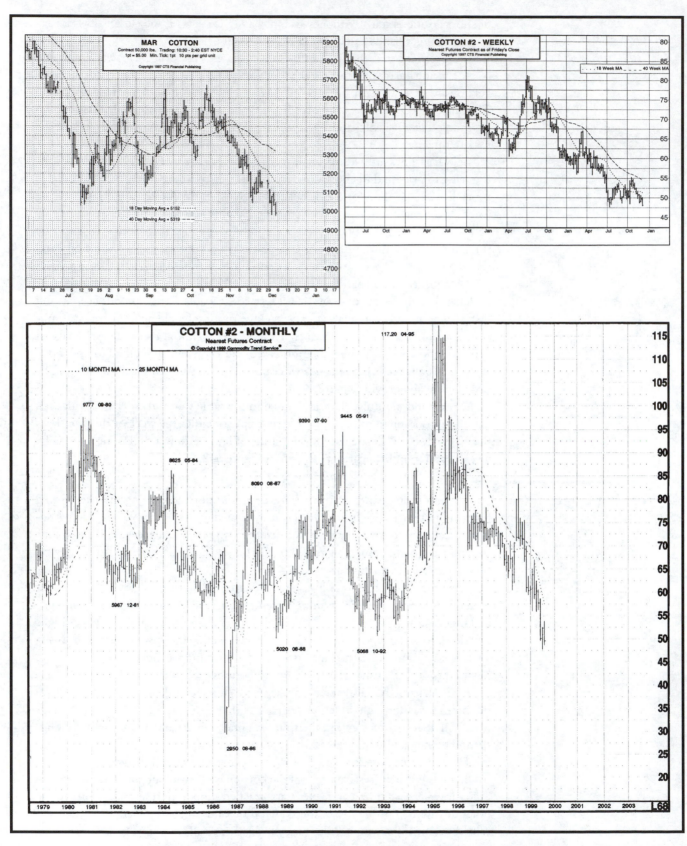

FIGURE 5–2: Daily, Weekly, and Monthly Charts of Futures Contracts.

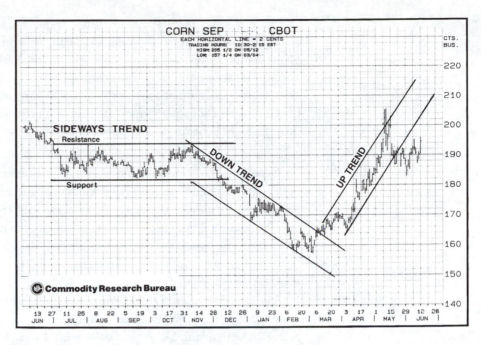

FIGURE 5–3: Illustration of Up, Down, and Sideways Trends, and Support and Resistance.

Psychology of a Trend[2]

Since no one is absolutely certain about the true value of the commodity in question, the relative buying or selling strength of the bulls or bears causes market price fluctuation. The fact that new buyers are willing to pay more for a commodity than the previous buyer in an uptrend is based upon the expectation that price will keep going up. At some point, however, not enough buyers are willing to pay a higher price.

At the top of a market, hedging (by selling short) makes a larger number of contracts available. When farmer-producers begin selling against their production, or elevator operators begin hedging their long cash positions, the market can turn around in response to their action. Speculators on the sidelines can be drawn into the action and also begin making sales. Prices drop. Previous buyers holding long positions have a natural inclination to hold their positions, expecting price to go back up. Some follow an old charting rule that states there should be two successive closes outside the trend line before expecting a change in trend. Even as prices erode further, many longs remain convinced that the dip is only temporary. They hang on, possibly buying more contracts offered by those who hold opposite convictions. Their confidence may stem the tide, possibly causing a recovery. But if buying is not strong and more contracts are offered for sale, the buyer's confidence is shaken. They begin to liquidate their positions to protect what profits remain or limit their losses. Their selling pressure adds to the momentum, and a downtrend is set into motion.

Eventually, prices decline to such a low point that both longs and shorts become convinced that prices cannot go lower. Price may be below cost of production. At some point traders with short positions begin covering to take profits. They must do so by buying. Their "buy" actions stem the downtrend and attract new buyers who expect a turnaround in price. At low price levels, farmer-producers are willing to assume price risks themselves rather than hedge at a loss. They do not offer short contracts. Farmer-buyers who want to lock in low-cost input prices for protein supplements, feed grains, or feeder livestock enter the market to hedge-buy their needs. Speculators seeing the rise beginning rush to get in the early action. The market moves up and an uptrend begins to form. Often, new fundamental information feeds into the market showing reduced supplies or increased demand. The psychology turns into a bull market and creates an uptrend.

Identifying a Trend

It is not easy to identify a trend until it is fairly well established. Many trend lines, both major and minor, may have to be drawn before a true trend can be ascertained. An uptrend is confirmed when the market closes above the top of the "short" covering rally (those holding short positions buy in order to liquidate their short position). Additional new buying reinforces the uptrend. Many bull markets are accompanied by a rapid and large buildup of open interest that shows new buyer willingness to get in on the action at a higher price.

A change from uptrend to downtrend is confirmed in a similar manner. At some point either a paucity of buyers or an influx of new sellers forces the market to close below the uptrend line. After an initial sell-off the market sometimes rallies at least to the downside breakout point. A close below the bottom of the first decline through the uptrend line confirms the downtrend. A decline in open interest further confirms the change in market direction as traders liquidate positions.

Commodity prices do not spend all their time in upward or downward trends. Often they move in sideways channels with no indication of which way they are headed. Typically, a "shock" from fundamental information is required to start a new trend direction. Sideways markets are usually at the bottom of the markets, seldom near the top. Uptrend markets usually take considerable time in building to their ultimate objectives. Although once a market tops, it can drop suddenly. Unless farmers are closely attuned to the market, good selling opportunities can appear on the chart and be gone before they have a chance to react.

Farmer hedge-sellers can use changes in uptrends as a signal to implement their sell decision. This is particularly important if topping action given by other formations such as gaps and reversals (defined later) have also appeared in the charts. Likewise, farmer hedge-buyers looking for market bottoms to lock in input costs should watch for chart confirmations that signal price bottoms. The same chart rules of change in trend apply in both up and down markets.

Brock and Martin listed 10 trend-following rules that give purpose to chart watching:[3]

1. Commodities spend a large part of their time in sideways trends.

2. A downtrend is valid until the market closes above a major downtrend.

3. An uptrend is valid until the market closes below the major upward trend.

4. An uptrend is confirmed when the market closes above the top of the first rally through the downtrend line.

5. A downtrend is confirmed when the market closes below the bottom of the first decline through the uptrend line.

6. A sideways (sometimes called trading) market exists until the market closes above the resistance line, signaling an uptrend, as in Rule #4, or below the support area, signaling a downtrend, as in Rule #5.

7. Sideways trends are generally at bottoms of a market—not tops.

8. Market bottoms often have narrow trading ranges and are dominated by overall bearish sentiment.

9. Downtrends generally go into a sideways trend, rather than immediately into an uptrend.

10. An uptrend often turns suddenly into a downtrend market.

Volume and Open Interest

Volume and open interest are important tools used in conjunction with bar charts. They provide measures of the extent of activity coming into the market or leaving it. Volume and open interest data are provided as part of many chart service reports, as shown in Figure 5–4.

Volume represents the number of trades for a particular commodity on a given day. As a barometer of trading activity, it measures the number of buyers and sellers participating in the market. Change in volume is often associated with price change in the same direction. For example, a gradual increase in volume during an uptrend suggests that more buyers are willing to come into the market expecting a continued price rise. However, rapidly accelerating volume, following a substantial upward price movement, often signals a major top and possible price reversal as traders liquidate positions.

Open interest is the number of open positions established, but not yet liquidated by delivery or offsetting trades. Open interest is not related to volume; there can be a high-volume day without a change in open interest. Open interest increases only when new purchases are matched by new sales. For example, if a buyer makes a new purchase of 5,000 bushels of July soybeans and the trade is taken by another person who initiates a new short position of 5,000 bushels of July soybeans, the net result is an increase in open interest of 5,000 bushels. If a new purchase is offset by the sale of a previously purchased contract, then there would be no change in open interest. Decreases in open interest occur when previous purchases are sold and matched by liquidation of previously sold positions. This means that both buyers and sellers are closing out previous commitments. Open interest figures, therefore, show whether new positions are being established or old ones are being canceled out.

Analysis of volume and open interest information can often detect market conditions that are not readily apparent as changes in supply and demand conditions of the actual commodity. For example, new export business may be reflected in futures markets before it is publicly announced. Increased open interest is generated by commercial exporters buying futures contracts of the commodity they plan to export to protect themselves against possible price increases.

It is important, however, to adjust changes in volume and open interest futures to account for seasonal activity before applying general rules to make marketing decisions. Open interest in grains normally increases during harvest seasons as elevators establish routine hedges to offset cash purchases. Later in the year a decline in open

FIGURE 5–4: Example of Volume and Open Interest Report.

interest can signal that country elevators are lifting hedges by buying back previously established positions once cash sales have been concluded. Their buying can temporarily result in higher prices. Furthermore, sharp declines in open interest may occur in soybeans after the first of January because cash sellers waiting for a new tax year are selling and lifting hedge positions. Seasonal patterns do not vary much between years. Therefore, changes due to seasonality can be anticipated fairly closely. It is the changes in volume and open interest outside seasonal influences that support analysis of price action in detecting future moves, as summarized in Table 5–1.

**TABLE 5–1: Volume and Open Interest Signals
Associated with Price Action**

When Prices Are Up	When Prices Are Down
Volume and Open Interest Up —Means STRONG MARKETS	Volume and Open Interest Up —Means WEAK MARKETS
Volume and Open Interest Down —Means WEAK MARKETS	Volume and Open Interest Down —Means STRONG MARKETS

Chart Formations

Psychology of Support and Resistance

Support and resistance psychology applied in trend analysis has broader application in the real world market. Any point in a chart that has been established by a price objective as the extreme of a price move, either up or down, creates an important psychological barrier to trading beyond the extreme. Resistance has several strong psychological barriers to continued market strength, or a rise in price. First, buyers are reluctant to enter new higher price territory where others have feared to tread. There is a shortage of buyers at resistance points. Second, many holders of long positions see rallies to old resistance levels as an opportunity to exit their positions. Those who bought before the rally have profits in their positions that they are willing to take. Those who bought near the top of a previous rally may want to get off the roller coaster ride. When the market went down after a period of strength, many longs refused to sell. They watched their positions decline in value and losses mount. Reluctance to realize actual dollar losses kept these traders from liquidating. When the market turns upward again, these longs are ready to liquidate when contracts reach the price they originally paid. This tendency to "let losses run" in hopes of breaking even on a rebound is characteristic of neophyte traders who fail to establish stop (limit) losses.

Resistance levels entice sidelined traders to initiate short positions. In the absence of a change in fundamentals required to send price higher, they think that a short position has less risk than a long one. New sellers enter the market. As price approaches an old high, there is an avalanche of sell offers from varied sources. Often, the market falls under the cumulative selling pressure, and another correction occurs.

The psychology operates in opposite fashion in downtrending markets. Support refers to a level on the chart where a price decline may be expected to stop. A support level could be stopping points of earlier down moves and can be expected to halt subsequent declines.

All past price levels at which a significant reversal in direction has occurred are prime targets for future resistance and support areas. Several chart formations develop from this phenomenon.[4]

Defining Price Objectives

Head and Shoulders Formation. The oldest, most popular, and possibly most reliable trend reversal pattern is called the Head and Shoulders Formation. It can signal a downtrend if it occurs at the top of the market, or an inverse head and shoulders can signal an upturn from a market bottom. An example is given in Figure 5–5.

Descriptively, the topping formation begins with a rally and a decline within an uptrend to form the left shoulder. A second rally, carrying beyond the first rally but falling back to the starting point, outlines the head. The right shoulder is formed by a third rally and a decline that extends to the two earlier stopping points. The two stopping points form a neckline. The head and shoulders formation is completed when the neckline is broken; the breakout confirms the change in trend.

A common measuring rule applied to head and shoulders formations is that once the neckline is broken, the subsequent move will at least equal the distance from the top of the head to the neckline. In the hog chart (Figure 5–5), the measure from

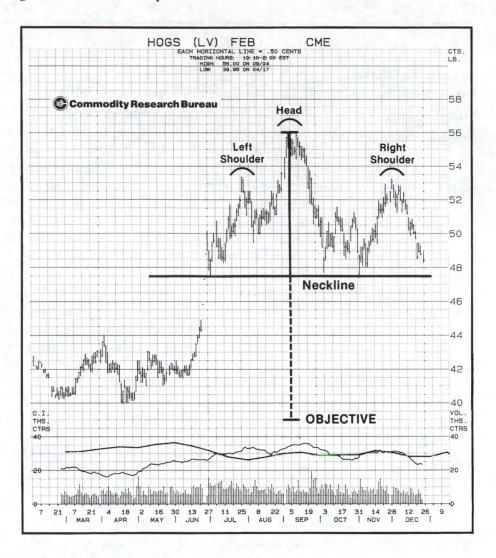

FIGURE 5–5: Example of Head and Shoulders Top.

$56.00 to $47.50 gave a downside objective of $39.00. The measuring rule applies conversely for uptrend signals generated by an inverted head and shoulders.

Flags and Pennants. Appropriately named, flags and pennants fly from a mass to symbolically foretell and lead major advances and declines. They have decorated many rapid and extensive commodity price moves.

Flags and pennants have a quick vertical price move that forms a flagpole to the body of the formation. A period of consolidation follows the initial move as traders await new information. Flags usually extend in the opposite direction from the major move, and when tightly constructed they are especially reliable indicators of direction and price objective. Loosely formed flags that point in the direction of the move may be unreliable. Normally, if flags and pennants are not completely formed in a month, they are likely to be part of another formation and should be viewed with suspicion.

In measuring the move to be expected from flags or pennants, one guide is to use half the length of the flagpole. The distance from the initial breakout to the center of the flag would be half the distance of the next expected move out of the formation. In other situations, the length of the flagpole, transposed on the lowest trading price prior to the breakout, establishes the price objective. As illustrated in Figure 5–6, price moved through resistance strongly upward from a level of $7.00 to form a pennant. Its center point of $8.05 suggested the next move would extend $1.05. The breakout occurred at $8.25, giving an objective of $9.30; the next up move actually reached $9.20. A bull flagpole was set at $8.25. Measuring from the breakout to the center of the flag at $8.75 suggested another objective of $9.25. The market eventually traded to almost $9.90 before declining significantly. The subsequent bear flag in the down move (from the $9.20 breakout to the $8.00 midpoint) projected a downside objective of $6.80. Flags and pennants can be used to measure objectives in either down moves or up moves.

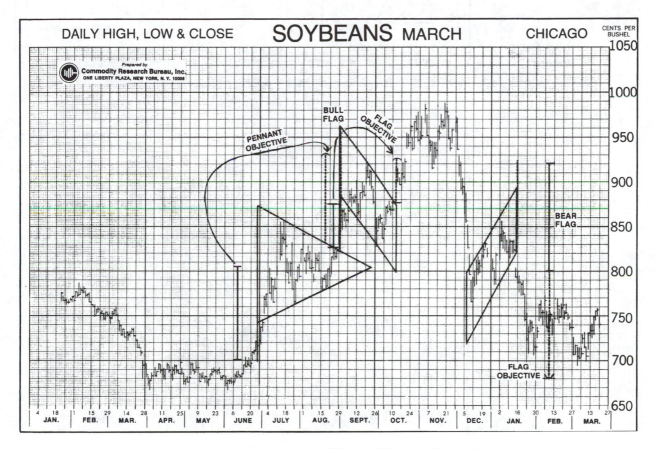

FIGURE 5–6: Examples of Flag and Pennant Formations.

Triangles. Triangle formations have been classified as ascending, descending, or symmetrical (Figure 5–7). The formation is also referred to as a coil. As trading activity approaches the apex, the balance between buying and selling gets narrower; the coil spring is wound tighter and tighter. In an ascending triangle, market bottoms edge higher and higher, but overhead resistance holds; this action points to an upward breakout. In a descending triangle, market tops move lower and lower, but support holds; this action points to a downward breakout. In a symmetrical triangle, tops become lower and bottoms are higher. They generally give no warning as to direction of breakout. Buying and selling become so closely balanced that a mini-

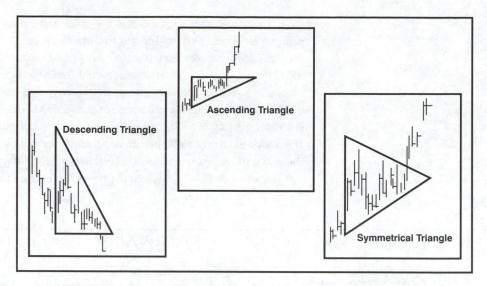

FIGURE 5–7: Illustrations of Three Types of Triangle Formations.

FIGURE 5–8: Example of Ascending Triangle Formation.

mum of bullish or bearish fundamentals will upset the balance and spring a move in either direction. Triangles figure prominently as signs of impending activity, but they are the most untrustworthy forms as far as their ability to predict trends. It is extremely difficult to anticipate the direction of price movement out of a triangle. Frequently, the breakout proves to be false and the trend is quickly reversed.

The measuring rule for triangles is that price will move a minimum of the vertical side of the triangle once the breakout occurs. Furthermore, the breakout generally occurs about two-thirds of the distance to the apex. Trading volume typically increases after the breakout. In the example of July cotton (Figure 5–8) the vertical side, measured from 33 cents to 52.5 cents, pointed to a move to 71 cents. Eventually, the price exceeded that level.

Gaps. Gaps are open spaces in charts where no trading occurs across a vertical distance. At times, changes in market fundamentals occur overnight and set off a significant move in market direction. Such changes often take place after consolidation in one of the previously mentioned formations. An up-gap develops when buying enthusiasm floods the market and causes the lowest price of a particular day to be higher than the highest price of the preceding day. A down-gap results from heavy selling pressure, causing the highest price on one day's trading to be lower than the lowest price of the preceding day.

Four types of gaps occur: common gap, breakaway gap, runaway gap or measuring gap, and exhaustion gap, as shown in Figure 5–9. A common gap exists temporarily. It has no special significance because it is typically filled a short time later as the trading range extends to overlap the opening. It seems that the market wants to tidy up unfinished business by moving back to fill previous gaps. Gaps become price objectives, since traders believe that all gaps are eventually filled. Their trading takes prices back to fill common gaps.

Breakaway gaps signal the completion of consolidation formations where the market has traded for some time without direction. Gaps are sometimes indicators of a dynamic move to follow. Often a fundamental shock stimulates buying enthusiasm by drawing high volume and open interest into the market as traders sense a strong uptrend. Trading pressure does not move the price back to cover the gap. Longs become bolder and add to their positions; shorts buy back their positions and go long. A little later, a flurry of buying snaps open the uptrend and forms the runaway gap. With astonishing reliability, this also becomes a measuring gap of half the distance of the remaining uptrend, calculated from the consolidation base, which produced the breakaway gap originally. Toward the top of the trend, the market is flooded with bullish news. Conventional wisdom convinces everyone that strong demand will outstrip limited supplies and the market is going to record highs. Two groups of buyers are attracted into the market; small speculators off the street and belabored short sellers. Small traders waited two or three weeks to be convinced the market was going up, but once convinced, they buy aggressively. The belabored short sellers buy in panic to cancel their positions as margin calls mount. This wave of buying creates an exhaustion gap, or "sucker surge." The commodity is priced well above what any user can translate into value. At this point, the market is overbought and the stage is set for correction. Often downside gaps occur in the correction to set downside objectives in the same fashion as in the uptrend.

In Figure 5–9, the breakaway gap occurred at $7.44 in early July. A measuring gap at $8.26 in early September signaled an objective of $9.08. The market moved to

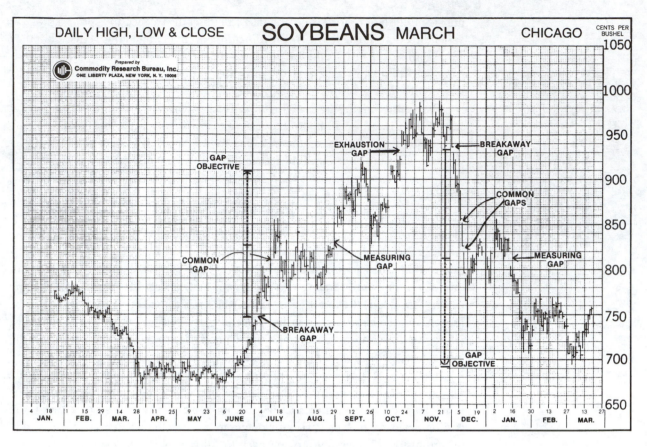

FIGURE 5-9: Examples of Gap Formations.

$9.20 later that month. In the downtrend after the top, breakaway and measuring gaps signaled a downside objective of $6.91.

Retracement. Markets often react to sudden and extreme changes in trend by bouncing back in the direction from where the move started. On its way to the top of a move, the market fades and selling increases. As the mood shifts, often a 50 percent retracement, or rebound, occurs, as in Figure 5–10. At other times, sentiment pulls the price back a little further to a 66 percent retracement. Following a sell-off in an uptrend, many buyers see the lower price as a bargain relative to recent market highs. Bullish sentiment, still hovering over the market, attracts new buyers who missed the previous upsurge. Their collective action bids price back toward the old high. Eventually, bearishness prevails and the change in trend does occur to send the market lower. Likewise, there can be a 50 percent or 66 percent retracement back toward a market low after the market has bounced off a bottom.

In the wheat example, a 50 percent retracement of its 92-cent move set a $2.70 objective, and a 66 percent retracement set an objective of $2.54. The July contract traded to $2.52 before expiring.

Identifying Market Tops and Bottoms

Reversals. An important chart signal is a reversal day. Its use is to identify a market extreme by spotting the day of a market turnaround. When a reversal day occurs after a relatively large move, and prices a start to retreat on subsequent days, the reversal may signal an important move in the opposite direction.

FIGURE 5–10: Example of Retracement.

Three kinds of reversals are key reversals, hook reversals, and island reversals, as shown in Figure 5–11. They occur at both market tops and bottoms.

A bearish key reversal at the top of a market has the following characteristics:

> A high higher than the previous day.
>
> A low lower than the previous day.
>
> A close that is lower than the previous day's low.

A bullish key reversal at market bottom has:

> A high higher than the previous day.
>
> A low lower than the previous day.
>
> A close that is higher than the previous day's high.

A key reversal provides a single day's picture of the battle between bulls and bears. Volume is unusually high, indicating considerable trading activity. In an up market, strong bullish buying pushes the high above the previous day's high. At the same time, the bears become heavy sellers to pull the low below the previous day. Often, the close foretells who will win the-tug-of-war. If it is below the previous low, the bears have won and the market is likely to turn down due to heavy selling pressure. At the bottom of a downtrend market, a close higher than the previous day's high means the bulls seem to have won and the market may continue up.

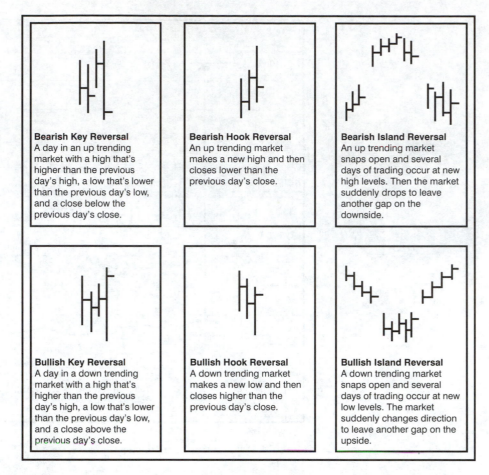

Bearish Key Reversal
A day in an up trending market with a high that's higher than the previous day's high, a low that's lower than the previous day's low, and a close below the previous day's close.

Bearish Hook Reversal
An up trending market makes a new high and then closes lower than the previous day's close.

Bearish Island Reversal
An up trending market snaps open and several days of trading occur at new high levels. Then the market suddenly drops to leave another gap on the downside.

Bullish Key Reversal
A day in a down trending market with a high that's higher than the previous day's high, a low that's lower than the previous day's low, and a close above the previous day's close.

Bullish Hook Reversal
A down trending market makes a new low and then closes higher than the previous day's close.

Bullish Island Reversal
A down trending market snaps open and several days of trading occur at new low levels. The market suddenly changes direction to leave another gap on the upside.

FIGURE 5–11: Examples of Reversals in Market Direction.

Hook reversals occur more often but are less reliable as indicators of change in market direction. Either the market high in an up market or a market low in a down market is exceeded, but the other end of the range is not exceeded. They key is the direction of the close. A lower close in an up market means market weakness, and a higher close in a downtrend market suggests market strength.

An island reversal consists of a few days of market action set off from the main body of prices by gaps on both the up and down side. These reversals should be considered as warnings of impending potentially large price moves.

Reversals are not perfect predictions of market bottoms or tops, though they are more reliable for tops. Their importance is enhanced if they are concurrent with other chart signals, such as exhaustion gaps, high volume, or price objectives defined by other formations.

Saucer or Bowl. The name saucer or bowl describes a rounded bottom or top formation, depending upon the depth and slope of the picture formed (Figure 5–12). Over a period of time, prices gradually curve up or down, depending upon whether they are establishing a bottom or top formation. They indicate the probable direction of the subsequent move. Round tops and bottoms are viewed as reliable trend indicators. They also imply that a major move is shaping up. The width of the saucer, determined by the time element taken to form it, roughly indicates the extent of the move

FIGURE 5–12: Examples of Saucer Top Formations.

to follow. When a saucer formation portends a lengthy move, a resting plateau, or platform, may develop between the saucer and the main part of the move.

Double Tops or Bottoms. Double bottoms occur at areas of major support at market lows, and double tops occur where major resistance halts price advances (Figure 5–13). The formation develops in an uptrend when price moves to the same level within a three-to-five-week period but there is not sufficient bullish sentiment to carry it to higher levels once a topping action has formed. Similarly, a second effort to drive prices lower in a down market that fails to penetrate the support established by the previous move causes a double bottom.

The outstanding implication of a double top or bottom is that a large subsequent price move in the opposite direction will follow. Real double top and bottom formations are rare. More often they become parts of other patterns.

Bar Chart Rules

Farmers can use a number of rules with bar charts to form guidelines for marketing decisions. Oster suggests several rules:[5]

1. Sell breakouts of uptrends. Watch for breakout of an uptrend as a selling opportunity. Initiate a short position after the second consecutive day the market closes below the trend line.

2. Buy breakouts of downtrends. Initiate a long position after the second consecutive day the market closes above the uptrend line.

3. Sell double or triple tops. Whenever a contract hits the same price area for the second or third time, it is a good time to sell. If price action cannot penetrate the same price after several tries, chances are that reduced demand for the moment will lead to a major price decline.

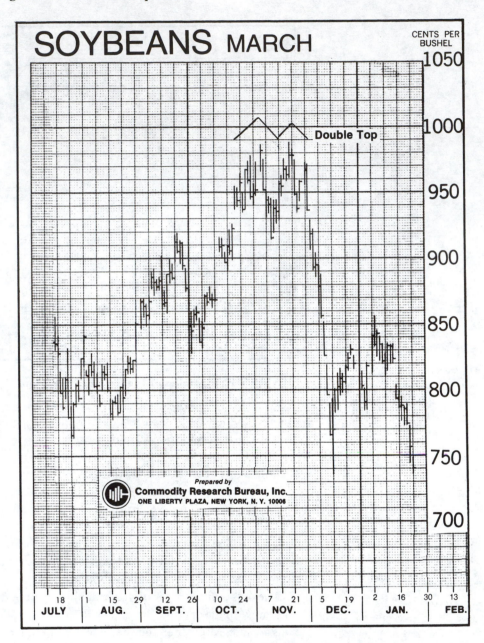

FIGURE 5–13: Example of Double Top Formation.

4. Buy double or triple bottoms. Commodity buyers looking for a buying opportunity can use a double or triple bottom as a signal that sellers trying to run prices lower cannot break support levels. This means that a price rise is probably imminent.

5. Sell when prices hit resistance areas. Areas of price congestion are future support or resistance levels. As prices move along in channels, the congestion areas become a hard level to penetrate. Many advisory letters suggest selling on rallies because reaction to resistance brings prices back up to support levels. If fundamentals dictate that the sideways channel is the right price level, the rallies are about the only sales opportunity to gain a few cents in price.

6. Buy when prices dip to support areas. Support areas set the floor to which prices fall. Unless there is a change in fundamentals, such as a larger supply that must be moved or a drop in demand levels, the traders will have a hard time forcing prices any lower. When the bottom is hit, an upward rebound is likely.

7. Sell a 50 percent retracement of a downtrend. If a commodity has dropped severely from a top formation and begins to rise again, it is likely that prices will move to 50 percent of what they were when the downtrend began. Producers who missed a selling opportunity at the high of the market often get another opportunity when the market has retraced 50 percent. If prices continue to rise, the next peak can be a 66 percent retracement. A scale-up sales technique works well under these conditions. Initiate the first sale at a 50 percent retracement level and double the amount sold at the 66 percent level.

8. Sell head and shoulder tops. This is not so easy, but when a head and shoulders occurs, confidence of a downtrend increases once the price breaks below the neckline after the second shoulder is formed.

9. Use price gaps as objectives of both corrections and extremes. After an upgap occurs, wait a few days for the price to return to fill the gap. If the gap does not fill, prices often move strongly away from the gap. Learn to read breakaway, measuring, and exhaustion gaps. They indicate how high prices are likely to move or the extent of a fall. Gaps are a major tool in determining price objectives. Price gaps are usually filled by later trading. It may take weeks or months (or years) to return to price levels left open, but it is nearly certain that a gap will not be left open eternally. Old gaps become price objectives eventually.

10. Use flags and pennants as measures for price objectives. Their midpoint is often half the anticipated move. Though questionably reliable, their use in conjunction with other measures to establish market objectives may point to market extremes both high and low.

Confirmation

Confirming a chart formation enhances confidence in its reliability. A combination of gaps, flags, and reversals on a single chart often confirms a similar price level. Several formations pointing to the same price objective can give more confidence that the objective will be reached. Further confirmation is determined by consulting different trading months of a particular commodity. For example, when attempting to determine whether a head and shoulders is forming in July corn, every other corn contract being traded should be studied for confirmation that the formation is occurring. Weekly and monthly continuation charts should also be consulted to confirm longer-range trends, support, and resistance levels. Objectives defined by formations on the daily charts may also appear in the longer-range charts. As a rule, formations should be regarded with suspicion when a majority of charts fail to confirm the formation, unless there are fundamental factors that suggest that divergence should occur.

Reliability

Technical analysis of chart patterns is more art than science, and allowance must be made for errors. Unexpected occurrences can change price trends suddenly and without warning. Rain or hail, overnight political decisions, wars, and any number of unpredictable events can invalidate forecasts made by interpretation of chart patterns.

Another factor limiting their use is the fact that chart formations are often very indecisive. Patterns are quite obvious after they have been formed; they are not as obvious during the formative stage. The example in Figure 5–14 shows three uptrend lines and projects in $3.46 objective price. Six days after this analysis, prices dropped below $3.00. Subsequently, the market rallied to $3.41. Chartists usually maintain a neutral attitude until patterns become clearly discernible.

No Sign of a Top in Corn

"Corn prices have climbed steadily since establishing lows at harvest. Solid uptrends have formed and only mild downward corrections have occurred. The latest correction started on Apr. 11 at $3.25 and ended Apr. 20 at $3.09. Since no bearish reversals or other signs of a top have occurred, this could prove to be a bull flag formation and the midpoint of a rally to $3.46 in the July futures.

The $3.46 objective is calculated by taking the 37–cent distance from the Mar. 14 low at $2.88 and the Apr. 11 high of $3.25 and adding it on to the Apr. 20 low of $3.09. Rallies to this swing objective will be indicated by a close above $3.21, a new rally high close,

The failure of prices to rally through the previous rally high at $3.25 would be bearish. If this occurs, a potential double top could form, which would offer tough resistance to futures price advancement attempts.

Should prices sell off from current levels and close below $3.10, a downtrend will form and declines to $2.95 in the July contract should be expected."
Quote from Top Farmer Intelligence. Used by permission.

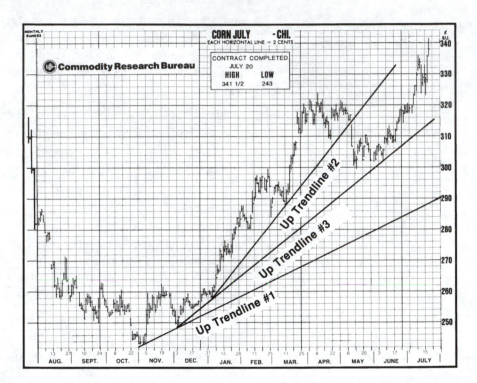

FIGURE 5–14: Example of Indecisive Analysis in Chart Reading.

Robert J. Taylor reported a Technical Reliability Index to evaluate the price-forecasting dependability of chart patterns.[6] The Index measures the percentage of time that a completed formation resulted in the expected price movement for that formation. Since technical formations occur with greater frequency in some commodities than in others, the evaluation was further refined to indicate frequency of occurrence of each formation in each commodity. Three symbols, used in connection with the Technical Reliability Index, where F = frequent, A = average, and I = infrequent. The Index and frequency are reported in Table 5–2. It shows, for example, that breakouts from sideways trends occur frequently for cattle prices and are more reliable (75 percent) than ascending triangles (47 percent) in detecting market bottoms. Likewise, head and shoulder tops occur frequently in soybeans with a high reliability (82 percent) that they are the true tops.

The dual problems of reliability and accurate detection of the formation in the development stages limit chart formations in their use as a marketing panacea. A topping or bottoming formation may occur once in a year. The farmer is left with what

TABLE 5–2: Technical Reliability Index

	BOTTOM FORMATIONS					
Commodity	Head and Shoulders	Double Top	Ascending Triangle	Island Bottom	Breakout from Sideways	Saucer Bottom
Cattle	A/71%	A/67%	A/47%	I	F/75%	F/73%
Cocoa	A/72	A/69	A/57	A/67%	F/71	F/73
Copper	A/73	A/68	A/67	A/71	F/77	A/69
Corn	A/70	F/73	A/46	I	F/68	F/75
Cotton	A/68	A/67	A/52	I	F/81	A/71
Eggs	A/70	F/74	A/59	A/71	A/73	I
Hogs	A/72	A/68	A/60	I	A/74	A/69
Lumber	A/76	A/71	A/68	I	A/73	A/74
Orange Juice	A/74	A/71	A/69	I	F/77	A/76
Plywood	A/81	A/72	A/68	I	A/70	A/73
Pork Bellies	F/85	F/72	A/41	A/74	A/57	I
Potatoes	A/70	A/69	A/62	I	A/73	A/71
Silver	A/73	A/70	A/61	I	A/76	I
Soybeans	F/70	F/76	F/38	I	A/66	A/62
Soybean Meal	A/70	A/74	F/71	I	F/77	A/72
Soybean Oil	A/68	A/71	F/75	I	F/77	A/70
Sugar	I	A/60	A/62	A/62	A/71	A/69
Wheat	A/75	A/74	F/41	I	F/71	F/64

	TOP FORMATIONS				CONSOLIDATIONS	
Commodity	Head and Shoulders	Double Top	Descending Triangle	Island Top	Pennant	Flag
Cattle	A/69%	A/62%	A/61%	I	A/72%	A/67%
Cocoa	A/78	A/65	A/62	A/72%	A/71	A/73
Copper	A/77	A/69	A/66	A/81	A/60	A/57
Corn	A/81	A/73	A/54	I	F/73	F/71
Cotton	A/70	A/72	A/52	I	F/76	A/70
Eggs	A/68	F/74	A/51	A/76	F/79	F/73
Hogs	A/72	A/71	A/55	I	F/72	F/70
Lumber	A/71	F/70	A/65	I	A/69	A/66
Orange Juice	A/70	A/66	A/67	I	A/72	A/69
Plywood	A/80	A/72	A/66	I	A/69	A/67
Pork Bellies	F/87	F/75	A/40	A/77	F/65	A/60
Potatoes	A/80	A/72	A/70	A/78	A/65	A/61
Silver	A/77	A/71	A/70	I	A/73	A/69
Soybeans	F/82	F/73	F/55	I	F/68	A/61
Soybean Meal	A/72	A/65	F/71	I	A/64	A/60
Soybean Oil	A/65	A/72	F/73	I	A/67	A/62
Sugar	A/60	A/61	F/61	A/72	A/55	A/52
Wheat	F/81	F/77	F/51	I	F/64	F/58

F = Frequently   A = Average   I = Infrequently

Source: Robert J. Taylor, "Technical Personalities of Major Commodities." Used by permission.

to do in the interim until another objective is reached, especially if he or she missed the extreme of a move. Bar charts, therefore, become means to complement other methods of analysis in establishing price objectives at which marketing decisions should be executed.

Moving Averages

Whereas bar charts are used to identify price objectives, another approach to making pricing decisions is based upon trend analysis. This method of analysis does not attempt to predict price movement but simply follows price trends to take advantage of the major moves a market might make during the course of the marketing year. It relies on signals from the market analysis to define price direction. While market prices may be moving up one day and down the next, there is often an underlying direction to the market. Trend-following analysis attempts to identify the direction of major trends while filtering out minor corrections.

A moving average is an arithmetic average of prices over time. The time span is constantly moving forward to reflect only the most recent prices. For example, Table 5–3 reports closing July wheat prices during the month of April and shows the 3-day

TABLE 5–3: Calculation of 3- and 10-Day Moving Price Averages, July Contract

Date	Closing Price	Moving Average 3-Day	10-Day
April 1	368.4		
4	368.4		
5	370.2	369.08	
6	370.6	369.83	
7	369.2	370.08	
8	372.4	370.83	
11	374.6	372.17	
12	366.0	371.08	
13	361.6	367.58	
14	361.4	363.17	368.38
15	362.0	361.83	367.73
18	359.4	361.17	366.83
19	356.2	361.08	365.43
20	352.2	359.41	363.58
21	352.6	357.16	361.93
22	357.0	357.41	360.38
25	359.2	359.83	358.83
26	362.0	362.90	358.43
27	363.4	365.08	358.6
28	361.4	365.83	358.6
29	361.2	365.58	358.57

and 10-day moving averages. To calculate the 3-day moving average for April 5, the closing prices for April 1, 4, and 5 are added and divided by 3 to produce the average of $369.08. The 10-day moving average is calculated in similar fashion, but it includes 10 days of prices rather than 3, and the sum is divided by 10. The respective calculations for the next day are made by deleting the first day's price from the calculation and including the fourth or eleventh day's price. By so doing, each average always includes 3 or 10 days of most recent closing prices.

A variation on the technique of calculating moving averages is a front-weighted procedure. It is based on the assumption that most recent market activity is more important in detecting market action than earlier price data. More weight is given to the most recent day's price as compared to past prices. Table 5–4 shows how to calculate a front-weighted three-day moving average of the wheat prices from the previous example. April 1 is the farthest day away and consequently receives a weight of 1.

TABLE 5–4: Calculation of Front Weighted 3-Day Average

Date	Price Closing	Weight	Weighted Value	Front Weighted 3-Day Moving Average
April 1	368.4	1	368.4	
4	368.4	2	737.0	
5	370.2	3	1110.75	369.38

The next day, April 4, receives a weight of 2. The most recent closing price of $3.702 is given a weighting index of 3. Multiplying each of the closing prices by its corresponding index weighting and summing the results produces a total of $1,110.75. Dividing the total units by 6 yields a front-weighted three-day moving average of $369.38 in comparison to $369.08 using the simple average method.

A 10-day weighted moving average can be calculated by assigning the most recent day's price a weight of 10 and multiplying the next oldest day's price by 9, the third oldest by 8, and so on to the tenth oldest by 1. The sum of the weighted values is divided by 55, the sum of the weighting units.

Weighting gives greater emphasis to more recent price action, but it is a more complicated calculation. Since calculations are completed each day with new information and can be very time-consuming, personal computer spreadsheets greatly simplify the procedure and reduce the time required for calculation.

Moving averages are plotted along with bar charts, with price on the vertical axis and time on the horizontal. Figure 5–15 illustrates price activity of corn and sugar markets, along with their moving average prices. Several observations can be made in reading the charts. First, the moving average smoothes out the price action. No one day's price can have an extreme effect on the average value, since the last day's price is always averaged in with several others. Using an average eliminates or reduces the distraction (market noise) caused by a sudden and extreme daily fluctuation. It enables the user to observe a smoother definition of trend rather than market reaction. This feature is the major reason for the widespread use of various types of averages as market analytical tools. Some chartists use 3-9-18-day averages; others use daily-10-40-day ones. The number of days for calculating the average depends upon market performance of which one works best, or possibly one's personal preference. Reasons for using different lengths of time are to improve sensitivity to changes in direction, but at the same time to eliminate false signals. The time period chosen should be sufficiently short to be sensitive to true trend changes, but sufficiently long to filter out market noise.

Farmers who decide to use them should be aware of the features that characterize moving averages:

1. They reflect only what has occurred in the past and do not attempt to project into the future;

2. They change direction after the prices from which they are derived change;

3. They smooth the price waves of market action;

4. The time lag that occurs before moving averages signal trend change depends on the time span covered by the average and on the velocity of price movement after the change occurs;

5. Averages are always less than the prices from which they are derived during advances and greater during declines;

6. The more days of prices included in the moving average, the longer the time required to realize crossovers;

7. During sideways price action, moving averages cannot clearly indicate trend direction, since no trend exists at such times.

Moving averages provide a method for detecting changes in price trend. When plotted as lines on charts, the crossing above and below one another provides signals of change in price direction. In using moving averages as a market decision tool, the supposition is that once a trend has been established, it will continue for some time.

The pattern in Figure 5–15 uses daily, 10-day, and 40-day moving averages. The crossover of the daily and 10-day average below the 40-day average signals a downtrend; likewise, a double crossover above the 40-day average signals an uptrend. In the sugar chart, a downtrend was in force at the beginning of the chart, and an uptrend detected on September 13 continued throughout the time period covered, even though all averages had turned down after February 1. The averages were very effective in detecting major trends. By comparison, the corn market bounced along in sideways action rather than staying on a prolonged trend. The market changed from down to up on September 11 but turned down again on October 27. It changed to an uptrend November 17, a downtrend on December 19, and an uptrend on January 11. In a sideways trading pattern, moving averages whipsaw in and out of uptrends and downtrends.

Incorporating moving averages into a marketing decision framework relies on crossovers to signal changes in trends. A change from uptrend to downtrend would signal farmer-producers to place hedges on their production and tell input buyers to exit long market positions. A change from downtrend to uptrend would signal input buyers to place long hedges on items they planned to buy and producers to lift hedges or delay taking market positions.

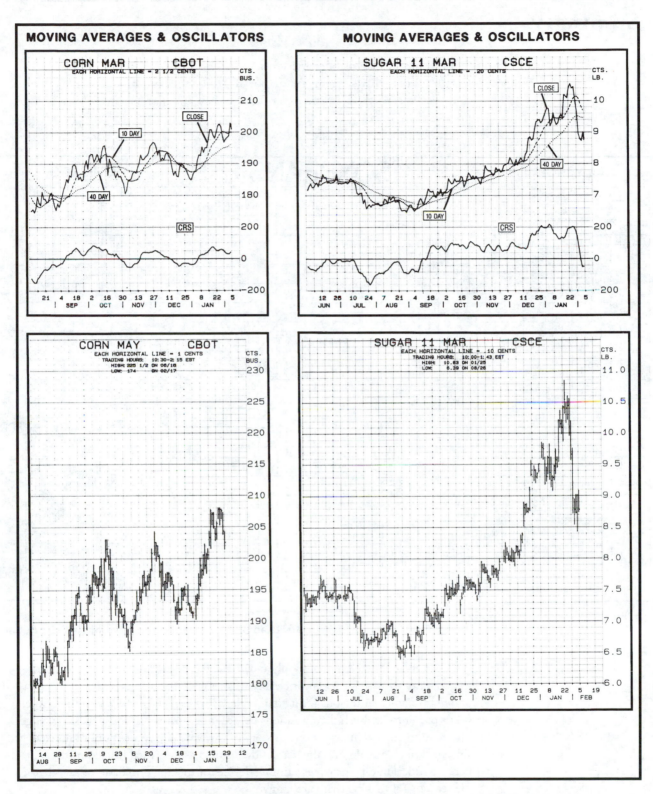

FIGURE 5–15: Moving Averages for Corn and Sugar, In Addition to Their Respective Daily Bar Charts.

Summary

Is technical analysis of charts a form of magic or witchcraft? (See Figure 5–16). Certainly the charts seem far removed from the fundamentals of production–weather, usage rates, and government policy–that farmers are familiar with as price-making forces.

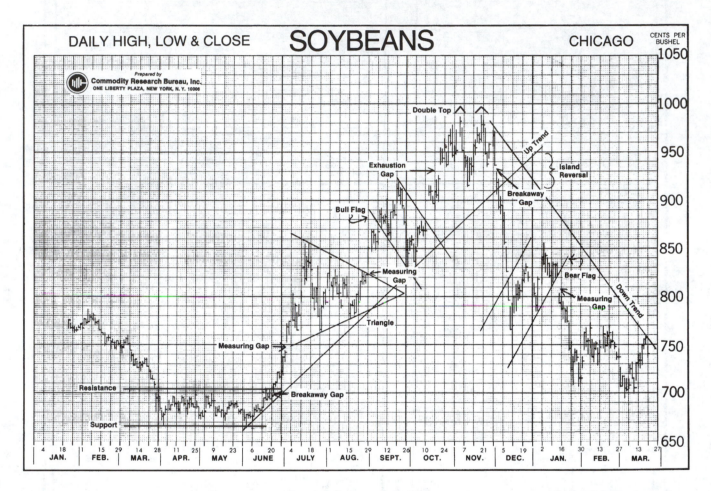

FIGURE 5–16: Magic or Witchcraft? Examples of Many Chart Formations in Price Action.

The fundamental forces of supply and demand come together in the market place to generate prices. But when prices start to move, charts often show the activity before fundamental reasons for it are revealed. Professionals who are fulfilling their needs prefer to establish their positions before the information is public knowledge. Often, fundamental information is history by the time it becomes public knowledge, since its effects are bid into the market by insiders as they established positions earlier. Technical trading provides a timing device for making purchases or sales before the reasons for making them are known.

Fundamentals change. As they do, the questions arise: How high will prices rise, or how low will they fall? Producers will sell all that is produced at some price, and price will ration supply to avoid empty pipelines. These two conditions create the boundaries for price level fluctuations. But the middle leaves considerable room for price negotiation. Technical analysis is a means to measure the negotiation process.

There is an old investment proverb that says, "It is not what is true; it is what investors believe to be true that counts." Technical market analysis is a method for giving market participants confidence in reading what other people believe to be true. Pure technicians ignore fundamental factors of supply and demand; they pay no attention to planted acres, harvest progress, weather conditions, number of animals on feed, or any other fundamental facts. They simply read the market, knowing that eventually all fundamentals will be registered there. In the end, fundamentals will determine final market value. A chart is simply a road map of how and when the prices set by fundamentals come to be realized.

There is no economic reason that price charts should establish price direction or price level. But so many technicians are watching the same pattern that their collective analysis causes self-fulfilling prophecies. Charts work because chartists make them work. Charts are important to farmers because they indicate how high or how low prices are, or may go. In an uptrend market, traders move the market up to a point where no buyers are left. In a downtrend, they move it to where there are no sellers. A producer with a perspective from charts can use his or her knowledge of fundamentals to know when the market has gone beyond logical reason. Overbought and over-sold markets create marketing opportunities. At those times, farmers who are familiar with chart analysis can "harvest" the speculators. If technicians have driven prices higher than logical levels determined by fundamentals, farmers and ranchers should take advantage of the circumstances and price their production. Likewise, when downward pressure drives prices to low value, or below cost of production, farmers and feeders should lock in low input costs. Speculators provide opportunities for both sellers (wanting high prices) and buyers (wanting low prices) to be satisfied by the market. Technical analysis signals when to cash in on these opportunities.

Questions for Study and Discussion:

1. What problems of fundamental information cause some market participants to turn to technical analysis of the market?
2. What is the logic behind use of technical market analysis?
3. Explain the three factors upon which technicians base their analysis.
4. What is the psychology of a trend in commodity prices?
5. Why are volume and open interest important technical factors?
6. What is the foundation of chart formations?
7. List and explain the chart formations that define price objectives. What are these rules of measurement?
8. List and explain chart signals of market extremes.
9. What are the rules to follow in bar charting for market decision-making?
10. What are confirmation and reliability in connection with chart reading? Why are they important?
11. Explain moving averages as a tool of technical analysis.
12. Is technical analysis magic or witchcraft?

Exercise: Pricing Charting

Bar charts are an example of the adage "A picture is worth a thousand words." Once a person learns how to read them, charts provide pictures of what is happening in the market.

Commodity charts should be kept on professional graph paper that can be purchased at any office supply store. Chart paper should have five divisions, one for each trading day with a heavy line to indicate start of the week. Preferably it should have room for 52 weeks of time on the horizontal axis. Prices are designated on the vertical axis. The increments selected should relate to price volatility of respective commodities in terms of both daily fluctuations and annual range of fluctuation. Soybeans can be plotted in 5-cent increments over a $3.00 range; corn, 2-cent increments over a $1.50 range; wheat, 2 cents and $1.50; cotton, 50 points and 30 cents per pound; cattle, 50 points and 30 cents; hogs, 50 points and 30 cents. Rules change as volatility and price level change.

Accuracy and scale in plotting charts cannot be overemphasized. Plotting the high, low, and close of each day's market activity in the exact price at which they occurred is of key importance. Inaccurate plotting can change the angle of trend lines, show false market signals, and in the final analysis give misleading forecasts of future prices.

Most beginners are well served by charting nearby contract options in commodities of most interest to them. As they gain experience and learn charting rules, they turn in time to subscribing to a charting service. Several firms publish weekly or monthly reports that have space for charting until the next report is issued. They supply charts on all options of the various commodities. They typically also provide long-range weekly and monthly charts for long-term analysis. A request to a couple of the following should secure a sample of what they offer.

Charting Services:

Commodity Price Charts, 250 South Wacker Drive, Chicago, IL 60606.
Commodity Trend Service, P.O. Box 32309, Palm Beach Gardens, FL 33420

Endnotes for Chapter 5:

1. Material for this section comes from the excellent work of one of the masters: William L. Jiler, *How Charts are Used in Commodity Price Forecasting* (New York: Commodity Research Publication, 1973). Used by permission.

2. Merrill J. Oster, *Commodity Futures for Profit* (Cedar Falls, Iowa: Investor Publications, 1979) 15/1.

3. Richard A. Brock and John F. Martin, *Charting Farm Markets* (Milwaukee: Top Farmer Publications, Jan., 1981) 14.

4. Jiler, *How Charts are Used in Commodity Price Forecasting.*

5. Oster, *Commodity Futures for Profit* 15/11-15/14.

6. Robert J. Taylor, "Technical Personalities of Major Commodities," *Commodities* (Cedar Falls, Iowa: Oster Communications, Inc., Aug., 1972).

Chapter 6

Pricing Strategy

Pricing Strategy

Modern farming is a complex production process and an equally complex business enterprise. It is inherently speculative, since the future is inherently filled with so many unknowns. Farmers thrive on speculation; the nature of their business requires them to be risk takers. In addition to insects, drought, and pestilence that create production risks, price risk has become more important in affecting profitability. As the 20th Century drew to a close, the trend toward freer markets and the demise of government regulations, production controls, and subsidy payments produced greater price uncertainty than experienced in the previous sixty years. Surplus production, economic deflation, and shrinking exports added fuel to the fire of price uncertainty. Volatile prices provided the opportunity for both high profits and disastrous losses.

Financial success depends upon the successful management of risk and uncertainty inherent in farming. Improved marketing is one way to manage risk. But two psychological attitudes prevent many producers from being successful marketers: Farmers are optimistic, and they like to speculate. Because of the nature of their business, farmers are the world's greatest optimists. They live by the rule that things must grow. No one else would borrow $150,000, scatter seed, fertilizer, and pesticides over 1,000 acres, and pray for rain. Farmers have an inbred psychology that a crop must grow higher and higher. That optimism extends from their production practices into their attitudes about prices. Prices are always too low and must go higher to give them a greater return tomorrow than the market offers today. Consequently, they often delay pricing decisions until forced to sell. The reality of the market is that prices go down as often as they go up, and often with much more vigor. Many farmers can accept only up, because higher prices mean so much to their well-being.

Farmers are price and production speculators, and they are risk takers. Crop farmers have no control over the most important factor determining production—weather. Furthermore, they have no control over the most important factor determining the price of their product—the market. The nature of their business requires that farmers live with production and price risks; the challenge is to control them as best they can.

Risk-bearing ability varies greatly among producers. Some people are risk takers, psychologically prepared to accept setbacks as offsets to opportunity for significant gains. Others have a risk aversion, preferring to minimize risk. Financial condition greatly affects attitude. A full owner with money in the bank can withstand more risk than a young farmer with high debt payments, but his or her reaction to risk may be just the opposite. An operation with high cash-flow requirements, relative to gross receipts, is more vulnerable than a high-fixed-cost operation. Lenders may place bounds on what they will finance due to risk. While excessive caution may limit financial growth, excessive risk taking may bankrupt the business.

Not all risk can, nor should, be avoided. The problem is to balance the probability of loss against its effects. Large potential losses and their disastrous effects should be avoided as much as possible, but small loss-large reward situations may generate profitable opportunities.

Farmer-producers may attack the problems of production and price speculation on two fronts. The first is to minimize the effects of unfavorable outcomes, and the second is to improve market pricing strategies. To practice the first, they may minimize risk by transferring it to others. This may be through insurance protection on crop and livestock production and on stored commodities. They may participate in government price support and disaster relief programs. They may build cash reserves or increase equity in the business, and they may use forward pricing methods to lock in prices before harvest. The second step is the development of a marketing strategy. Instead of being price takers, controlled by emotions and financial requirements, farmers should develop a price-maker philosophy regarding the products they sell. It takes more time and work, but can be very rewarding. A farmer's marketing strategy consists of a set of objectives, plans, and tactics that guide production and marketing decisions. Since the market place presents the farmers with a wide variety of choices, opportunities, and risks, a marketing plan is the blueprint for making decisions and exploiting the market.

Strategies for Mitigating Risk

Harwood[1] lists several strategies for mitigating risk. He notes that: Farmers have many options for managing the types of risks they face. For example, producers may 1) plant short-season crop varieties that mature earlier in the season to beat the threat of an early frost; 2) install supplemental irrigation in an area where rainfall is inadequate or unreliable; or 3) use custom machine services or contract/hired labor to plant and harvest quickly during peak periods.

Most producers use a combination of strategies and tools, because they address different elements of risk or the same risk in a different way. Following are some of the more widely used strategies.

Enterprise diversification—assumes returns from enterprises do not move up or down in lockstep, so low returns from some activities would likely be offset by higher returns from other activities. Diversification can also even out cash flow. According to USDA data, cotton farmers are among the most diversified in the United States, while poultry producers, with poultry products accounting for 96 percent of the value of their production, are the least diversified.

Vertical integration—generally decreases the risk associated with the quantity and quality of inputs (or outputs) because the vertically integrated firm retains ownership control of a commodity across two or more levels of activity. Vertical integration also diversifies profit sources across two or more production processes. In farming, vertical integration is common for turkeys, eggs, and certain specialty crops.

Production contracts—guarantee market access, improve efficiency, ensure access to capital, and lower startup costs and income risk. Production contracts usually detail inputs to be supplied by the contractor, the quality and quantity of the commodity to be delivered, and compensation to be paid to the grower. The contractor typically provides and retains ownership of the commodity (usually livestock) and has considerable control over the production process. On the downside, production contracting can limit the entrepreneurial capacity of growers, and contracts can be terminated, or not renewed, on short notice.

Marketing contracts— set a price (or pricing mechanism), quality requirements, and the delivery date for a commodity before harvest or before the commodity is ready to be marketed. The grower generally retains ownership of the commodity until delivery and makes management decisions. Farmers generally are advised to forward price less than 100 percent of their crop until yields are well assured to avoid a shortfall that would have to be made up by purchases in the open market.

Futures contracts—shift the risk from a party that desires less risk (the hedger) to one who is willing to accept risk in exchange for an expected profit (the speculator). Farmers who hedge must pay commissions and forego interest or higher earning potential on money placed in margin [good faith] deposits. Generally, the effectiveness of hedging in reducing risk diminishes as yield variability increases and the relationship (correlation) between price and yield becomes more negative. Hedging can reduce, but never eliminate, income risk.

Options contracts—give the holder the right, but not the obligation, to take a futures position at a specified price before a specified date. The value of an option reflects the expected return from exercising this right before it expires and disposing of the futures position obtained. Options provide protection against adverse price movements, while allowing the option holder to gain from favorable movements in the cash price. In this sense, options provide protection against unfavorable events similar to that provided by insurance policies. To gain this protection, the hedger in an options contract must pay a premium, as one would pay for insurance.

Liquidity—involves the farmer's ability to generate cash quickly in order to meet financial obligations. Some of the methods that farmers use to manage liquidity, and hence financial risk, include managing the pace of investments (which may involve postponing machinery purchases), selling assets (particularly in crisis situations), and holding liquid credit reserves (such as access to additional capital from lenders through an open line of credit).

Crop yield insurance—provides payments to crop producers when realized yield falls below the producer's insured yield level. Coverage may be through private hail insurance or federally subsidized multi-peril crop insurance. Risk protection is greatest when crop insurance (yield risk protection) is combined with forward pricing or hedging (price risk protection).

Crop revenue insurance—pays indemnities to farmers based on revenue shortfalls instead of yield or price shortfalls. As of 1998, three revenue insurance programs (Crop Coverage, Income Protection, and Revenue Assurance) were offered to producers in selected locations. All three are subsidized and reinsured by USDA's Risk Management Agency.

Household off-farm employment—may provide a stream of income to the farm-operator household that is more reliable and steady than returns from farming. In essence, having household members work off-the-farm is a form of diversification. In 1996, according to USDA's ARMS data, 82 percent of all farm households reported off-farm income exceeding farm income. In every sales class (including very large farms), at least 28 percent of the associated farm households had off-farm income greater than farm income.

The purpose of this chapter is to set the framework for developing pricing strategies and to review several price enhancement and risk reduction techniques.

Risk Management[2]

Reduced government intervention in the markets for program crops (wheat, corn, cotton, and other selected field crops) under the 1996 Freedom to Farm Act heightened producers' uneasiness about price risk and income variability. The age-old problem of yield variability still had to be reckoned with. What are the tools available to producers?

Natural Hedge

An important concept that must be understood is the **"natural hedge."** This is the tendency for yields and price to change in opposite directions and thereby helps to stabilize gross farm income over time. In drought years of low total production, prices rise. In abundant years, prices fall in response to higher production. A 100-bushel corn crop selling for $3.00 per bushel generates the same gross revenue as a 150-bushel crop at $2.00 per bushel. In each case, the gross income is $300.00 per acre. The farmer's marketing challenge is to extend the high price level into years of high production levels. Selling 150 bushels for $3.00 raises gross income to $450.00 per acre.

The prevalence of natural hedges for corn, soybeans, and wheat is illustrated in a series of maps included as Figure 6–1. The more negative the correlation between annual yield and annual price, the greater the natural hedge.

In the case of corn, the natural hedge is concentrated in Illinois, Indiana, Iowa, western Ohio, southern Minnesota, northern Missouri, and eastern South Dakota, Nebraska, and Colorado. However, Illinois, Indiana, and Iowa are often referred to as the "heartland" and for good reason, that is where corn is produced. When Illinois has a bumper crop, corn prices plummet. When a drought hits, Illinois yields are severely impacted, production nose-dives, and prices soar to the extent that the negative correlation (relationship between price and yield) is fairly robust: below –0.40. The northeast, southeast, West Coast, and mountain states produce relatively insignificant quantities of corn and therefore, have little, if any, impact on the level of corn prices.

Soybeans experience a comparable scenario. By checking crop prospects in central Illinois, southern Iowa, northern Missouri, and eastern Nebraska and Kansas, one can develop some insight into expected soybean harvest prices.

The correlation is far from perfect (e.g., –1.0), but it is there. Furthermore, –0.40 for corn is much stronger correlation than –0.25 for soybeans. The –0.07 for Kansas winter wheat, means that there is virtually no relationship.

Developing Alternatives for Managing Risk

The late 1990's produced a proliferation of market-based mechanisms available to agricultural producers for managing yield, price, and revenue risks. Risks associated with high yield variability and resulting income variability can be mitigated by

Offsetting Price-Yield Relationship, a Key Factor in the Farm Risk Environment, Varies by Region and Commodity

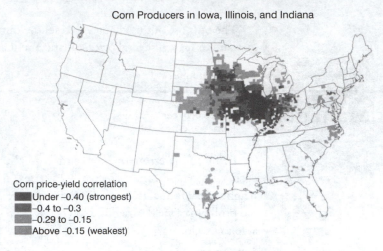

Corn Producers in Iowa, Illinois, and Indiana

Corn price-yield correlation
- Under –0.40 (strongest)
- –0.4 to –0.3
- –0.29 to –0.15
- Above –0.15 (weakest)

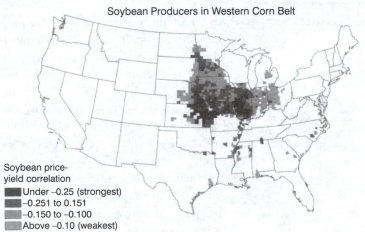

Soybean Producers in Western Corn Belt

Soybean price-yield correlation
- Under –0.25 (strongest)
- –0.251 to 0.151
- –0.150 to –0.100
- Above –0.10 (weakest)

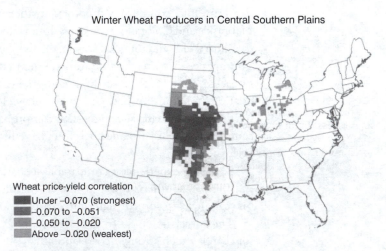

Winter Wheat Producers in Central Southern Plains

Wheat price-yield correlation
- Under –0.070 (strongest)
- –0.070 to –0.051
- –0.050 to –0.020
- Above –0.020 (weakest)

Price-yield correlation indicates strength of offsetting relationship between price and yield movements—the more negative, the better the "natural hedge" works to stabilize revenue. Based on annual county-level data, 1974–94.

Economic Research Service, USDA. Agricultural Outlook/April 1999.

FIGURE 6–1: Offsetting Price-Yield Relationship, a Key Farm-Risk Environment, Varies by Region and Commodity.

programs like Federal Crop Insurance. Despite the growing complexity of agricultural insurance programs, the majority of policies sold can be reasonably represented by two generic types: standard yield-based crop insurance and revenue insurance.

The largest share of farm coverage continues to be traditional yield-based crop insurance, although revenue insurance coverage is gaining wide acceptance. Traditional yield-based insurance, referred to as multiple peril crop insurance (MPCI), includes both minimum catastrophic coverage (CAT), which insures against severe losses and whose premiums are fully subsidized by the federal government, and higher levels of coverage called "buy-up" coverage that has partially subsidized premiums. Revenue insurance policies include income protection, revenue assurance, and crop revenue coverage.

Two time periods are relevant in calculating insurance program prices. The first is planting time, when a *Projected Price* is used to set insurance premiums, select price coverage, and set value coverage levels. The second period is harvest time, when the *Harvest-time Futures Price* is used to value the farm's production, whether sold or stored.

For yield-based insurance purposes, the USDA's Risk Management Agency (RMA) establishes a Projected Price about three months before the insurance sign up for each commodity. The yield-based-insurance version of the Projected Price is not derived solely from a futures market price average, but from a forecast of the season-average price that incorporates additional market information.

For revenue insurance valuation, the Projected Price is the average of the daily settlement prices of the harvest-time futures contract during the month preceding program sign up. For the price at harvest time, the average closing price of the harvest-time futures contract during the month prior to the contract's expiration is used. For example, the Projected Price for a corn revenue insurance contract is the February average closing price of the Chicago Board of Trade's (CBOT's) December corn contract. And the harvest-time futures price for the December corn contract is the average daily settlement price during November.

Yield-based crop insurance (MPCI) pays the operator an indemnity if the actual yield falls below a yield guarantee, but multiple peril crop insurance (MPCI) does not offer price protection. Under MPCI the producer pays a processing fee for the minimum catastrophic coverage (CAT) and a premium for buy-up coverage to obtain partial protection against yield loss only. The *yield guarantee* is determined by multiplying the producer's average historical yield, referred to as the actual production history (APH), by the coverage level. Coverage levels range from 50 to 75 percent (expanded to 85 percent in some areas for 1999) of the APH yield, and from 60 to 100 percent of the Projected Price.

For example, suppose a corn producer has an APH yield of 150 bushels per acre, the Projected Price is $2.50 per bushel, and the producer selects 75-percent APH coverage, referred to as the *elected price*. The producer's yield guarantee is 112.5 bushels per acre (75 percent of 150 bushels). An actual yield below 112.5 bushels will result in an indemnity payment to the producer equal to the elected price of $2.50 times the difference between the yield guarantee and the actual yield, even if the harvest-time price rises above the Projected Price. However, if the actual yield does not fall below the yield guarantee, even if the harvest-time price falls below the Projected Price, the farmer receives no indemnity. Thus, the multi peril coverage (MPCI) partially insures against production risk, but does not insure against price risk.

Revenue insurance in the form of income protection and the standard revenue assurance program protects farmers against reductions in gross income when a crop's price and/or yield decline from early-season expectations. The *revenue guarantee* equals the product of the farmer's APH yield, the Projected Price, and the coverage level selected by the producer. A farmer receives an indemnity when the actual yield, multiplied by the harvest-time futures price, falls below the revenue guarantee. Since revenue insurance coverage is generally available at a maximum of 75 percent, it provides only partial protection against both price and yield risk, and is less effective at reducing risk when the natural hedge is strong.

Revenue insurance with replacement coverage protection is available to farmers via the Crop Revenue Coverage program or the Revenue Assurance program when purchased with an increased price guarantee option. The added replacement coverage protection (RCP) feature offers a revenue guarantee that depends on the higher of the price elected at sign-up or the harvest-time futures price. Thus, the grower's revenue guarantee may increase over the crop season, allowing the producer to purchase "replacement" bushels if yields are low and prices increase during the season. Replacement coverage complements forward contracting or hedging by partially ensuring that the farmer can buy back futures contracts when yields are low and harvest-time prices are high. Growers are still subject to basis risk, and only partial coverage is available. Perhaps most important, this is not a free lunch since producers must pay substantial annual payments and if the programs are highly subsidized, the taxpayers will bear the burden.

In general, the revenue guarantee of RCP equals the product of the farmer's APH yield, the coverage level selected, and the higher of the early-season Projected Price or the harvest-time futures price. Indemnity payments are triggered when the harvest-time revenue falls below the revenue guarantee. Thus, revenue insurance with RCP also provides only partial protection against yield and price risk, and it is less effective when the natural hedge is strong, because high prices offset low yields and revenue is more likely to stay above the guarantee.

The premium for revenue insurance with replacement coverage is more expensive because replacement cost protection provides greater price protection. In addition, premium differentials increase when producers are permitted to subdivide their acreage into "units," sections, or irrigated/nonirrigated rather than total acreage (as under Income Protection).

With a 75-percent coverage rate, the standard revenue insurance guarantee for a corn producer with an APH yield of 150 bushels and a projected harvest-time price of $2.50 is $281.25 per acre. A revenue insurance policy with RCP (less than 75 percent coverage) has $281.25 as an initial minimum revenue guarantee, but this guarantee may increase if market prices rise during the growing season. If a low harvest-time price causes the market value of the crop to fall below the revenue guarantee, revenue insurance policies, with or without RCP, will pay the same indemnity. However, if the low yield is accompanied by a high harvest-time price, revenue insurance with RCP will pay an indemnity, while policies without RCP will pay a lower or no indemnity.

The combination of hedging and insurance can function together to enhance farm income. Hedging through forward pricing involves setting the price, or establishing a lower limit on price, for a commodity to be delivered at some time in the future. Forward pricing strategies include contracts like: cash forward, futures, basis, minimum (maximum for feed purchases) price, and options. Regardless of which tools are chosen, farmers face the challenge of when to commit to the offers presented in the marketplace.

Commodity Pricing Problems

Farmers have two commodity pricing problems. The first involves the basis market that dictates when to move the physical commodity in a cash sale. The second applies to the price level decision of when to commit to the price offered by the market.

The first task is relatively straightforward with storable commodities. Farmers weigh the costs of holding grain in storage against the expected cash prices later in the marketing year. Demand for food and fiber is fairly stable from month to month throughout the year, so inventory must be carried forward after harvest. If the price were constant month to month, any producer would have the option of selling the inventory this month, or holding until next month and receiving the same price, while incurring the cost of carrying the product for a month. Rational producers would sell now rather than incur the carrying charge if next month's price were not sufficiently higher to pay storage charges. The larger supply coming onto the immediate market would depress the current price relative to next month. Thus, cash price tends to trend upward over time. If the up trend in price equaled cost of carrying inventory, the producers would be indifferent about which month they made their sale. If late season prices were expected to exceed cost of carry, producers would hold inventory for late season sale. The month-to-month spread in prices tends to settle at levels sufficiently wide to cover the costs of carrying the inventory from month to month. The gradual rise in price is the method of charging customers for the service of holding the inventory. The specific details of basis level decisions in a market economy will be discussed in the next chapter.

The second task in pricing is picking the day to set price level decisions. As illustrated in Figure 6–2, farmer pricing has one objective—"Top the Market." As the market price level fluctuates up and down throughout the year, farm profit rises and falls relative to a fairly constant cost of production. Only at the highest price level is profit maximized on products being produced. Similarly, a farm operator purchasing feed and feeder animals seeks to buy inputs at the lowest price level of the year. His or her livestock cost-of-production is minimized with low-cost inputs.

There is a wide array of possible events that influence price on any given date. A wheat farmer being offered a current price of $3.00 per bushel realizes that the following possibilities exist:

1. Crop failure, which could be minimal or extensive, resulting in price of $4.25 or up to $6.00, with smaller probability of extensive failure and extreme prices.

2. A future bumper crop, also of variable size, with a corresponding price decline from $3.00 down to $2.00.

3. A business recession, leading to a decline in demand and lower price.

4. A war, leading to reduced output and a higher price.

5. A prospect for either increased or decreased export demand.

6. A shift in consumer taste away from bread toward meat.

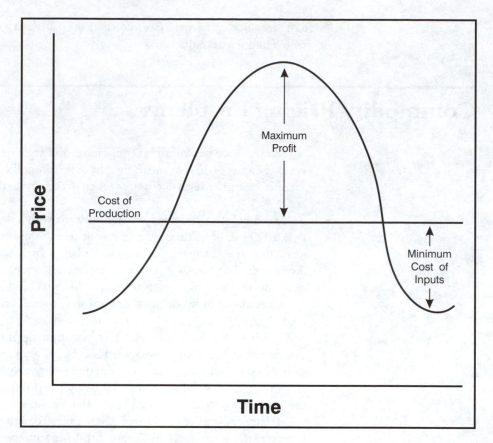

FIGURE 6–2: Farmer Pricing Challenge—Top the Market.

Each of these forces will affect the available supply and its rate of use. The role of price is to provide an adequate but not excessive inventory carryover. The expected increase or decrease in carryover sets the tone of price level in the market. As forces from increased demand or reduced supply feed into the market, the price level rises; as decreased demand or increased supply enters, the price level falls. The only thing a holder of wheat can be certain of is that Murphy's Law will prevail and something unusual will happen to change price from its current level.

In making pricing decisions, each market participant, in effect, estimates the probabilities of various conditions of supply and demand at a future date and their effects upon price. These likely estimates may be assembled into a frequency distribution of various prices, such as:

Possibilities		Expected Price of Occurrence		Probability Product
$4.00	×	0.05	=	$0.20
$3.50	×	0.10	=	$0.35
$3.25	×	0.20	=	$0.65
$3.00	×	0.35	=	$1.05
$2.80	×	0.20	=	$0.56
$2.60	×	0.10	=	$0.26
		Expected Price		$3.07

The weighted average expected price is derived by multiplying each probable price by its respective probability of occurrence and adding the results, giving, in this case, $3.07. The producer of wheat would make the decision to continue speculating on price or accept the market offer, depending on the current offer relative to expected average. The producer will sell if the market bid is greater than $3.07 or continue speculating if it is less.

The speculative game of pricing based upon fundamentals and price probabilities is to delay making a price commitment when current pricing opportunities are too low. As price rises, the other part of the game is to attempt to identify mistakes bid into the market and to profit from them by making pricing decisions when high price opportunities occur. Some use technical analysis, such as bar charts, in an effort to identify when over-bought and over-sold opportunities exist.

Each seller or buyer has a different set of price expectations and different probabilities attached to them. Even though he or she may not go through the rigors of the example in calculating probabilities, in the end, the producer is aware that market price will react to events to prevent the accumulation of an excessive carryover or, likewise, prevent depletion of stock before year end. Thus, the second pricing decision for a commodity producer depends upon **expectations** of PRICE LEVEL changes. The time for making sales decisions depends upon one's assessment of whether events will add strength or weakness to price as determined by fundamental analysis. Whether one's timing turns out to be correct, or not, depends upon one's skill in accurately assessing future events.

Price projections based on fundamentals and technical analysis might go as follows:

2000 Soybean Forecast

We look for a possible sideways trading pattern to develop over the next three months, with prices trading in the 50-to-60-cent range from the current $5.00 level in the first quarter. Based on technical chart patterns, July beans could hit $5.50 in November, and potentially, could reach $6.00 in January, at which point they could encounter overhead resistance. Failure to penetrate that resistance will mean a setback of some magnitude. Furthermore, if no serious problems develop in Brazil, July could easily test the $4.50 level. However, in view of falling interest rates and a weakening dollar, any serious problems in China might test the $7.00 and usher in a run at $9.00 by July. Expectations for a long-term up trend are contingent upon too many factors that are still uncertain at this time.

In saying that price level could be $9.00 a bushel or $4.50, the analysis says nothing. It simply shows the frustration of trying to base a pricing plan upon fundamental analysis of uncertain events. Nevertheless, a farmer must make pricing decisions at some time during the production or storage period. The producer did not grow the crop to hold it forever.

Pricing Choices

Farmers are faced with three choices in making pricing decisions:

1. *Price Takers* who deliver their product and ask, "What is the price today?"

2. *Routine Hedgers* who initiate pricing in advanced production without analyzing price; or they deposit grain in storage, hedge against price change, and hold until time of sale.

3. *Selective Pricers* (or selective speculators) who proceed with production decisions in hopes of making profitable pricing decisions sometime during the year.

Price Takers

Price takers sell when bills must be paid, loans are due, or storage bins are needed for the next crop. They make no pricing decisions but take what is offered when some circumstance forces them to make a sale.

The other alternatives require a more active role in the timing of pricing decisions. The lure of increased profits encourages many producers to pursue those alternatives.

Routine Hedgers

For routine hedging in advance of production, fixing prices is the final decision criterion in the production plan. Land productivity, equipment, facilities, past production methods, and future production plans limit the alternatives in enterprise selection. Credit availability, cash-flow requirements, and risk-bearing ability further limit the number of production alternatives. From among many combinations of livestock and crop possibilities, a few viable alternatives remain for any producer. A farmer can switch planting among a few crops and raise livestock or leave lots empty. To make the production decision, the farmer prepares budgets to determine which production alternative offers the highest net return between variable costs and gross receipts. Routine hedgers rely on price signals during planning periods before production commitments are made and decide what to produce based on relative profits. They fix prices by pre-pricing production. Market price quotes are fulfilling one of their important roles in signaling farmers what should be produced. In effect, currently available price quotes determine the production plans that farmers select.

The Pricing Ahead of Production strategy dictates that a producer either set profitable hedges before initiating production or avoid production. Few producers practice such a strategy, for a variety of reasons. Seldom are producers satisfied with pre-production pricing opportunities; they hope for higher profit opportunities and are willing to speculate on that hope. Many are reluctant to price what they do not have for fear that production problems will occur. Drought or disease could wipe out production committed to forward contracting. Often a single crop production alternative is not distinctly profitable. Futures prices among several related commodities tend to rise and fall together. High soybean prices encourage high prices in competing crops, such as corn in the Midwest and cotton in the South. Seldom do profits of one alternative greatly out distance others when all limitations are weighed. Finally, agricultural production is characterized by high fixed costs relative to total costs. Production decisions are typically based solely on variable costs and enterprise net margin. What is at risk is their return to labor, management, and capital. Consequently, producers prefer to practice selective pricing, or conversely selective speculation, by going forward with production plans hoping that a profitable opportunity will present itself before the cash crop must be sold.

The common mistake of pre-production pricing is to read futures prices, base production plans on them, but avoid making pricing commitments ahead of planting crops or filling feedlots. High futures prices elicit increased production that in turn produce low market prices when production is ready for the cash market. Producers feel betrayed. Their production decision was correct, but they did not complete the strategy by fixing price. They failed to fully appreciate the role of futures prices as presenting a pricing opportunity. Instead, they incorrectly chose to view futures quotes as a price prediction, which they are not.

Farmers who delay pricing until the crop is in the bin could use routine hedging to fix a future price for their inventory held in storage. However, most farmers do not hedge stored production. They store grain to speculate that price will improve. Sometimes they are correct and profit accordingly; often they are wrong and sell stored grain late in the season for less than what it was worth at harvest, after incurring the costs associated with carrying commodities.

Selective Pricers/ Selective Speculators

Producers of grain and livestock who do not pre-price make a pricing decision every day that markets are open for trading. Most days, their decision is to not make a pricing commitment and wait until the next day. Planting seed, filling feedlots, or storing unpriced grain create speculation on price until some form of pricing commitment is made. With a selective speculation/selective pricing strategy, the decision, as to when to commit to pricing commodities for sale is a decision based on controlled speculation. The challenge is to manage price speculation in the way most advantageous to the individual producer.

To initiate a selective pricing strategy, objectives must be established and restrictions placed on the scope of permissible speculation. Every day does not necessarily offer a profitable pricing opportunity, because price often falls below total production cost. Some days offer more profitable prices than others. One producer may set an objective to sell at the high net price of the year or buy inputs at the lowest price. If he or she is successful, the strategy will yield maximum income. But, such an objective involves high risk because of the difficulty in picking highest or lowest price and unwillingness to commit 100 percent of production. There is only one way to pick the high price of the year. It is called: *The 365-Day Hindsight Method*—look at prices over the past year, pick the high day, and say, "That's when I should have sold." Unfortunately, that system of pricing does not work in the real world, since the opportunity is history and hindsight, not foresight, is 20/20.

Price uncertainty affects each producer differently in forming a pricing strategy. Some avoid making pricing forecasts and submit to the Greed-Hope-Fear strategy. GREED keeps them from committing to high profit opportunities when they arise. They refuse to sell at bonanza high price levels because they want only the top price. After a market slump, they refuse to make sales at profitable prices because they HOPE prices will return to their previous levels. Eventually, they do sell at a low price because they FEAR prices will go even lower. Or they must sell because a loan payment is due.

Another objective might be to get the average price of the year. They sell equal amounts out of storage every Wednesday of the marketing year, or equal amounts on the 15th of every month. This seems rather dull to many operators, but, in actuality, they often fail to achieve comparable results.

A different objective might be called income averaging, whereby the objective is to reach the same income level year to year. When price in a marketing year reaches the level of previous years, pricing commitments are made. But there is no guarantee that prices will be the same year to year. Some years prices are high; other years they are substantially lower.

In order to better understand the process of selective pricing/selective speculation, we need to delve into farmer psychology toward pricing production and into the results of various strategies.

Farmers Store Grain to Speculate

At harvest, farmers store corn, soybeans, and wheat for one reason—they expect the price to rise. No one would store grain if he or she expected the price to fall lower throughout the storage year. Nor would selling at the same price next July, following eight months of storage, make sense. During that period, the grain producer could have had money in the bank earning interest rather than tied up in grain that did not rise in value. Thus, farmers store grain to speculate on a price rise. Common sense tells them that the price is too low at harvest and, therefore, it can go only one way— UP! This is true whether corn price is $1.50 or $4.50 per bushel. Less-timid farmers store grain and shoot for the top cash price of the year. Occasionally these storage speculators succeed with part of their production.

Not to be outdone by these coffee-shop top price pickers, we found the **net** central Illinois High Cash Price—**the top cash price of each year adjusted for the cost of storing grain**—for 28 years (Table 6–1); 20/20 hindsight is a powerful tool in this search.

We chose the central Illinois cash price because that area is the heartland of corn production and other markets relate to it. The one-day storage-year high cash price (column 3) was adjusted for carrying charges incurred after December 1 until the *net effective high cash price* day arrived (column 6). Carrying costs include interest on the money tied up while grain is in the bin. We assumed a 12-percent per annum interest rate over the entire period. This is low for some years and high for others, but the discrepancies do not materially affect the conclusions. In addition, we calculated the cost of management, storage-bin ownership, and risk to equal 2 cents per bushel per month. These interest and carrying costs were deducted from the top cash price (column 3) to adjust harvest-time price to obtain the *Net effective Illinois high cash price* (column 3 – column 5 = column 6).

The highest cash price and the net effective high cash price do not necessarily occur the same day. For example, the 1998-harvest price was $2.14. The highest cash price, $2.18, occurred in mid-March 1999. Since it cost 16 cents to hold corn from December to March, the March net price was only $2.02. Table 6–1 indicates that the crop should have been sold at, or close to, harvest. Holding grain until March would have produced a 4-cent per bushel larger corn check, but with 16-cent holding costs, the net realized cash price would have been reduced to $2.02 per bushel.

Capturing the top net cash price each year produced average storage profits of 40 cents per bushel per year (average **high cash price** of $2.99 *minus* the average December 1 harvest price of $2.38, and *less* the average 21 cents per bushel storage cost for an average holding period of 19 weeks). Not bad! The obvious conclusion is "store corn at harvest and sell on the HIGH PRICE DAY of the year." The only obstacle is knowing when the top price day occurs. AND THAT IS THE PROBLEM!

TABLE 6–1: Results from Corn Storage Speculation, 1971 to 1999 (cents per bushel)

1	2	3	4	5	6	7	8	9
Crop Year	Illinois Harvest Price Dec. 1	Storage Year High Price	Weeks of Storage Until High	Storage Cost★	Net Effective Illinois High Cash	Change in Price After High	No. of Weeks for Change to Occur	Average Net Price Sell Every Wednesday
1971	105.00	126.00	6	4.60	121.40	−15	11	106
1972	132.00	340.00	41	38.50	301.50	−103	3	179
1973	232.00	396.00	41	48.20	347.80	−52	3	277
1974	364.00	365.00	0	0.00	365.00	−113	11	274
1975	258.00	314.00	32	35.90	278.10	−19	3	245
1976	232.00	263.00	11	11.80	251.20	−116	13	200
1977	205.00	270.00	29	29.10	240.90	−64	10	205
1978	231.00	310.00	33	36.40	273.60	−36	6	271
1979	250.00	354.00	42	50.20	303.80	−21	4	258
1980	350.00	369.00	4	10.40	358.60	−33	10	297
1981	255.00	265.00	0	0.00	265.00	−70	14	211
1982	210.00	377.00	41	47.20	329.80	−52	9	266
1983	338.00	346.00	0	0.00	346.00	−57	14	289
1984	260.00	262.00	1	1.10	260.90	−13	5	228
1985	211.00	240.00	4	4.30	235.70	−69	26	225
1986	150.00	160.00	0	0.00	160.00	−20	14	163
1987	182.00	324.00	33	31.70	292.30	−84	5	203
1988	248.00	273.00	6	6.00	267.00	−63	42	224
1989	229.00	284.00	23	23.90	260.10	−74	13	221
1990	219.00	253.00	18	18.58	234.42	−32	11	212
1991	237.00	269.00	15	10.56	258.44	−24	8	226
1992	204.00	272.50	51	54.00	218.50	0	0	198
1993	272.50	294.00	7	14.00	280.00	−110	40	221
1994	210.00	332.00	48	52.00	280.00	−20	2	242
1995	315.00	525.00	32	44.40	480.60	−275	17	365
1996	264.00	299.50	17	27.10	272.40	−69	14	235
1997	269.00	269.00	0	0.00	269.00	−104	39	202
1998	214.00	214.00	0	0.00	214.00	−56	33	171
Ave.	237.38	298.79	19	21.43	277.36	−63	14	229

★Storage Cost = 2 cents per bushel per month plus 1% interest per month on the average cash value of corn.

Cash prices are for central Illinois beginning in 1982.

The 1998 corn crop reached a high of 218 in Mar. 1999, but it cost 16 cents to hold corn that long. Thus, the net price was only 202.

That problem is compounded by the knowledge that the optimum holding period is either very short (less than 2 months) for 13 out of 28 years, or very long (more than 6 months) in 11 years. These observations may be confirmed by reviewing the data in column 4 of Table 6–1.

Table 6–1 provides one other instructive insight. Column 7 of the table shows how closely those who speculate on grain prices flirt with disaster—and, like the moth, are often consumed by the flame. It shows the drop in price AFTER the high day occurred. Column 8 shows how long it took AFTER the highs were in for disaster to strike. To top the 1993 corn crop market, corn should have been stored for seven weeks until the end of January 1994 and sold for $2.94 per bushel (less 14.0 cents storage cost). Those who waited until early October 1994 watched prices plunge 110 cents per bushel in 40 weeks. After the high, sharp price breaks are the rule, not the exception. During this 28-year period, prices dropped an average of 63 cents per bushel within 14 weeks after the highs had been made. In 10 years, price dropped more than 37 cents per bushel within 2 months of harvest and by more than $1.00 per bushel in 6 storage years.

Storage speculators hold grain in hopes of capturing 40 cents per bushel. Those who continue to watch see 63 cents evaporate in an additional three months of waiting, hoping, and praying. Without 20/20 hindsight, or a reliable crystal ball, their chance of catching the top of the market is about 1 out of 300 (the number of days elevators are open). There are folks who also try to fill inside straights at the poker table. Their odds of 1 in 11 are better than the grain speculator's 1 in 300. As the man said: "Filling inside straights is fun, but don't try to make a living at it." Trying to catch the top of the cash market is fun—but farmers should not try to make a living at it, because the odds are against them.

As noted above, the most profitable holding period is either very short OR fairly long. Equally important, crops that are held for a long time are often followed by crops that are held for a short time. This reinforces the adage "A short crop has a long tail." When corn prices were at $5.00 in 1996, many farmers put grain in the bin in hopes of getting $6.00—many finally settled for $2.75. When their gamble did not pay off, they complained about low prices.

The table also tells us something about "Summer Rallies." Summer price rallies after June 1 (26 weeks after Dec. 1 harvest) that were big enough to pay storage costs occurred in only 11 out of 28 years.

As an alternative, "scheduled selling" suggests dividing the crop into 50 parts and selling one-fiftieth each week. Column 9 shows the average adjusted price from 50 equal volume sales each year. Equal amounts were sold every Wednesday and the price adjusted for storage costs incurred on that portion up to the time of the sale. The overall average net price of $2.29 was 8 cents per bushel LESS than the average price achieved by _always_ selling at harvest.

Farmers store grain to speculate on price. Like the gambler, they occasionally fill an inside straight, but more often than not, they are beaten by the odds. Unless storage speculators are able to come close to the top of the market (and very few are able to do so consistently), the profits from storage speculation are zero to none.

To succeed at the storage game, farmers need to forget about storing to speculate on price level and learn to speculate on the basis (the difference between their local cash price and a selected futures price). Invariably, a strong basis occurs when price level is low. Farmers are then reluctant sellers because they feel downside price risk is minimal. Their reluctance creates the strong basis. When prices are high, buyers find ready farmer-sellers, and they can get all the grain they need to satisfy demand. Therefore, the basis is weak; there is no reason for buyers to bid a premium. Farmers can become storage space merchandisers by (1) storing when the basis is weak, (2) hedging away price level change, and (3) watching the basis narrow to cover the storage cost. Then storage becomes profitable. The sole purpose should be to capture basis gain.

Farmers should sell cash grain when the basis market says "move it," whether price level is high or low. If they believe that price level is going higher, there is a better way to speculate on improved price level. Grain need not be held in bins to speculate. Replacing the cash position with a long futures position is the better way to speculate on price level. Futures and cash market prices tend to move up and down together. If cash market speculation is profitable, paper speculation will also be profitable. Furthermore, the cost of paper speculation is only about two cents per bushel per year: 1-cent brokerage commission and 1-cent interest on margin money. The total cost of speculating on paper is 2 cents per bushel per year versus the 52 cents per bushel per year holding costs for grain stored in a bin for speculation.

Price may go down and create a loss in the long futures position held for price speculation. However, grain stored in a bin loses value just as rapidly as a futures position in a down-trending market. Unfortunately, farmers are less aware of inventory value erosion. No neon sign on the bin flashes daily losses in value, but the daily futures market margin calls capture their attention.

Many farmers (and their lenders) object to speculating in the futures market. Nonetheless, they borrow thousands of dollars to build storage bins to SPECULATE that the price of corn will go up 40 cents per year—every year. If the books are properly tallied, the cost of making a mistake speculating with the physical grain is far more significant than the SAME mistake made speculating with grain on paper.

Minimum/Maximum Pricing with Options [3]

One way to achieve protection from the possibility of falling prices or rising costs is to purchase PUT or CALL options, respectively. A PUT option provides a sturdy price floor, but allows the producer to take advantage of rising prices, should they occur. The CALL option provides a price ceiling on feed grains, but allows the buyer to enjoy lower costs in declining markets. The price of this market flexibility is the option premium that is paid in full at the time the option is purchased.

This explanation will focus on the analysis of establishing a hedge for a corn producer buying PUT options. It must be remembered, however, that a pricing strategy concurrently requires selecting the timing of purchasing the option and determining an acceptable target or floor price. The target price chosen could be the total cost of producing the commodity, the minimum price set by government programs, the minimum price assured by purchasing crop insurance, or the bonanza high prices that occur periodically on the long-term price charts. The producer/marketer must be aware that option premiums are usually higher in volatile markets trading at price extremes in the trading range.

The producer/marketer might utilize the December corn futures price during the January-February planning period as the pricing criterion. She might establish the option to protect the anticipated spring planted crop. As harvest is completed in November, the farmer would either offset the option, exercise the option to create a short futures position, or let the option expire and enjoy the higher cash price offered by the market.

After harvest, with the grain stored in the bin, she could continue to speculate on possible higher prices for the stored crop by buying a March PUT option to establish a floor price, but hoping that it expires worthless. Or by buying a July option, our producer/speculator hopes that the July futures is above the strike price so that the option expires "in the money," during the summer price rally.

A negative scenario results from using options when high premiums must be paid several times to achieve the desired results. All too often, large multi-premiums reduce the effective floor price below acceptable levels.

The expected price of a PUT option is: **PUT Strike +(–)Expected Basis –PUT Premium = Expected Floor Price**. The ultimate value of the crop depends on the actual basis at the time of sale. Stronger basis results in a higher selling price and vise-versa. Options establish the price level but leave the basis open to speculation.

Whenever the market strategy triggers the producer/marketer to execute an option, further analysis is required to select the most advantageous strike price. Suppose that during October, the December corn futures trade at the following prices:

Date	December Corn Futures
October 15	$2.62
October 16	$2.65
October 17	$2.68
October 18	$2.69

If she sells a December futures contract at $2.69 and realizes her expected basis of –$0.23 in December, her final selling price is $2.46. While she wants to fix a floor price, she believes (or hopes) that prices may still move higher. She wants price protection but does not want to forego the benefits of higher prices. On October 18, she finds that December PUTS are quoted as noted below:

December PUT Strike Prices	December PUT Premiums
$2.50	$0.08
$2.60	$0.12
$2.70	$0.17
$2.80	$0.23
$2.90	$0.31

With this information, she determines which strike price offers the best return. She calculates what her expected floor price would be for each strike using the formula outlined above. She finds that:

PUT Strike	+(−) Expected Basis	−	Premium	=	Expected Price Floor
$2.50	−$0.23	−	$0.08	=	$2.19
$2.60	−$0.23	−	$0.12	=	$2.25
$2.70	−$0.23	−	$0.17	=	$2.30
$2.80	−$0.23	−	$0.23	=	$2.34
$2.90	−$0.23	−	$0.31	=	$2.36

She may select the highest floor price ($2.36) or analyze which cost/price increment gives her the best value. With 100,000 bushels of corn harvested, the total premium for a $2.50 strike is $8,000; the $2.90 strike option will cost $31,000. The $2.60 versus the $2.50 strike costs 4 cents more to raise the floor price a net of 6 cents. Selecting the $2.70 strike increases cost by 5 cents and nets an additional 5 cents; $2.80 strikes cost an additional 6 cents to raise the net 4 cents. The last increment costs an additional 8 cents to realize a 2-cent higher net. In each case, the net has been adjusted for the premium per bushel. In her risk/reward analysis, she believes that the $2.70 PUT best meets her market objective. A $2.80 strike costs an additional 6 cents to raise the price 10 cents, but the net only increases 4 cents and she determines that is not worth it.

Producers with different outlooks and goals will choose a different strike price. A farmer who expects the market to rise, but wants some price protection might choose the cheaper $2.50 strike. A producer who has a price goal of $2.33 will buy the $2.80 option. Individual goals and expectations dominate the pricing decision.

What are the possible consequences of purchasing a $2.70 PUT for 17 cents and a realized basis of 23 cents under the December futures at time of expiration in mid-November?

At Sale December Futures	+	Actual Basis	=	Local Cash Price	+	Option Profit(-Loss)★	=	Final Value
Price falls to $2.25	+	$−0.23	=	$2.02	+	$0.28	=	$2.30
Price stable at $2.68	+	$−0.23	=	$2.45	−	$0.15	=	$2.30
Price rises to $3.20	+	$−0.23	=	$2.97	−	$0.17	=	$2.80

No remaining time value.

★To calculate option profit (loss) = strike price − futures price (at delivery/offset time) − premium paid

As shown in the summary above, the final corn crop value is **at least $2.30** as long as the basis is 23 cents under the December futures. The crop value increases if the futures price exceeds the $2.70 PUT strike price.

Premiums represent the cost of price protection. Premiums reflect the market's assessment of the level of risk option sellers assume. The premiums for hurricane insurance are much higher in Miami than they are in Maine. Insurance is not available if the eye of the storm is directly overhead. When markets are highly volatile at price extremes, the cost of insurance (premiums) may be prohibitively high.

Building a Marketing Strategy

In order to improve on cash market speculation, two important concepts provide a framework for building marketing strategy: the decision period for pricing and the role of futures markets.

Decision Period

The decision period for pricing should be extended beyond one-day selling. The marketing tools reviewed in Chapter 2 allow decisions to be effected long before planting or well after harvest. For corn, soybeans, wheat, and cotton, the marketing period extends from the time of production planning until the end of the storage season. This is generally a period of about 20 months for a single crop. Because the storage year overlaps the next year's growing season, a crop grower is concerned with marketing **two** crops most of the time—*one in storage and one growing in the field.*

Fortunately, the futures markets trade contracts for pricing 12 to 18 months into the future. They permit forward pricing of planned or growing production, or sales of stored commodities. As one contract month approaches maturity, a new contract for the same month in the next production year is opened for trading. Crop producers are not limited to harvest-time prices to make pricing decisions.

For livestock, the marketing year is not as distinctively defined, since cattle and hogs often involve a more continuous production process. Key production decisions on hog farms relate to increasing or decreasing the number of sows kept for farrowing. A market-ready gilt can be either sold for slaughter on the current market or retained to increase production in about 10 months. Hog futures are traded 12 months ahead of maturity. With both production and pricing decisions extending a year ahead, the marketing year should correspondingly be at least 12 months long. In cattle feeding, the production period extends from the time of purchasing feeders and feed until the finished cattle are sold. Trading in live cattle futures extends about a year. For feeder cattle, futures trading spans about nine months. With the time overlap in contracts, inputs of feeders and feed and production of finished cattle can be contracted about six months before the feeding enterprise is physically started and extended about six months into finishing. Thus, our cattle marketing period is also one year. In this way, futures contracts serve to extend the period of price and production decision-making.

Role of Futures Prices

Futures prices are today's price offers for a future business transaction. The price agreed upon today is what buyers and sellers **think** is the proper future value of commodities based upon currently available fundamental information. But futures prices are poor predictors of cash price when the future becomes the present. Many events can and do occur in the interim to cause great changes in price. Within any one year, futures may fluctuate up and down by several cents or several dollars per unit of production. Futures prices tend to overshoot their targets; they are too high in anticipation of shortages and too low in anticipation of surpluses. It is logical to question why they are such poor predictors. Like any prediction of future events, they tend to be self-defeating. High futures prices, before planting or breeding, entice producers to shift acreage to crops with the highest profit alternatives or to breed more females. The resulting increase in supply depresses cash price when the production reaches the market. Likewise, low futures prices discourage production, and prices eventually rise to ration a reduced supply when it reaches the market. In one sense, futures prices are providing a social benefit by fluctuating to extremes. They tend to blunt realized effects by modifying shortfalls in production and discouraging over production.

A bigger problem in using a futures price as a cash price predictor is the unexpected event that affects price change. A widespread hailstorm, embargo, political turmoil, or export buying often occurs without warning. Uncertainty of events causes constantly changing forecasts, as reflected by constantly changing futures prices. An example is given in Figure 6–3. The chart shows fluctuating soybean futures in response to the newsletter account of changing fundamental factors.

> Last month's letter on the soybean market described a very bearish supply/demand outlook but called for a short-covering rally in beans, a short-covering rally in the meal versus oil relationship, and widening new crop bean spreads. Those rallies occurred, but the overwhelming bearish fundamental picture was altered in the process. First, the USDA estimated soybean-planted acreage at 63.3 million acres, which was more than 1 million acres below the lowest trade estimate. Then, weather changed. Abundant subsoil moisture and continuous rainfall through the first half of June suddenly changed to above normal temperatures and far below normal rainfall for virtually the entire belt. The trade has tended to focus attention on weather in the western corn belt, but this time around, the danger may lie in the north-central and eastern corn belt and into the mid-southern bean belt. This letter will concentrate on the prospects and effects of a weather market —a dangerous proposition in any year.[4]

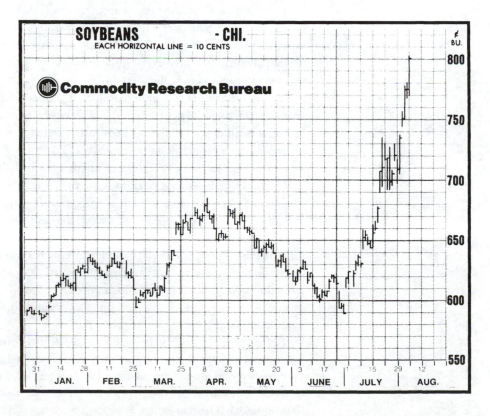

FIGURE 6–3: Illustration of Changing Fundamental Conditions and Resulting Changes in Futures Price.

The role of futures prices is to provide a quoted price to make pricing decisions today. Producers and buyers can use futures contracts to hedge away changes from today's price. Because so many things can and do change, futures prices cannot be relied upon to predict what cash prices will be at a later date. The pricing decision is whether to accept pricing offers day by day as time passes.

Futures Market— Workable, but not Designed for Farmers

The futures market provides a means for eliminating price level fluctuation and its associated cash market speculation. It was designed for grain merchants, not farmers. It allows grain merchants to hedge away the effect of price level change and to speculate on the basis and profit from the cash price—futures price relationship to earn storage profits. Hedging serves them well on both counts.

Grain merchant hedging, with a short futures position equal in size to the merchant's cash position, eliminates the effects of price level change. The parallel fluctuation in the value of cash grain held in inventory and of futures prices creates offsetting gains and losses. A rise in price level produces a deficit in the futures account but an equal gain in value of the inventoried grain, in a perfect hedge. Conversely, a drop in price reduces grain inventory value, but a gain in the futures position offsets the inventory loss.

Grain trade intermediaries use hedging to avoid the economic effects of inventory value fluctuation. If a Cincinnati grain merchant can buy corn 10 cents under a futures price and sell it in New Orleans for 15 cents over the futures, the merchant has a 25-cent margin. With a cost of operation of 20 cents per bushel, the grain merchant has a 5-cent profit margin. The possibility of the grain dropping 30 cents in value during the four days of the barge trip is too great a risk, given the small operating margin. The grain company hedges away the effects of a cash price change by taking an offsetting futures position. Should the cash grain price fall 30 cents in New Orleans before the barge arrives, the futures account will accumulate 30 cents of price protection to offset the loss.

As a grain handler, the grain company is not concerned whether the corn price level is $2.00 or $4.00 per bushel. It is only concerned with earning 5 cents for each bushel it floats down the river. The futures market serves the grain company well. It provides a vehicle to eliminate the effects of inventory fluctuation, while at the same time permitting it to collect an operating margin for handling grain.

Similarly, grain-storage operators, buying corn, wheat, or soybeans from farmers at harvest, focus their attention on earning storage profits. The storage operators need only calculate the price difference between the cash cost of the grain and some distant futures option in order to estimate their storage return. They know that the opportunity to deliver cash grain to satisfy the futures contract in the month of maturity will force the two prices to be approximately equal at delivery points. No matter how wide the basis at the time of the cash purchase, there is reason for basis to converge toward zero over time. Thus, hedging allows grain merchants to speculate on basis while eliminating price level speculation.

Armed with the knowledge that basis differences converge toward zero in delivery months at delivery points, grain merchants have a handle on what to expect from their basis speculation. They are left with their negotiating skills to buy grain at maximum discounts from the future price (weak basis) and sell it at a maximum premium (strong basis). Their objective is to earn a storage return to compensate them for holding grain from harvest until later use. As long as carrying costs are less than the amount of basis convergence, grain merchants profit from storing grain.

In both situations, the grain merchants' role is to trade the basis. Whether the futures, or cash, price level is high or low makes no difference to the merchants. Hedging allows them to eliminate the effects of price-level change and allows them to concentrate on basis trading.

In direct contrast, the farmer is vitally concerned with price level. To a grain producer, $4.00 corn returns significant profits and $2.00 corn generates significant losses, if the cost of production is $2.80 per bushel. The maximum difference between selling price and cost of production maximizes the crop farmer's profit. On the other hand, the livestock farmer buying feed seeks to capture low-priced corn and protein supplement. His or her production costs are reduced by using low-cost inputs.

The farmer who hedges eliminates price level change, just like the grain merchant. A farmer-hedger who establishes a fixed hedge position eliminates both lower prices AND higher prices. But since the farmer is most satisfied when capturing the highest price of the year, the producer is content only with executing a hedge on the one high price day of the year. With growing crop costs fixed for the year, that price extreme maximizes profit.

The farmer who forward prices production or inputs becomes, by necessity, a speculative hedger in comparison to the grain merchant who is a routine hedger. The elevator manager hedges to eliminate price change, but the crop farmer hedges in hope of capturing the ONE high price of the year on products to be sold. In similar fashion, a livestock farmer buying corn, feeder animals, and soybean meal attempts to hedge at the lowest price level of the year to capture low-cost inputs. As a buyer of commodities, the farmer achieves maximum profit when he or she can lock in the lowest input costs. As a speculative hedger, the farmer proceeds with unpriced production until he or she thinks the price extreme day occurs, and then establishes the hedge.

There is always the possibility that unpredictable events (weather, political decisions, unanticipated demand) might move prices higher or lower than what they were on the day that one makes a pricing commitment. Consequently, the farmer-hedger stands beside the market day after day gazing into various crystal balls and praying for divine guidance in selecting THE day to hedge. With analysis and dumb luck, he or she might sell on the one high day or buy inputs on the one low day—once in a lifetime.

Furthermore, the market does not rise day after day to a peak and then decline day by day to a bottom. The market moves in fits and starts, up and down, as fundamental and technical forces pull and tug at it. Clouding the fundamental forces driving the market in one direction or another is a lot of noise incorporated in daily and weekly price fluctuations.

In the meantime, the economic rewards from trying to capture high selling prices or low buying prices attract farmers to the futures market. But since the futures market was not designed for farmer pricing, achieving their objectives is not easy. Knowing when to be in a market position and when to be out of it is the challenge for farmers using the futures market. Very few have used the futures market successfully.

Farmer marketing plans and pricing strategies must be based on the fact that futures markets were NOT designed for farmers. But the baling wire/fodder twine/cutting torch inventiveness of farmers can change nonworkable designs to make them functional. By accepting the limitation of futures markets, farmers can develop selective hedging/selective speculation programs that offer hope for controlling cash market speculation. A properly developed marketing plan, executed with strict discipline, has the potential for serving farmers better than the usual practice of cash market speculation.

Farmer Speculation in the Futures Market

Use of the futures markets is touted as a price risk-management tool. Many farmers believe what they hear and venture into futures trading without understanding the difference between hedging and speculating. The previously cited Commodity Futures Trading Commission study showed that 22 percent of futures market participants were farmers pursuing price speculation.[5] They neither owned nor intended to own the commodities they were trading. Or they placed hedges and lifted them; they traded in and out of the market on whim and emotion, without plan or discipline, and became grist for the commodity market mill. Goaded by brokers who collect a commission on every trade, farmers who could be hedgers drift into small trader speculation. All too often they get in and out of the positions on rumors, emotions, and market excitement. Inevitably, they join the ranks of losers.

In addition to their production risk from drought, locusts, and pestilence, some farmers use the futures markets to increase speculation through trading the markets. Their speculative nature takes over to tempt them into doubling up on price risk rather than minimizing it. Whenever farmers buy somebody else's price risk, they are adding to risk already inherent in their business. Most farmers insist they are hedging. That is what they tell their spouses; that is what they tell their bankers. But deep down in their hearts they know that by trading coffee, gold, and pork bellies they are really speculators and not hedgers.

There is a clear distinction between hedging and speculation, as outlined in Figure 6–4. Hedgers own or intend to own the product; speculators do not. Hedgers seek to shift the price risk associated with owning a commodity; speculators purchase risk. Speculators do not own commodities and assume risk only when they take a futures position. They participate in the futures market and hope to profit by accepting the price risk that hedgers are trying to avoid.

But farmers have no place to hide. Price risk is with them as long as they farm—farmers always have a cash position, whether it is a growing crop or grain in the bin. The only way for them to avoid price risk is to sell the farm. Of necessity, farmers

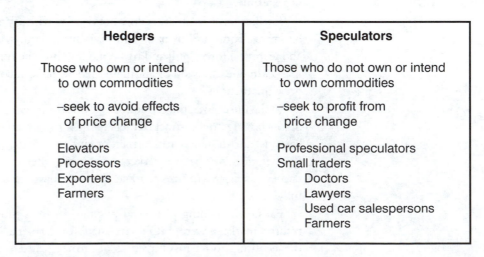

Hedgers	**Speculators**
Those who own or intend to own commodities	Those who do not own or intend to own commodities
–seek to avoid effects of price change	–seek to profit from price change
Elevators Processors Exporters Farmers	Professional speculators Small traders Doctors Lawyers Used car salespersons Farmers

FIGURE 6–4: Characteristics of Hedgers and Speculators.

must be selective hedgers/selective speculators on the value of their production; fixed hedging in its strict definition does not work for them as it does for the grain merchant. Therefore, their only viable alternative is to learn to manage that ever-present price risk with a marketing plan and pricing strategy.

Dealing with Brokers and Brokerage Accounts

The role of the account executive, or broker, is to fill orders and maintain records of the account. Brokers have contacts on the floors of the various exchanges who execute orders quickly and accurately. The brokerage office procedures should provide timely and accurate statements. There should be little difference between various firms, since their practices must conform to exchange and government regulations. However, as in all business, some execute their responsibilities better than others.

Brokers often provide market information and have trading plans for clients who request them. But their objectives might differ from those of a hedger. Many brokers are biased toward market trading for speculative clients. A hedger with a marketing plan and pricing strategy does not need marketing advice from a broker. He or she is usually better off following the plan rather than mixing several sources of information. Discount firms that do not have large overhead in market research and information services typically charge lower brokerage fees than big firms. Their service is adequate for the producer who has a well-defined marketing plan.

A farmer should understand that basic conflicts in objectives may exist between a hedger and a broker. The farmer prefers as few trades as possible to achieve marketing objectives. The broker may prefer as many trades as possible to generate commissions. The broker collects the commission fee when the position is closed out, rather than when it is initiated. Brokers are therefore anxious to have positions close out as soon as possible. The farmer-hedger should be concerned with long-term, major market trends; the broker watches minute-by-minute price change and wants customers to trade short-term action by reaction. Finally, the farmer should understand that a broker has both buyers and sellers as clients. He must have a reason to reinforce each of his or her decisions, as first one, then the other calls to place an order.

In daily chats with a broker, a farmer may get involved in the dynamics of market action and stray from his or her marketing plan into "playing the market." Few have the time or discipline to be successful traders, as confirmed by studies of small traders. Farmers must keep in mind that the broker does not know what direction the market is going to take. If brokers did know, they could make larger profits trading for themselves than taking market orders from clients. Farmers should develop their own trading plan and follow it. The broker should be an order taker, not an advisor. The conflicts in objectives between the two are so great that few brokers can be successful trading advisors for farmers.

A farmer-producer should deposit his or her initial brokerage margins as U.S. Treasury Bills, rather than a cash balance with the broker. The interest income from the T-bills will be credited to the client's account and generate income. Otherwise, the brokerage firm will invest the idle balance and gain the return.

Account surpluses produced by favorable futures positions should be transferred to interest-bearing accounts either with the brokerage firm or with the producer's lender. Properly structured savings accounts with a bank offer additional security by the Federal Deposit Insurance Corporation.

Pricing Strategies

Many strategies are used by farmers to make pricing decisions. Some involve analysis; some do not. Fundamental analysis and technical strategies were discussed in previous chapters. Other strategies include:

1. Calendar Selling,
2. Cost-of-Production Models,
3. Scale-Up Selling,
4. Scale-Down Buying,
5. Seasonal Pricing, and
6. Trend-Following Models

Calendar Selling

One method, with several variations, is scheduled selling. A portion of production is priced according to the calendar. Producers may consistently sell at harvest and avoid storage costs, or they may pick July as their sales month, hoping to earn storage profits. Others sell a portion of production each week to achieve an average price. They may vary the schedule by selling larger amounts during the December-January and June-July periods to avoid harvest periods in both the Northern and Southern hemispheres. Those who are comfortable with forward selling may price one-fourth of their anticipated production before planting, another fourth in late summer, another fourth about January 1, and the final fourth in the spring. Livestock producers often close their eyes to fluctuating prices by selling at regular intervals (monthly, weekly). They produce an equal number of hogs or cattle for sale each period. Scheduled selling strategies eliminate the need for making forecasts, while yielding close to an average price for the year, less holding costs.

Cost-of-Production Models

Livestock Sales. A more active program bases pricing decisions on the cost of production. Outlined below is a hog production cost analysis and pricing decision procedure. The procedure emphasizes the various "cost of production" figures that can be calculated. When asked, "What is your cost of production?" a farmer should answer, "It depends." Cost depends upon how the calculation is made and what is assumed in estimating fixed costs. This example analysis is based on both price of the product and the major feed cost item, corn. It essentially uses the hog-corn price ratio as a decision criterion to initiate hedges on a larger and larger portion of anticipated production. It includes locking in feed cost as well as product price. Editorial comments in using the procedure are included.

Hog Pricing Decision Analysis

Today, before breeding gilts, go through these steps:

1. Call elevator to get corn price contracted over next nine months.

   $_____/bu.   Contract price on purchased corn for next nine months, OR

$_____$/bu.   Elevator bids if corn produced on the farm will be used for feed and grower is willing to wait nine months to receive those prices.

2. $_____$/cwt.   Buyers offer to buy market hogs in nine months.

3. Calculate your total cost of production coefficient (CP5) in Table 6–2.

4. Decide.

5. Act.

6. Forget about prices or costs: the profit locked in by the technique will be achieved.

TABLE 6–2: Budget Calculations to Determine Decision Costs of Production (CP's)

	Example Farm[1]	Producer's Farm[2]
	($ per cwt. hogs sold)	
Purchased Inputs		_____
Supplies/meds	$ 1.29	_____
Utilities	.66	_____
Boar Replacement	.43	_____
Misc.	.46	_____
CP1	$ 2.84	$_____
Supplement: premix	13.45	_____
meal	5.00	_____
CP2	$21.29	$_____
Corn		_____
5.43 bu./cwt. sold @ $2.00/bu.	$10.86	
CP3	$32.15	$_____
Labor ($20,788 total)	4.26	_____
CP4	$36.41	$_____
Interest		_____
Breeding Herd $50,000 investment × 12% = $6,000 total	1.37	_____
Buildings & Equipment $165,000 investment × 20% = $33,000 total	6.75	
CP5	$44.53	$_____

[1]One year cost data from Wilmington College Commercial Swine herd, 120 sows, farrowing house. Nursery, partially slotted finishing floor. 4,884 cwt. produced. 18.95 pigs/sow/year.

[2]Fill in the blanks on your farm from records or 1040F tax forms.

The decision when the buyer's offer is less than CP5 (i.e., price does not cover total costs): Individual producers have the freedom and the responsibility to decide when they will produce and when they will not produce. An economic principle says that farmers will continue to produce as long as the price received covers average variable cost (AVC). From the BUDGET CALCULATIONS, each producer can determine what his or her individual AVC is. Depending upon individual circumstances, it may be CP1, CP2, CP3, CP4, or CP5.

If price falls below CP2, all producers will quit production. At this price level, they would be using the buildings free, working free, and giving their corn away, but they would refuse to incur cash production costs and give them away.

Other operators will cease production when price falls to CP3. They are willing to use the buildings for free and work for free, but they refuse to sell their corn through hogs for less than the cash corn market will pay.

Some operators depend upon hired labor. Their quitting price is CP4, since they must pay production cost, plus feed cost, plus labor cost. They will use the buildings without a return because there is no other use for that fixed resource. When the buildings must be replaced, these operators will be forced to quit, rather than rebuild.

Finally, an investor who is considering setting up a pork-producing enterprise does so depending upon whether there is the expectation that price will exceed CP5 (total costs) for the lifetime of the investment. The investor should leave the money in the bank to earn interest rather than put it into hog buildings and earn zero return on it.

Farmers who do not develop and use a marketing system to guarantee a satisfactory profit level continue to produce during nonprofit periods, waiting for their neighbors to quit. Eventually, a sheriff's auction eliminates the least efficient producers. But there is a better way.

LEARN what to do.
Overcome FEAR of making a nine-month commitment.
And remember, GREED tempts many producers into losses.

Calculation of Satisfactory Sales Prices

CP1 $_____
CP2 $_____
CP3 $_____
CP4 $_____
CP5 $_____

Buyer's offer $_____/cwt. for delivery in nine months.
Buyer's offer of $_____ equals _____.
(Which CP?)

Are you willing to accept the returns being offered?

YES Then call your hog buyer and contract to sell _____ (number of) market hogs, nine months from today. Next, call your supplement supplier and contract a nine-month supply of supplements. Finally, call your grain supplier and contract purchase of your corn needs for the next nine months, or agree that the price of your corn as entered in the budget is what you are willing to take for it and that you are willing to wait nine months for the income.

NO. Select Alternative A or B.

A. I expect higher prices for hogs or lower prices for feed. I will take the risk of being naked in the market in hopes of a higher net return.

What do you expect hog price to be? $_____ Why?

What do you expect corn price to be? $_____ Why?

What do you expect supplement price to be? $_____ Why?

However, based upon my risk-bearing ability, I want to guarantee that percent my production cost of $_____/cwt. is covered. Make the above calls to contract hog sales and feed costs representing that percentage.

B. I will not operate if I cannot be assured that I will cover cost of production: $_____ (Which CP?)

DO NOT BREED SOWS

1. If price prospect is looking up, or costs are down, delay breeding until next cycle and recalculate CP's.

2. If price is looking weak, or costs are up, take sows to market.

Total Cost + 10 percent pure profit

Calculation: CP5 × 1.10 = $_____/cwt.

Buyer's offer for delivery in nine months $_____/cwt.

Are you satisfied with covering total costs plus a 10 percent profit?

YES. Then call your hog buyer and feed supplier, and contract hog sales and feed inputs.

NO. Select Alternative A or B

A. I expect higher prices and will take the risk of being naked in the market in hopes of higher price. However, based upon my risk-bearing ability, I want a guarantee that _____ percent of my production covers total cost plus 10 percent pure profit. Call and make contracts to cover that percentage of your output and feed costs.

B. I will not operate if I cannot cover total cost plus earn a 10 percent pure profit.

DO NOT BREED SOWS

Take care, however, that the green-eyed monster, GREED, does not tempt you into being forced to sell at a loss nine months down the road, when your market hogs weigh 220 lbs.

Total Cost + 30 percent pure profit.

Calculation: CP5 × 1.30 = $_____/cwt.

Buyer's offer for delivery in nine months $_____/cwt.

Are you satisfied with covering total costs plus 30 percent pure profit?

YES. Then call your hog buyer and feed supplier, and contract sales and feed inputs.

NO. Select Alternative A or B.

A. I expect higher prices and will take the risk of being naked in the market in hopes of higher price. However, based upon my risk-bearing ability, I want to guarantee that _____ percent of my production covers total cost plus 30 percent pure profit. Call and make contracts to cover that percentage of your output and feed costs.

B. Hog production is profitable at this price level; I plan to expand production _____ percent. I feel I must guarantee that _____ percent of this expanded production earns this profit level, however.

Remember, at times in the past, a 5 percent expansion in aggregate pork production has led to a 21 percent decline in hog prices.

Total Cost + 50 percent pure profit.

Calculation: CP5 × 1.50 = $_____/cwt.

Buyer's offer for delivery in nine months $_____/cwt.

Are you satisfied with covering total costs plus 50 percent pure profit?

YES. Then call your hog buyer and feed supplier and contract sales and feed inputs.

NO. Select Alternative A or B.

A. I expect higher prices and will take the risk of being naked in the market in hopes of higher price. However, based upon my risk-bearing ability, I want to guarantee that _____ percent of my production covers total cost plus 50 percent pure profit. Call and make contracts to cover that percentage of your output and feed costs.

B. Hog production is profitable at this level. I plan to expand production _____ percent. I feel I must guarantee that _____ percent of this expanded production earns this profit level, however.

Remember, in past hog cycles a 10 percent expansion in pork production has sometimes led to a 42 percent decline in hog prices. If GREED convinces you not to contract, what do you expect hog prices to be in nine months?

Refinements to the technique:

1. Decide how often to make calculations: annually, monthly, daily, before each breeding, before each feed purchase. Individual interest in marketing and desire to refine expected profits will dictate how often to make calculations.

2. Calculate grinding and storage costs of corn. If you use a fall harvest price on home-grown corn, add storage and interest costs to grinding costs to determine as-fed value of corn when it is converted to cash as hog sales.

3. Hedge cost of feed supplements and live hogs as separate activity in futures market rather than cash contracting.

The National Pork Producers Council developed a comparable analytical procedure called: *Step-by-Step Pricing*.[6] The spreadsheet program identifies various costs of production to permit managers to relate to market prices being offered. Please see Figure 6–5.

The cost of goods sold includes purchased livestock and all feed. Production expenses include supplies, medicine, and incidentals. The middle steps of the "climb to profits" adds the cost of interest, depreciation on fully owned assets, labor, and loan

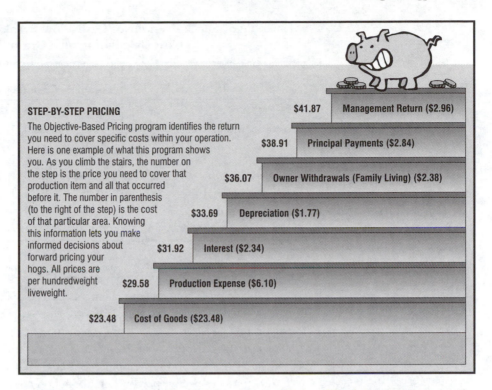

STEP-BY-STEP PRICING

The Objective-Based Pricing program identifies the return you need to cover specific costs within your operation. Here is one example of what this program shows you. As you climb the stairs, the number on the step is the price you need to cover that production item and all that occurred before it. The number in parenthesis (to the right of the step) is the cost of that particular area. Knowing this information lets you make informed decisions about forward pricing your hogs. All prices are per hundredweight liveweight.

$41.87 — Management Return ($2.96)
$38.91 — Principal Payments ($2.84)
$36.07 — Owner Withdrawals (Family Living) ($2.38)
$33.69 — Depreciation ($1.77)
$31.92 — Interest ($2.34)
$29.58 — Production Expense ($6.10)
$23.48 — Cost of Goods ($23.48)

FIGURE 6–5: Step-by-Step Hog Pricing. Source: National Pork Producers Council.

principal payments (in lieu of depreciation). The final step to profit is the payment to management. The total costs of production, in this case, are almost $42.00 per cwt.

Grain Sales. A similar method of cost analysis for grain production can be developed from university budget estimates.[7] Separate costs into appropriate variable and fixed cost categories, as illustrated in the hog model, to calculate the various CP's (costs of production) for your operation. Determine which costs must be covered or how much profit is required before hedges are placed.

Another method of cost production is to calculate cash-flow break-even requirements.[8] A positive cash flow requires that sufficient cash must flow into the business from the sale of grain to meet cash expenditures. The procedure is outlined in Table 6–3. The first expenditure item is cash production expense for seed, pesticides, fertilized, insurance, fuel, etc. Production expense for 200 acres of corn and 200 acres of soybeans is given as approximately $30,400 and $19,400, respectively. With yields of 125 bushels per acre for corn and 40 bushels per acre for soybeans, variable production costs are $1.22 for corn and $2.43 for soybeans. As long as price level exceeds these costs, production will be initiated; larger losses would result by leaving resources idle.

Prices must be higher to meet the other cash-flow requirements. One other cash expenditure is the annual payment on a machinery loan. The payment of $10,800 includes interest and a portion of the unpaid principal. It amounts to 22 cents per bushel of corn and 68 cents per bushel of soybeans. The third cash expenditure item includes all the costs of controlling land, such as cash rent payments, real-estate taxes, or the payments on land debt. In the example farm, it was assumed that 200 acres were owned free and clear, 100 acres were cash rented, and 100 acres were recently purchased with 100 percent financing. The land expenditure includes cash rent on

TABLE 6–3: Example of Cash-Flow Break-Even Method of Cost Calculation on a Grain Farm

Item	Total $	Corn (200 acres)		Soybeans (200 acres)	
		($per acre)	*($per bu.)*	*($per acre)*	*($per bu.)*
Direct Production Costs		152.00	1.22	97.00	2.43
Corn	30,400				
Soybeans	19,400				
Machinery payments	10,800	27.00	0.22	27.00	0.68
Land payments	50,100	125.25	1.00	125.25	3.13
Family living & taxes	17,150	42.88	0.34	42.88	1.07
CASH-FLOW REQUIREMENTS:	127,850	347.13	2.78	292.13	7.31

100 acres, principal and interest payments on 100 acres recently purchased with 100 percent financing. The total land payments exceed $50,000. Spread evenly over all 400 acres, that expenditure is $125.25 per acre, or $1.00 per bushel of corn and $3.13 for soybeans.

The last cash-flow expenditure incurred covers family living needs. The $17,150 also includes income and self-employment taxes. This labor income adds 34 cents and $1.07 to the respective per-bushel production costs. Total cash expenditures are $347 per acre of corn and $292 per acre of soybeans. The figures translate into cash-flow expenditures per bushel of $2.78 for corn and $7.31 for soybeans.

If producers can price their grain at these cost levels, cash inflow will exactly equal cash outflow. Higher prices would produce a surplus to increase cash reserve, retire debt, expand the operation, or increase living expenditures. If the grain were sold at prices below these expenditure levels, a cash deficit would occur. The deficit would have to be covered by drawing upon previously accumulated cash reserves, increasing indebtedness, decreasing family living expenses, or selling assets.

Highly leveraged farmers may well use this type of cash-flow analysis to calculate break-even price objectives. It gives them a reference point on the minimum price level that would keep them out of financial difficulty. Note that depreciation charges are not accounted for, and principal payments are included. Thus, the method does not reflect the total cost of production; it only accounts for cash-flow requirements.

These examples provide two methods out of many to calculate cost-of-production price decision criteria. Other methods may better serve individual producers in their particular situations. However, all cost-of-production models have serious logic problems in contributing to the development of grain and livestock pricing strategy.

Shortcomings of Cost-of-Production Pricing. The cliché "Know Your Costs of Production" has been quoted so often to farmers that some believe cost of production can be legitimate marketing strategy. It cannot! And those who have practiced it know that it can lead to bankruptcy instead of guaranteed profit.

First, **the market does not care what it costs a farmer to produce corn, hogs, or anything else!** If the market were sensitive to costs, why would it offer farmers $1.50 for corn (when costs were $2.78) or $18.00 for hogs (with $45.00 per hundredweight costs of production) in 1999? The market is oblivious to farmers' production costs; it has more important roles. In one situation, the market's role is to get prices high enough to ration a short supply among highest-return users to avoid a diminished supply from being prematurely exhausted. Conversely, its role is to drop prices low enough in the face of oversupply to encourage increased usage and discourage further production. Those are legitimate roles of the market; it does not fret about farmers' cost of production.

Secondly, basing a marketing plan on the cost of production will win practitioners a quick trip to the poorhouse because of the way it functions. Assume that the total production costs for corn are $2.75 per bushel (excluding land charge).* What happens if the market rises to $2.70 and then falls to $1.50? The cost of production strategy never signaled a sale. The producer slides along with the market, holding unpriced grain, into financial difficulty. On the other hand, suppose the market rallies to $2.80. "Time to price" rings the strategy. "You can't go broke taking a profit," even if the profit is only a nickel per bushel. Then the market rallies to $3.80 (as it did in 1983 and 1988 and once more, with vigor, in 1996). The price move creates lots of margin calls. And the salve of "selling at a profit" wears thin when hedgers calculate what might have been. Their problems are often compounded further in high-priced markets by pricing too early. Widespread summer drought characteristically creates bonanza high harvest-time prices; droughts also reduce yield. Few situations are worse than having 125-bushel yields hedged at $2.80, but the harvest produces only 70 bushels per acre and a price level of $3.80. The textbook hedge definitions of offsetting gains and losses do not match because of diminished production levels.

Over time, a cost-of-production marketing strategy is guaranteed to create financial problems. The mechanics of the strategy cause farmers to sell too late (or never) in down-trending markets or too early in up-trending markets. Cost-of-production hedgers miss out on bonanza high prices by selling too early and they go unpriced in low, declining markets. They suffer from $1.50 corn prices and eliminate $3.50 prices. Their long-term average price will typically be less than cash market speculation that captures the benefits of a "natural hedge" of prices moving higher in response to low production.

Scale-Up Selling, Scale-Down Buying

Producers can use scale-up selling in conjunction with other analysis to provide decision points for committing to pricing opportunities. The objective is to capture a higher than average price. Scale-up selling involves setting a price objective at which the first sale will be made and then committing larger and larger amounts at selected price increments. Scale-down buying involves making larger and larger purchase commitments at successively lower prices. Both price increment and percentage of production to hedge at each increment must be selected along with the initial entry price.

Individuals may use cost-of-production estimates, chart objectives, or historical prices to determine their first sale price. Pricing increments should be based on the price volatility of each commodity. They may be 30 cents in corn and wheat, 40 cents in soybeans, and 2 cents in cotton. The percent of production at each sale point dictates how many sales decisions will be made. One producer choosing large price increments might price larger percentages of production each time. Smaller price

---

*The calculation is based on variable production costs of $150.00 (labor–$50.00; machinery–$100.00) and overhead ($30.00 per acre). Total: $330.00 with a 120-bushel yield.

increments would fit with smaller percentages each time. The procedure is outlined in Table 6–4 with two combinations of increments and production percentages.

Two problems plague scale-up selling (or scale-down buying). First, market price may never reach the first sale point, and if it does, it may not follow through to signal sales of the remainder of the production. The method guarantees the first sale is wrong if the price does happen to rise further. There will have to be liquidity to carry hedges to the top of the market and sufficient fortitude to suffer the agony of having sold a portion too early. Second, the original selling point may be set too low so that all sales will have been made long before a strong price move finally exhausts itself. Should bonanza high prices occur, the producer will have no remaining production to commit at the very high levels. An average price of $3.40 looks good unless the market reaches $4.00. At that point regret sets in.

A variation to scale-up selling is to use a scale-down selling procedure. The first sale is made after a market has peaked and has made a predetermined decline. Larger commitments are made on subsequent declines. Should the market rally before passing through all the decision points, the same procedure is used on the next market decline after a second peak. This method has the advantage of causing the seller to wait until a major market thrust has run its course and reduced the chance of selling too soon. It suffers from the dual problems of being able to detect market tops and farmer psychology against selling in a down market. The agony of knowing they missed the top makes it extremely difficult for many producers to make pricing decisions at lower levels. Hope of a rebound and capturing market tops springs eternal in many producers.

TABLE 6–4: Illustration of Scale-Up Hedging at Two Percentage Rates and Price Increments

Percent of Production		Pricing Points		Results
		($/bu.)		
Option I				
5 initial	×	3.00	=	0.15
10	×	3.15	=	0.32
20	×	3.30	=	0.66
30	×	3.45	=	1.03
35 last	×	3.60	=	1.26
100			Average Price	$3.42
Option II				
20 initial	×	3.00	=	0.60
30	×	3.30	=	0.99
50 last	×	3.60	=	1.80
100			Average Price	$3.39

Seasonal Pricing

Seasonal pricing has formed the foundation of many market pricing plans. An analysis of cash price indicates that there are times of the year when higher and lower prices can be expected. Logic would suggest that the end months of crop years should have higher prices than harvest time due to the reduced availability of supply. Furthermore, on a particular day, grain futures prices are typically higher in distant contracts than in nearby months.

Seasonal price variations within a crop or marketing year are due to seasonality of production, demand factors, or marketing patterns. Fall holiday demand for turkeys is a classic example of demand-induced seasonal price effects. Reduced sow farrowings during winter months lead to reduced supplies in late summer and produce supply-induced seasonal price effects in hog prices. The seasonal change in livestock prices is due largely to traditional production practices and seasonal eating habits. Seasonality of crop prices originates from the dual effects of short harvest periods and the costs of storage over longer consumption periods. For most crops, prices are lowest at harvest, when the supply is most abundant. Charts of seasonal tendencies of cash prices are presented in Figure 6–6.

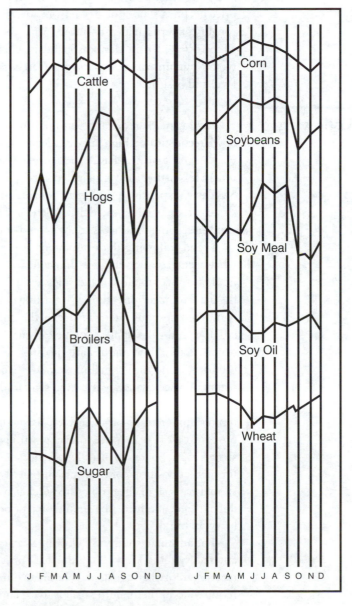

FIGURE 6–6: Cash Price Seasonal Tendencies.

Producers who use a seasonal pricing strategy find that it often does not work. They store grain at harvest, waiting for a summer rally, only to sell grain at a lower price in July than existed at harvest, while incurring storage costs for the period. Livestock producers who put animals on a feeding floor on a schedule to finish them at historically high seasonal periods find that the market forgot to read the history book when all too often the anticipated seasonal price rise fails to materialize.

An analysis of historical cash grain prices leads one to conclude that there was a consistent improvement in the late-season price compared to harvest price. This relationship has been dubbed "the summer rally" in grains. During the time period of the futures price level was fairly constant due to government price support and storage programs. The historical rally in price was basis level gain of cash price relative to futures price level. To emphasize the point, the 13 years of recent futures prices depicted in Figure 6–7 discount the existence of either a summer corn rally or spring hog price rally. While some years did show such rallies, about an equal number had price declines. The conclusion from analysis of futures prices is that there are no seasonality patterns in level of futures price upon which to base a pricing plan. The effect all too often identified as seasonality is actually basis level gain. Seasonality pricing is typically an unworkable marketing plan because it mixes the two pricing effects, price level and basis level.

Trend-Following Models

Picking market tops and bottoms can be an exercise in futility. Preconceived ideas of what constitute a high price or a low price can be harmful to a successful marketing plan. As most farmers have experienced, prices can, and often do, move in both directions a lot further than anticipated. Cheap prices have tendencies to become cheaper. Similarly, expensive prices may become more expensive. To incorporate a pricing strategy more sophisticated than calendar scheduling and not as frustrating as picking extremes, some producers adopt a trend-following technique.

Trend following does not attempt to predict price action; it merely interprets what occurred in trading pits and contains rules for reacting to signals generated. As markets move up, down, and sideways, the purpose of trend following is to analyze the current major direction. Is it turning down from up, or simply settling back for a continued upswing later? To decide the question, various methods of trend analysis are used and rules applied for a clear-cut decision about when changes in direction occur. Popular techniques are moving averages, price channels, linear regression, oscillators, and directional movement. Each one has its criterion to determine the extent of price change required to signal a change in direction.

The change in trend dictates market action. A change from up trend to down trend signals farmer-producers to place hedges on their production and signals input buyers to avoid or remove market positions. A change from down trend to up trend signals input buyers to place long hedges on items they plan to buy and producers to lift market positions or delay placing them. As a pricing strategy, a farmer-producer should have all of the crop or livestock production hedged or sold during a primary down trend, and he or she should own or be long on all the production during a primary up trend.

The advantage of trend following as a pricing decision technique is that trades are entered systematically. Trend signals provide an automatic method for dictating reaction to market signals and eliminate human judgement and indecision. All that is required to implement the plan is to execute the appropriate position whenever signals are generated.

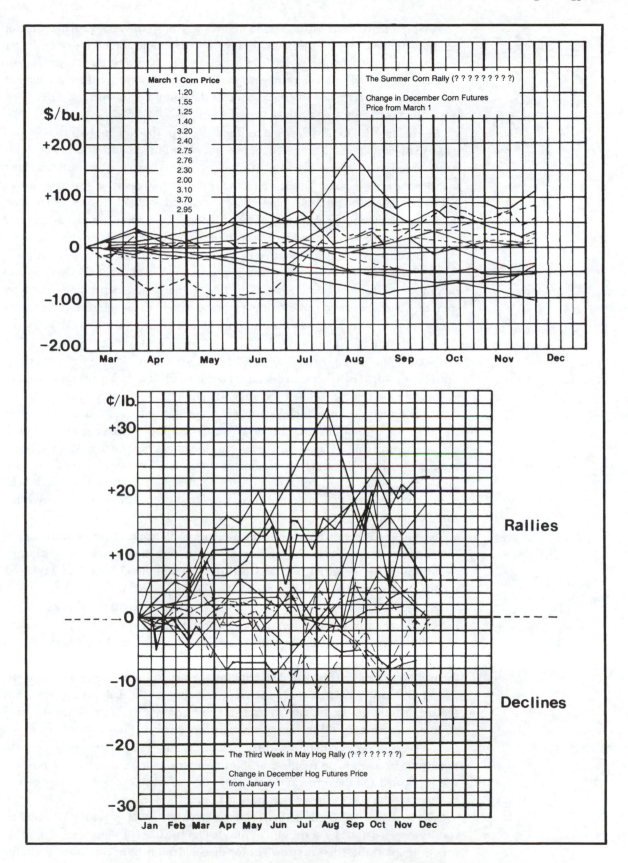

FIGURE 6–7: Intra-year Changes in Corn and Hog Futures Prices.

An important characteristic of trend following is that it will not signal initiation of positions at market tops nor market bottoms. By definition, a trend is detected after the extremes have occurred; the change in direction is confirmed by the signals generated. With a trend-following strategy, producers will never price their production at the top market price nor buy their inputs at the bottom of the market. They must be satisfied with capturing a part of the major moves.

A problem with trend-following techniques is that they often produce false signals. An up trend might be signaled, but within a short time a down trend signal follows. An input buyer would have set buy hedges only to be signaled a short time later that the market was going lower. The same situation arises as falsely signaled down trends fail to follow through. These false signals are called *whipsaws*. They arise because the criterion for detection of trend changes is set too close to market price action and corrections within a trend touch the signal criterion. Then the true trend again exerts itself. The action dictated by a whipsaw is to lift the original hedge position and replace it when another change in trend is detected. The canceled position may, however, have generated a loss in the futures transaction, depending upon how soon the switch occurred and extent of price movement in the interim. If the whipsaw occurs within a short period without much price change, the loss can equal the width of the criterion boundary.

No trend-following system can avoid false signals. The success of these systems must be measured by relative frequency of correct calls and the profits they produce as compared to number of incorrect calls and amount of losses. An analogy can be drawn to flipping a coin, which has a 50/50 probability ratio of producing heads or tails. If heads yields a 20-cent gain and tails a 10-cent loss, a farmer would spend all day tossing the coin because the overall result would be profitable. With a 2-to-1 reward-to-risk ratio, a participant would need be correct only one-third of the time to break even. Thus, for an evaluation of trend following as a market decision technique, the bottom line is the number of correct calls times their profitability compared to the incorrect calls times their losses.

We analyzed three types of moving averages to illustrate the procedure and to search for a workable one for farmers to use in executing a marketing strategy. Results of two of the analyses are attached at the end of this chapter, and Table 6–5 reports the results of a popular three-way (4-9-18-day) moving average.

A simple, one-variable moving average consists of M days of prices being included in the average, where M can be any number of days. For example, to calculate a 10-day moving average, we summed the closing prices of the previous 10 days and divided by 10. On subsequent days, the oldest closing price was subtracted and the newest closing price added. The criterion for analysis of the trend was the comparison of today's price relative to the average. As long as the daily closing price was greater than the 10-day moving average, an up trend was in effect. When the daily price was less than the 10-day average, the market was in the down trend.

To execute a marketing plan with a moving average pricing strategy, long positions were initiated in up trends, and short positions in down trends. One position was liquidated and the other initiated upon signal of trend change. A long or short position was held in the nearby contract at all times.

We analyzed farm commodity prices over two 10-year periods. The trading results reported in tables at the end of the chapter included all simple moving averages from 2 days to 28 days in span. The top diagonal numbers of each table report the result of each respective moving average analyzed.

Few positive results occurred, and many large losses did occur. In soybeans, for example, the best simple average for the first 10-year period was a 22-day average. It produced a total profit of $65,145. But in the second 5-year period, the 22-day average yielded a $15,160 loss. Only the 20-day average yielded positive results for the second 5-year period.

We concluded from these results that a daily price versus a simple moving average was unsuitable for detecting market trends in executing a marketing strategy.

The second analysis consisted of the crossover of daily price and two moving averages, defined as follows:

1. The average of M days' closes and N days' closes were calculated, where the number of days in the M average (the faster-moving average) was less than the number of days in N (the slower-moving average). For example, if M = 10 and N = 20, we analyzed a 10-20-day combination.

2. If the short-term (M-day) average was greater than the long-term (N-day) average, and the daily market closed above the M-day average, a long position was established on the opening the next day.

3. If the short-term (M-day) average was below the long-term (N-day) average, and the daily market closed below the M-day average, then a short position was established on the open the next day.

4. If the high and low of the next day were equal, it was assumed the market was locked the limit and no trade was made. The trade was made the first day the market was not locked the limit.

5. The model was always in the market in selected nearby contracts with either a long or short position.

6. The model was tested on November or December, March, and July grains and cotton contracts; February, July, and December hog contracts; and February, June, October, and December cattle contracts.

The tables at the end of the chapter report the results of trading one contract in farm commodities over two 10-year periods. They show total results of all combinations of the faster-moving M-day average (from 2 to 28 days) against the slower-moving N-day average (From 2 to 60 days). The intersecting cell of the respective row and column reports the trading results of each combination. At the bottom of each table, the return from the most profitable combination is reported. For example, trading corn in the first 10-year period, a 12-60-day combination yielded the highest return. It totaled $24,065 per contract ($4.81 per bushel) over the period (almost 44 cents per bushel per year). Over the second 5-year period, however, the 24-30-day combination was best for corn, with total returns of $8,240 per contract. In the second 5-year period, the 12-60-day combination yielded only $970 per contract.

We concluded that trading profits from a two-crossover moving average model were very elusive. Furthermore, the best combination in one period did not guarantee that it would be the best over the next period. Some other combination of moving averages worked better in every second period tested.

Optimization of a pricing model in one test period does not guarantee that it will perform satisfactorily next year, or over the next several years. The difference in results between periods in our test should alert farmers who are offered marketing schemes to ask how consistently the scheme performed over two or three different marketing periods. A marketing strategy is of no value if it works only one year in five. And fantastic average results from a 10-year test period may turn to disastrous losses in the eleventh year of actual market trading. The farmer who expects to follow a marketing plan should look at both the average results of the test period AND the extremes in profits and losses for the period.

The results of the third moving average model are presented in Table 6–5. This one uses the popular 4-9-18-day averages. When the 4- and 9-day averages exceed the 18-day average, an up trend is in effect; when the 4 and 9 averages are less than the 18, a down trend is in effect. The crossover is signaled when both the 4 and 9 averages are either greater or less than the 18. In every commodity and in both periods tested, returns from the 4-9-18 scheme were less than the best of the 2-day moving averages. A simple 4-9-18-day average thus proved to be unsatisfactory as well. We leave the problem to researchers to find a more market-sensitive method for calculating moving averages for farmers to use in a pricing strategy.

TABLE 6–5: Trading Results of a Three-Way (4-9-18-Day) Moving Average, Major Farm Commodities

Commodity	Time Periods	
	First ten-years	Second five-years
	(total dollar returns over each ten-year period)	
Corn	$ 6,180	$ −770
Soybeans	68,785	14,420
Wheat	25,085	−4,580
Cotton	39,240	12,875
Soymeal	28,580	−1,840
Live Cattle	−6,912	−8,776
Hogs	2,237	15,156

Trend following is a systematic, technical marketing plan that rules out human judgement in making decisions to let profits run while cutting losses short. As long as the number of profitable positions and their amount of profit exceeds the number of loss positions and their amount of loss, the overall result will be profitable. The problem is to find a consistently profitable moving average combination.

Keeping Score

Every farmer likes to brag about making a $12.00 soybean sale, a $60.00 hog sale, or a 95-cent cotton sale. Each year every coffee shop has its high-selling hero. But because prices realized at different times of the year do not have the same value, there must be some procedure to calculate price equivalency, irrespective of when sales are made during the year. A procedure for keeping score is suggested as follows. Use late harvest-time price as the base price. Since wheat harvest in Oklahoma, South Dakota, and Montana is completed on different dates, the base month will be different depending on location. But that makes no difference in calculating the results for a particular farm. The objective is to have a specific initial pricing date year to year. The second step is to add profits from forward sale commitments to the harvest price. If a contract price for fall corn delivery was set at $3.50 early in the season and harvest price was $2.85, the forward pricing profit was 65 cents per bushel on that portion of production so committed, say one-third. Next, add storage returns to the harvest base price as sales are made during the year, but deduct costs of storage up to time of sale. Assume another third was sold in January at $3.00, and the final third was sold in June at $3.40. They represent additions of 15 cents and 55 cents to the base harvest price. However, storage costs must be deducted from the storage sales. Using the carrying cost of 7 cents per bushel per month, the January gain must be reduced by 14 cents and the June sale by 56 cents on the portion of sales made on each date. The summary calculation follows:

Price per Bushel

	$2.85	Base Nov. 1 Cash Price
+	.22	Pre-harvest Hedge (65 cents × 33% of production)
+	.05	January gain over Nov. 1 base (15 cents × 33%)
+	.18	June gain over Nov. 1 base (55 cents × 33%)
−	.05	Storage Cost to January (14 cents × 33%)
−	.18	Storage Cost to June (56 cents × 33%)
	$3.07	Effective Price

This method produces an effective net price of $3.07 per bushel. This compares to an average of $3.63 from the three sales when carrying charges are not accounted for.

This method can be used to evaluate the bottom line of various marketing strategies suggested in this discussion or offered by various advisors. It was used in the analysis of marketing strategies cited at the end of this chapter and in the analysis of 730-Day Marketing Plan explained in Chapter 10.

Profit Rather Than Price

Specialized production in farming means that many producers buy commodities as production inputs and sell finished products. Poultry, hog, and cattle producers buy corn and soybean meal as feed ingredients. Feedlot operators also buy feeder livestock. To the extent production inputs are traded on commodity exchanges, these producers can hedge their input purchases as well as sell contracts to cover sales of the product.

A viable strategy is to hedge in major production cost items while simultaneously selling against production. This technique is effectively a futures spread that focuses on relative prices rather than price level. Product prices do not have to be high to lock in profits if input prices are low. The objective is to lock in a profit spread rather than being concerned with price level per se. Since feed costs are 60 to 70 percent of costs of hog production, the pork producer should focus on the hog-corn price ratio as the decision criterion. In cattle feeding, the *feeder spread* is relevant.

To produce finished cattle, two major inputs are required, feeder animals and feed, primarily corn. Other cost items are supplements, interest, veterinary expenses, death loss, annual costs for buildings and equipment, labor, management, and a profit margin. Most costs are stable or fixed; feeders and feed are the most variable. The objective is to maximize the difference between the two major variable costs and price of finished cattle.

Balancing the number of contracts among the three items is essential for an accurate spread. Feeder cattle contracts, trade on the Chicago Mercantile Exchange, in lots of 50,000 pounds of feeder cattle per contract. Using an average weight of 700 pounds, each contract represents 72 head. Live cattle contracts, also traded on the CME, represent 40,000 pounds of live weight, or about 36 head of 1,100-pound cattle. Corn is traded on the Chicago Board of Trade in 5,000-bushel contracts. Producing 400 pounds of gain with a conversation rate of 7.5 pounds of corn per pound of gain requires 53.6 bushels of corn per animal. Putting these figures together in the approximately correct proportions produces a trading portfolio of one feeder cattle and one corn contract (1,100 lbs. of surplus feed) on the input side and two live cattle contracts on the output side. This combination represents 72 head. Smaller-size contracts on the Mid-America Exchange allow other combinations or four 1,000-bushel corn contracts would eliminate most of the feed surplus in one 5,000 bushel contract. To calculate the proportions for any size volume of operation, first determine the number of head of finished cattle to be sold and their weight (72 head at 1,100 lbs. = 80,000). Divide by the size of the contract to determine the number of live cattle contracts to sell (80,000 divided by 40,000 = 2 contracts). Next decide the weight of feeders to be bought and the number of contracts to cover them (72 head times 700 lbs. = 50,400 divided by 50,000 pounds per contract = 1 contract). To calculate how much corn is required for the gain on the animals, subtract purchase weight from selling weight and multiply by the number of pounds of corn per pound of gain and number of animals (1,100 lbs. – 700 lbs. = 400 lbs. of gain per animal; 400 lbs. of gain per animal times 7.5 pounds of corn per pound of gain = 3,000 lbs. of corn per head. Multiply the 3,000 lbs. of corn per head times 72 head equals 216,000 pounds of corn total). Finally, divide the corn weight by 56 pounds per bushel to calculate the number of bushels and number of corn contracts required (216,000 lbs. of corn divided by 56 lbs. per bushel = 3,857 bushels of corn, or about one 5,000 bushel CBOT contract or four 1,000 bushel Mid-America contracts).

Using closing prices (rounded) on November 1, 1999 for the respective contracts permits calculation of the profitability of the spread. December corn futures were $2.00, November feeders were $81.00 per cwt., and April live cattle (five months later) are $70.00 per cwt., the dollar differential is:

Fats	70 cents × 40,000 × 2	**sell**	=	$56,000
Less Feeders	81 cents × 50,000 × 1	**buy**	=	–40,500
Less Corn	$2.00/bu × 4,000 × 1	**buy**	=	– 8,000
		Spread Margin		7,500
Margin per head (72 head)	For other costs and profit			$104.17

If this margin is sufficient to cover other production costs, the feed lot operator can set the hedge to lock in this spread. If it is not satisfactory, he or she can delay setting the hedges and/or delay initiating the enterprise until profitable opportunities exist.

When market conditions do not allow profitable feeder spreads to be established, placing opposite futures positions in each respective contract permits a feeder to speculate on the possibility of improved price relationships without placing cattle in the lot. It is called a *reverse feeder spread.* The opposite action is to sell feeder and corn contracts and buy live cattle contracts. If a feeder fills his or her lots when the spread profit is abnormally low, that farmer-speculator is betting that the spread will widen as the cattle mature. It may not do so, but setting a reverse feeder spread in futures is no more speculative than placing cattle in the lot and hoping relative prices will improve.

Setting feeder spreads when it is profitable to do so eliminates the effects of price level change and reduces feeder and corn buying and fat cattle selling to basis considerations. The feeder who can buy corn and feeders on basis below the futures contract prices and sell fats at a premium to futures will improve on the results of the spread. But cash buying and selling at poor basis can eliminate profits hedged into the spread. In effect, futures contracting eliminates price-level speculation while allowing basis level speculation.

Government Programs Affect Pricing

Government programs exert an influence on market pricing. The U.S. Food and Agricultural Act was instituted in 1977 and was revised periodically thereafter until 1996. This legislation utilized reserves, storage, and non-recourse loans. The program encouraged farmers to put grain in the farmer-owned reserve when stocks were abundant and market prices were low. When market prices became relatively high, disincentives were applied to encourage farmers to remove the grain from storage and market it to place a damper on prices. If total grain stocks became too large, production was reduced through cropland set-aside and diversion programs. If grain prices were low, producers received deficiency payments to maintain income levels. The major justifications for these programs were to protect farmers from severe price declines and to assure consumers of adequate food supplies without sharp increases in food costs.

A major change in agricultural policy attitudes occurred in the early 1990's. Taxpayers, unhappy about national budget deficits, demanded budget cuts. Farmers were dissatisfied with government controls restricting agricultural production. Disgruntled citizens urged the Congress to enact major changes in public policy. The Federal Agricultural Improvement and Reform Act of 1996, became known as the Freedom to Farm Act. It removed government restrictions on what and how much farmers could plant. Pundits who understood the surplus production capacity of American grain farmers dubbed the 1996 Act as: "The Freedom To Go Broke Act." These naysayers anticipated surplus production, burgeoning carry-over stocks, and low prices

Despite weather-related mediocre yields in 1997 and 1998, corn carry-over stocks almost quadrupled (486 million to 1,600 million bushels) in that period. An anemic

Japanese economy only exacerbated the dilemma. Between 1995 and 2000, corn prices declined from $5.25 to $1.60 per bushel. Soybean stocks grew from 132 million to 738 million bushels. Soybean prices plunged from $8.85 to $3.95 per bushel. Surpluses grew as wheat stocks more than doubled (376 million to 885 million bushels) and prices plummeted from $6.35 to $1.92 per bushel.

The 1996 Act did contain safety-net provisions to protect farm income. Fixed, but time dependant, declining transition payments were added to non-recourse loans and crop loan deficiency payments on grain, oilseeds, cotton, and rice for seven years from 1996 to 2002. The transition payments were a high percentage of the subsidy payment farmers had previously received in the 1970's and 80's.

Non-recourse loans allowed farmers to store the eligible crops and receive a cash flow without being forced to make commitments at "fire-sale" prices. As prices improved, growers could sell their crops and repay the loans. Farmers also had the option of delivering grain to the Commodity Credit Corporation as full-payment for the loan at maturity. The non-recourse provision meant that if the grain was worth less than the loan, the deficiency would be forgiven. The loan rate effectively set a local price floor irrespective of global conditions. The 1998 loan deficiency provision (LDP) allowed commodities to flow into the market at whatever price they would command, rather than holding surpluses in storage. If market-clearing prices were lower than the loan rate, the LDP payment equaled the deficiency, thereby effectively stabilizing the price floor at the loan rate. These provisions allowed grains to continue to flow into the market irrespective of market prices.

No doubt, the new millennium farm bill will significantly rearrange the playing field and the rules of the 1996 "game." Due to continually changing government programs, it behooves farmers to stay up-to-date with these policies and adjust their production and marketing plans accordingly.

Results of Pricing Strategies

An analysis of grain marketing/pricing strategies by Kentucky researchers provides some factual results of various marketing strategies for corn and soybeans.[9] Nine grain marketing strategies were defined with respect to pricing mechanisms that triggered sales, timing of sales, quantity sold, and the delivery period. The strategies were separated into two categories: (1) five, which required no market information, only timing; and (2) four, which required information. The producer who required no information delivered and cash priced production in a specific and predetermined harvest or post-harvest month; he or she made no attempt to forward price. Strategies requiring market information instituted sales based upon price action in the futures markets. In each instance where applicable, the variable costs of marketing and storage were net price in terms of a harvest season equivalent. Definitions of the various strategy alternatives:

Those Requiring No Market Information:

1. Entire crop priced and delivered at harvest. No carrying or storage costs were incurred with this strategy.

2. At harvest, grain was stored and equal amounts (1/12) of production were priced and delivered for cash each month. With this strategy, a producer received the season's average price, net of carrying charges.

3. Grain stored at harvest with half of total production delivered and cash priced in historically high price months, August for corn and June for soybeans. The remaining half of production was sold equally in each of the remaining 11 months beginning at harvest.

4. Grain stored at harvest and one-third of total production priced and delivered for cash in June, July, and August. With this strategy, the producer attempted to capture the season's high gross price by making sales in three historically high cash price months.

5. Grain stored at harvest and delivered for cash in December after the harvest glut was safe in storage and cash basis had improved.

Those Requiring Market Information:

1. Scale-up selling. Grain was stored at harvest, and 25 percent of the production was sold each time the average monthly price rose a selected percentage above the projected season's average price. Corn sales were made at 5, 10, 15, and 20 percent above the projected average; soybean sales at 10, 20, 30, and 40 percent levels. If sales had not been triggered by the targeted prices by September following harvest, the remaining inventory was sold at September's average price.

2. Trend analysis. The producer was assumed to sell the commodity and/or hedge when the closing price for two consecutive days fell below an up trend line as defined by chart analysis. Grain was stored at harvest, and one-half of the remaining inventory was sold for cash each time a sell signal (as defined above) occurred in the futures market.

3. Pre-harvest hedging of fall sales. The producer hedged one-third of expected production in March, July, and September for harvest delivery by using December corn and November soybean futures contracts. The producer bought back the contracts in October and delivered the total production for cash at harvest. Timing of the sales decisions during the selected three months depended upon the trend analysis and use of sales signals as defined in option 7. If the producer did not price during any one specified month due to a lack of sell signals, the cumulative quantity was priced in the next scheduled pricing month that contained selling signals. Any portion not hedged by September was sold for cash at harvest.

4. Pre-harvest hedging for sales from storage. The producer hedged one-third of expected production in April, July, and September according to the same signals used in option 7 for delivery in February. The March contracts were used to establish the hedges.

Average results for a 10-year period are presented in Table 6–6. Few strategies produced better results than selling at harvest. Alternatives 5, 7, and 9 produced substantially better results in corn. In soybeans, strategies 4, 7, and 9 were significantly better than harvest sales.

TABLE 6–6: Average Net Prices Produced by Nine Alternative Corn and Soybean Marketing/Pricing Strategies, Kentucky, 10 Year Average

Marketing/Pricing Strategy	Average Net Price[1]	
	Corn	Soybeans
	($/bu.)	($/bu.)
Calendar sales:		
1. Delivering and pricing crop at harvest.	2.37	5.80
2. Selling stored grain in equal amounts each month.	2.36	5.80
3. Half of crop sold in historically high-priced months (Aug. corn, July soybeans); other half sold 1/11 per month.	2.37	5.88
4. Thirds of stored crop sold in June, July, and August.	2.36	5.92
5. Stored at harvest and sold for cash in December.	2.45	5.87
Scale-up selling:		
6. Sales when monthly price exceeded projected season's average price by defined percentages.	2.14	5.58
Trend analysis:		
7. Grain stored at harvest and ½ of remaining volume sold on trend violation signals.	2.47	5.95
8. Hedging ⅓ of production in March, July, and September on trend violation signals for fall delivery.	2.38	5.86
9. Hedging ⅓ of production in April, July, and September on trend violation signals for Feb. delivery.	2.57	6.18

Source: Riggins, Skees, and Reed 61.

[1] Net of carrying cost

The analysis emphasizes the importance of storage and carrying costs in affecting net return. Selling one-twelfth each month, hitting both high- and low-priced months, produced no better results than selling at harvest after deducting carry charges on stored grain. Nor did selling in historically high-priced months improve on harvest sales. It appears that early season basis gain may be more important in improving cash sale results than trying to pick high prices.

Using the Options Market

Using the options market for price protection appears to offer an attractive alternative to use of the futures market. Farmers dread the possibility of margin calls, and they are reluctant to lock in a price when prices could go higher. With options, a minimum price level can be established for a premium payment, without eliminating the potential of higher prices. Options supposedly eliminate the threat of margin calls.

The hog market provides an example of use of the options market. Various prices bid into the market on November 1, 1999 are summarized in Table 6–7. A $45.00

December PUT gives the hog producer the right, but not the obligation, to market his or her hogs at $45.00 (less the basis and option premium) at any time prior to the option expiration date in mid-November. The option "insurance" premium for the right to sell $45.00 hogs was $1.73 per cwt. The $48.00 option was $3.35, while the $43.00 option was "only" $1.00. Respective net prices "insured" by the options were $44.65, $43.27, and $42.00, less local basis on the day of delivery and interest on the premium.

TABLE 6–7: Hog Futures and PUT Option Prices, November 1, 1999

December Futures (400 cwt.)	Options Strike Price	Terminology	PUT Options Premium	Price Level Insured by Options[1]
	($/cwt.)			
	$48.00	In the Money	$3.35	$44.65
$45.00	45.00	At the money	1.73	43.27
	43.00	Out of the money	1.00	42.00

[1] Less local basis on day of delivery. The producer faces the same basis risk as in the futures market.

A farmer who purchased a $45.00 December PUT paid $1.73 per cwt. for the right to sell hogs in a $45.00 market at any time prior to expiration of the option in mid-November. If hog prices rally to $50.00 at expiration, the right to sell hogs for $45.00 is worthless. The farmer's net price is $48.27 ($50.00 cash less $1.73 option premium, less the basis). On the other hand, if prices fall to $43.00 by mid-November, the right to sell $45.00 hogs makes the option worth approximately $2.00 per cwt.* The farmer's net price would be about $43.27 ($43.00 cash market, plus $2.00 option value, less $1.73 premium, less the basis).

The $45.00 December PUT was "At the money" since the December futures price also traded at the same price on November 1. The entire $1.73 premium was thus attributed to the "time value" inherent in the remaining two weeks of life before the option expired.

The $48.00 December PUT is "In the money," since the price level locked in exceeds the $45.00 futures price. The right to sell $48.00 hogs in a $45.00 market is worth $3.35; $3.00 of the premium constitutes "intrinsic value" (the difference between $48.00 and $45.00); the remaining $0.35 ($3.35 total value minus $3.00 "intrinsic value") is attributed to "time value" of the option's remaining two weeks of life. The $43.00 December PUT is "Out of the money" since the option strike price is below the futures close. The $1.00 premium is all "time value."

Were the $3.35, $1.73, and $1.00 respective premiums high or low? The answer depends upon how much value the producer places on the $44.65, $43.27, and $42.00

*For the option to be worth $10.00 in a market price move of $10.00, the delta must be 1.0 that is, a 1:1 change in option value versus change in market price. A constant 1.0 delta is not a valid assumption. Deltas of 0.50, a 50 cent move in the options for a $1.00 move in the futures, is more realistic. Thus, it usually takes two options to get the same price protection as one futures contract.

minimum price levels guaranteed by the respective options. The producer eliminates the possibility of lower prices but does not eliminate the opportunity of higher prices by purchasing options. Cost and value are relative.

Some hog producers concluded that the $3.35 per cwt. premium on $45.00 options was an exorbitant price to pay for two weeks of time. Others felt that the $1.00 premium on $43.00 options was a better buy because it was cheap. We are confident in these assertions since the number of PUT options outstanding on November 1, 1999 was 11,552; representing an inconsequential 210,000 hogs. A $43.00 option was cheap, but it was not inexpensive. In addition to the $1.00 premium, the $43.00 option cost $2.00 per cwt. of value foregone at the onset because the strike price was $2.00 per cwt. under the Chicago futures market. The purchaser was guaranteed a price level of only $42.00 per cwt. ($43.00 strike price less $1.00 premium). In order to buy low-cost disaster insurance, the purchaser gave up $3.00 per cwt. from the beginning ($2.00 deductible and $1.00 premium). Out-of-the-money options are like insurance with a deductible clause. The $43.00 PUT had a $800.00 "deductible clause" (400 cwt. times $2.00 per cwt.). Therefore it had a low-cost premium of "only" $400.00 (400 cwt. times $1.00 per cwt.). By comparison, the $45.00 At-the-money PUT covered all the downside price risk (no deductible clause) at a cost of $692.00 (400 cwt. times $1.73). If the $43.00 option were bought and hog prices fell to $43.00, the protected price is below $42.00 per cwt. The option buyer regretted not having purchased the $48.00 option that netted $44.65 in a $43.00 hog market. Hindsight is always 20/20.

Options are often touted as having a distinct advantage over the futures market because there are no margin calls. That claim is largely erroneous when options must be carried over time in order to guarantee a long-term price level. A comparison of futures and options supports the point. For example, if a December hog futures contract is sold at $45.00, the 40,000-lb. hog contract requires an initial margin deposit of $600.00. If the price level goes up, each $1.00 per cwt. increase requires an additional $400.00 to maintain the hedge position. By comparison, an option with a $1.73 per cwt. premium costs $692.00 initially. (Actually, the premium cost is $1,384.00 if the delta is 0.50 and two options are purchased for full coverage.) Futures prices require at least a $1.73 move against the position before margin calls exceed the initial purchase cost of the comparable option. Furthermore, if hog prices are constant or increase by the mid-November option expiration date, the $45.00 option expires worthless. The $692.00 ($1384.00) "insurance" premium would not be collected upon and would become an out-of-pocket expense. Thus, in November, the producer is faced with another decision. In order to again lock in a price and guard against lower prices, an At-the-money $45.00 February PUT option is purchased for, say, $3.00 per cwt. The premium is now higher because the February option has a two-month span. Should prices remain constant, or higher, until mid-January, the February option would also expire worthless. In mid-January, the decision of whether or not to buy additional price protection must be made once again. A $45.00 April At-the-money PUT might cost $3.00. In mid-March, the decision to continue extending price protection might be made once again, if the anticipated price break has not yet occurred. Use of a June $45.00 At-the-money option might cost $3.00.

If by mid-May hog price falls to $35.00, the producer will have achieved the long sought-after objective—price protection—but at what cost? The four options for eight months of price protection on one month's worth of production cost $10.73 per cwt. ($1.73 December + $3.00 February + $3.00 April + $3.00 June). Hogs marketed in June realize a price level of $34.27 ($45.00 June option strike price less $10.73 cost of options). That price is $0.73 per cwt. less than the cash price. During the time period, the options buyer enjoyed an average cash price of $43.66 for all hogs marketed during the eight months of options protection ($45.00/cwt. × 8 months − $10.73 /8) = $43.66 (less interest on the option money). In addition, the options hedger enjoyed peace of mind knowing that he or she had locked in minimum prices in case of price disaster. Neighbors who were cash market speculators enjoyed the same average prices on cash hog sales plus the $10.73 less cost for hogs marketed in June. They did not have the peace of mind associated with knowing what minimum price to expect when the price drop materialized. The annual insurance premium for peace of mind cost $10.73 per cwt. in this example. Is it worth it? Ask the options buyer.

The point is that the assertion that there are no margin calls with options is not necessarily true. If the market does not "cooperate" by falling significantly lower shortly after the first option is purchased, additional options (equivalent to additional "margins") must be purchased to maintain the price protection.

Buying options for price protection is like buying fire or life insurance. The best time to buy insurance is the day before it is needed. The worst case is to cancel the insurance the day before it is needed. Not buying the June PUT in the above example was the worst possible case. It is not until then that the price protection was needed. By the same token, the reader may rightly question the wisdom of buying a December option rather than a February when there was only two weeks left. Also, why buy a February option rather than a July? One of the major problems with options is open interest—the number of contracts outstanding and hence liquidity. This is particularly troublesome for hogs; for as noted above, the open interest was only 54,000 contracts. One-half were in December, 25 percent were in February, and 1,100 were in July. Purchasers of July options were very lonely because they had little company.

The alternative form of price protection is to sell a hog futures contract. If the market follows the same pattern noted above, the hedger enjoys the same cash market prices and the same price protection at a cost of $0.40 per cwt. ($60.00 commission* for a 40,000-lb. futures contract plus $60.00 interest for the $600.00 margin money at 12 percent). Additional margin is required if the futures market moves higher than the $45.00 hedged price. However, cash hog prices will usually go higher as well to offset the futures loss. Higher prices often generate consternation when margin calls have to be met. What has really happened is that the hedger has developed regret. What started as a $45.00 hog market can become a $60.00 market. A $45.00 hedged price level loses its attraction in a $60.00 market—when regret sets in. Then the hedger laments: "If I had just waited." The problem with waiting is that $60.00 possibilities can quickly turn into $30.00 realities.

Options appear to solve the problems caused by futures. But PUT options are not cheap, and they are no panacea. Unless the buyer is purchasing PUT options that are well in the money and, therefore, have a lot of intrinsic value (e.g., $55.00 hog PUTS in a $45.00 market), he or she is paying a premium price for time. But time is being given away for nothing in the futures market, thus one does not need the options

---

*Commissions total $240.00 if the hedge stays in the near month and rolls from December to February to April to June.

market. If one cannot trade the futures market, the options market may not be all that it has been publicized to be.

Summary

Pricing decisions have become more important than production decisions in their effects upon farming profitability. Economic effects of price variation have become relatively more severe than production variation in items within control of the farmer to decide. Unstable prices force operators to develop risk management and pricing strategies. Some strategies are active; others are passive. Some producers intend to capture market extremes; others are content with averages.

The purpose of this discussion was to emphasize the time period available for price decision making, the role of futures prices in pricing and production decisions, the choices available for mixing speculation with hedging, and techniques used to make pricing decisions. We suggested a method for keeping score of the results of a pricing plan and suggested it could be used to evaluate proposed strategies. Finally, we introduced the concept of locking in profits irrespective of price level whenever differences between prices of input and output items presented profitable opportunities.

We reviewed the features of current agricultural price legislation. Rather than removing uncertainty, current agricultural policy has added another element of uncertainty because of its impact on total supply, available supply, and potential changes in the program.

These ideas were intended to form a foundation of thinking to help an individual define a marketing plan suitable to his or her situation. A Kentucky study provided results of average prices for comparing various strategies for making price level decisions. And an effort was made to place agricultural options in perspective.

Questions for Study and Discussion:

1. How may a farmer attack the problems of price risk?
2. Discuss the factors that should be considered in framing a market strategy.
3. Are futures prices accurate predictions of cash price?
4. What are the two problems of commodity pricing?
5. What choices do agricultural producers have in making pricing decisions?
6. Explain selective hedging/selective speculations.
7. List and explain several pricing strategies.
8. Why is cost-of-production strategy inadequate as a pricing strategy?
9. Give examples of scale-up selling and scale-down buying.
10. Why did seasonal selling appear to be a viable strategy prior to 1972, and why has it proven unsatisfactory since 1972?
11. What is the reasoning behind a trend-following market strategy?
12. Explain a method for "keeping score" so various pricing strategies may be evaluated on a common basis.

13. Explain the concept of "profit rather than price."

14. Do government programs increase or reduce market uncertainty?

15. Explain non-recourse loans and their effect on price level.

Exercise: Making Pricing Decisions

A group exercise in committing to pricing decisions can be conducted by using playing cards to generate changing prices. Students can be put into a position of having to make pricing decisions by assuming that they have 100,000 bushels of commodity in storage and are planning production of an additional 100,000 bushels in the new crop year. They can pick any day to price all or a portion of each crop. The first day's price should be the current local price.

Face cards are removed from two decks of playing cards. The cards are shuffled, and each draw represents weekly change in price. A red card represents decline in price equal to the number on the card; a black card is a price increase, again equal to the number on it. Jokers represent limit day price moves in the same direction as the previous color drawn. (The instructor can bias the price trend, which will evolve by removing either some black or red cards. This will also serve to confound "card counters.")

The exercise begins when the instructor asks participants whether they want to price any of their inventory or planned production at the current market price. Those who do should write down the amount and the price. A card is drawn, and price rises or falls according to the card's color and number. Students are again offered the opportunity to price their production by duly noting quantity committed and the new price.

The exercise continues with successive card draws and time lags for students to record their decisions as they wish to make them. After about 40 draws, one marketing year of time for making pricing commitments will have lapsed. The students should average together their individual chosen prices and respective quantity commitments. The high- and low-price sellers might be recognized.

The exercise is intended to emphasize the difficulty in committing to price offers as prices randomly change. It emphasizes the necessity of developing a pricing strategy based upon some foundation rather than having to react to changing market prices.

Endnotes for Chapter 6:

1. Joy Harwood, et.al., *Agricultural Outlook,* USDA-ERS, Washington, DC, March 1999.

2. Randy Schnepf, et.al., *Agricultural Outlook,* USDA-ERS, Washington, DC, April 1999.

3. _____, *Strategies for Selling Crops with Options* (Chicago: Chicago Board of Trade, 1997).

4. Quoted from "Monthly Grain Report" (Chicago: Stotler & Co., July, 1983).

5. Blair Stewart, *An Analysis of Speculative Trading in Grain Futures*, Tech. Bulletin No. 1001 (Washington, D.C.: Commodity Futures Trading Commission, Oct., 1949).

6. Steve Meyer. "Climb the Stairways to Profit", *Pork,* Vol.19, No.12, Madison, WI: December 1999.

7. Gary Schnitkey, et al., *Ohio Crop Enterprise Budgets* (Columbus, OH: The Ohio State University, OSU Cooperative Extension Service, 1995).

8. Gene Futrell and Robert N. Wisner, ets., *Marketing for Farmers,* St. Louis, MO 63146: Doane Information Services, 11701 Borman Drive, St. Louis, Mo. 63146, 1982, 27.

9. Steven K. Riggins, Jerry R. Skees, and Michael R. Reed, "Analysis of Grain Marketing/Pricing Strategies," *Journal of the American Society Farm Managers and Rural Appraisers*, Vol. 50, No. 1 Denver: ASFM and RA, April, 1986.

**Corn: First Ten Years: Total Futures Profits from
Trading Various Moving Average Combinations, Long and Short**

No. of days	Faster Moving Average													
	2	4	6	8	10	12	14	16	18	20	22	24	26	28
2	−100835													
4	−78045	−55225												
6	−47475	−46705	−44515											
8	−39315	−27695	−24645	−26705										
10	−29875	−15555	−7905	−20095	−16675									
12	−19545	−7315	565	−11125	−13735	−11985								
14	−6145	−7465	2095	2885	885	−11155	−13085							
16	−6005	135	5575	9415	−5335	−9285	−13495	−13795						
18	−3495	1545	7275	11795	4725	1215	−5405	−11335	−1195					
20	95	3525	5325	9985	5535	2305	−7555	−4535	−13865	−435				
22	−1065	5095	8285	8435	7745	7255	1315	6425	745	−4765	−7175			
24	−7175	4635	5065	13945	13145	15465	9025	9285	4105	−75	−9805	−4905		
26	735	6995	11205	15575	11585	8015	11875	9775	4585	915	2955	−4655	2725	
28	3755	6925	8185	12145	12795	9855	5325	7125	1295	6215	3445	−1605	−11305	1555
30	9395	9355	8585	9385	14495	9455	8355	6285	1515	1185	4465	−3935	−5295	−5295
32	5975	14495	11485	14105	12825	7745	10715	9205	3335	7395	2055	−4025	−1675	−125
34	9055	9915	11715	12975	11625	12875	12505	8125	6855	11775	6505	3675	8575	−105
36	8825	6015	16285	15375	14095	15885	13325	12635	12025	8425	4785	4475	8255	2635
38	9815	8605	20325	15875	14495	17805	16615	13015	15185	11715	10305	10905	6105	6445
40	11705	14675	17485	16345	15825	18185	19215	20575	15675	14335	8815	10155	3395	3475
42	11955	15585	17625	12865	12655	17885	17185	50645	15015	12625	8705	7775	7075	8475
44	11875	16755	11835	9845	10195	18865	13285	16845	9345	11785	11685	9305	11345	6835
46	15595	14245	14305	9275	5955	14645	12915	11245	12365	15785	11365	9655	10295	9625
48	14955	16035	12255	10555	13235	16685	16645	15725	13875	15435	9655	10545	9715	8715
50	14085	13125	11105	12305	16485	17765	14245	11265	14425	12505	12325	10575	13735	11375
52	15085	15085	12915	15915	20895	16105	18725	18045	16895	16235	11885	13335	11965	15365
54	11265	14655	15755	18995	22065	20065	18995	17315	16185	19725	12835	11635	13765	15355
56	12245	17275	16505	22655	22915	22685	20665	20295	17845	16985	15975	11835	12885	15685
58	11705	16765	19035	23875	22005	22035	23215	20255	19225	21145	18905	18925	13325	14535
60	12255	18795	18245	23395	22375	24065	22195	21325	19655	20195	19205	21105	16855	17115

(Row label at left: Slower Moving Ave)

Max. profit = $24065 @ 12–60 days

**Corn: Second Five Years: Total Futures Profits from
Trading Various Moving Average Combinations, Long and Short**

No. of days	Faster Moving Average													
	2	4	6	8	10	12	14	16	18	20	22	24	26	28
2	−51550													
4	−39570	−28430												
6	−32107	−32350	−31250											
8	−27650	−25810	−25370	−16040										
10	−20970	−17050	−12580	−10410	−7510									
12	−17510	−7210	−5670	−2910	−10130	−7790								
14	−13550	−8830	−470	−1910	−2630	−7150	−12150							
16	−10010	−10910	−5190	−4850	−5410	−11150	−13050	−8890						
18	−7270	−9570	−5210	−2650	−4470	−4390	−6870	−10590	−10170					
20	−5390	−5310	−6550	−2130	−2010	−2980	−3610	−2870	−3750	−4010				
22	−4890	−7130	−4910	−1750	1610	−710	−2750	−1690	210	−2270	−5310			
24	−4830	−5210	−3550	350	−170	−90	350	310	−950	−1610	−9470	−4870		
26	−2550	−4970	−1370	910	30	390	−890	670	−2550	1010	110	−6730	−150	
28	−4130	−4410	1430	870	4310	2330	−710	1810	−1330	2150	3990	3250	2090	2670
30	−2210	−3090	3590	4330	2170	4290	1290	2410	−1570	1150	4530	8230	6090	1350
32	−2550	−1550	2490	2590	530	2790	−330	670	−490	4410	7230	6910	5290	610
34	−2290	−190	2810	3130	1730	50	1750	1770	770	5070	7150	6450	4450	1650
36	1810	1890	6430	6730	3130	1710	1810	3090	3250	3750	6770	6270	2190	530
38	−1450	5090	7270	6410	4190	2610	2050	2450	2390	2990	5110	2290	310	2350
40	−1590	6310	5370	4990	2710	2870	1230	2510	1530	1530	2730	1910	1070	2250
42	−1830	3790	6330	6290	2150	990	830	2610	3390	870	890	−210	−770	1570
44	−770	4650	4170	5370	2970	2090	1210	2530	930	1630	2190	−590	190	430
46	−2070	3290	5690	4970	2010	690	−390	610	1250	2490	1550	1350	1490	870
48	−890	2930	4830	3270	490	−990	−1050	570	2450	3190	1710	2130	930	2150
50	−1430	1690	4430	2010	−290	−570	−230	1070	2170	2550	3330	2450	2550	1630
52	−2570	2290	3390	210	1010	−1370	−1830	−830	2170	3410	2890	3070	2250	2370
54	−430	2520	3250	1250	910	−1230	−530	670	4610	3410	3930	4110	4090	1570
56	−1950	3490	4050	2630	−70	−2390	−410	2510	2650	3550	4470	3850	2470	1090
58	−430	2150	4110	1650	1370	−1910	570	3750	3050	3730	4370	3710	1390	2610
60	1470	2350	3930	2930	1710	970	1490	4630	4450	4370	3810	2070	1170	910

(Row label at left: Slower Moving Ave)

Max. profit = $8230 @ 24–30 days

Soybeans: First Ten Years: Total Futures Profits from Trading Various Moving Average Combinations, Long and Short

No. of days	Faster Moving Average													
	2	4	6	8	10	12	14	16	18	20	22	24	26	28
2	−74490													
4	−61115	22040												
6	−865	14805	26300											
8	7645	44465	1355	22305										
10	−575	60075	12915	26165	23845									
12	17365	45045	14045	18585	12425	42630								
14	28365	57685	18575	20696	19575	49445	49370							
16	61195	50345	64105	80845	33865	76245	75655	59005						
18	53255	80335	55745	71745	65435	106265	85795	34785	57925					
20	66165	82685	74635	77425	84285	72575	61385	34795	28265	54240				
22	55355	89895	77755	73535	80005	77105	63785	53245	46055	52185	65145			
24	67395	85585	94795	91465	71905	70125	55245	53495	45185	39085	29475	64935		
26	80955	91525	81325	86945	59025	61555	52595	42925	42235	48725	36555	55165	51765	
28	82665	84545	78655	76415	52985	47485	45585	46365	47655	36215	44625	51125	29665	53900
30	85765	76905	81275	64745	56835	55215	53155	43095	23505	37725	47055	67315	58795	41495
32	77295	67895	70575	75375	64245	62635	62395	46735	57695	67295	53995	52005	43135	43295
34	83505	71125	64215	72575	62025	68355	62405	56615	69045	70515	75065	62355	56395	53235
36	68645	73405	62855	70345	67755	69135	65215	69705	77075	82475	77585	74695	64895	55385
38	66325	77595	74635	81595	75345	77515	77355	71555	80545	86975	90915	85965	59985	47745
40	63395	75625	74265	77295	79615	75985	71205	73815	81755	82165	89385	66855	48235	50975
42	68965	75055	66785	86075	84465	76355	67905	86195	78465	100055	78685	69125	50565	55025
44	58105	76325	73895	85925	82145	79425	79045	80375	92745	81635	73195	61075	58345	50765
46	50535	75705	70855	76125	82525	83215	81045	89155	88875	74995	62025	63365	53685	57075
48	62015	67665	70235	71765	85435	78875	80215	87095	77605	76505	61775	58695	60255	59005
50	75065	68395	73775	74675	76115	79455	83135	83165	67235	78875	62475	59955	65605	57315
52	77065	81675	79515	88305	85995	72915	77885	83465	80275	66995	73765	61325	70465	56835
54	76295	84855	92565	100365	85785	86005	87565	69355	75735	60255	47965	64795	61835	66425
56	91945	93325	85135	96105	91435	83455	85415	67505	65535	61215	57885	72165	73305	67205
58	100495	86775	89455	86635	77405	83435	81785	71195	58595	59605	66015	76345	66535	68715
60	92275	86005	86625	82855	81235	84065	75045	71345	70285	95615	73435	83195	81065	72285

Max. profit = $106265 @ 12–18 days

Soybeans: Second Five Years: Total Futures Profits From Trading Various Moving Average Combinations, Long and Short

No. of days	Faster Moving Average													
	2	4	6	8	10	12	14	16	18	20	22	24	26	28
2	−72720													
4	−51700	−18840												
6	−24380	−26400	−46970											
8	−25180	−40740	−34520	−28240										
10	−17620	−25980	−19260	−3760	−6900									
12	−13460	−23340	−11100	9560	7160	−2120								
14	−14520	−12160	−2080	19180	11040	1680	−8600							
16	−6780	−9960	−4560	2520	7400	−11640	−15020	−8120						
18	−2080	−3240	120	10360	−9900	−20980	−11360	−24720	−3900					
20	3900	−1500	3280	3240	−880	−14760	−8360	−6840	−8120	10				
22	−2460	−3320	9920	4390	1360	−5320	−5280	4180	1840	−5180	−15160			
24	−2060	−7080	8800	4480	6060	−3930	−10920	6440	13520	−940	−8480	−4540		
26	−2780	−5320	1820	6800	1740	−5960	−10620	−1460	11780	−200	−1420	−14120	−13680	
28	−2020	−3980	7360	10440	−3480	−15780	−5600	−2560	10780	−1520	−8840	−15300	−26260	−3240
30	160	−840	10220	5120	−8360	−15080	460	−6660	60	−7340	−10620	−21280	−19020	−26540
32	−7760	−7640	11660	1680	−12440	−6320	−4860	1700	−5720	−17740	−11440	−18100	−21860	−21960
34	−5000	−3960	8800	−1740	−10000	−6720	900	−7840	−12860	−11120	−11360	−11540	−10040	−10440
36	−6860	−3980	−160	−5640	−4180	−5100	−4780	−9460	−12040	−16740	−12080	−3560	−3680	−12620
38	−3100	−4880	−1820	−8100	−4400	−7540	180	−16660	−22160	−19480	−6080	880	−8820	−7860
40	−200	−7000	1160	−5900	−8960	−12080	−6300	−15040	−22460	−8940	−960	2220	−8160	−8780
42	−1260	−3580	−3440	−6680	−7280	−7440	−11960	−16500	−10380	−8260	−3440	−3920	−6120	−5460
44	5960	4140	−6880	−9120	−5760	−13020	−21900	−14040	−8080	−1620	5060	−3260	−940	−9960
46	3080	1000	−1120	−6760	−4880	−10740	−16340	−12620	−2260	−200	3860	1280	−6840	−8880
48	4740	4100	10280	−1980	−5660	−8720	−11760	−10980	−1440	2400	4820	120	−10120	−14500
50	17140	9160	13280	4320	−2200	−8940	−13360	−7820	5060	5300	1020	−5000	−7500	−10580
52	11300	7900	11580	3300	1540	−7260	−5240	−1360	7080	9000	3420	940	−2580	−5280
54	11400	13980	14720	7640	580	−4100	−12200	−3120	10140	8140	2520	1340	−3380	−5420
56	13120	14120	13140	9640	2160	−3320	−13420	680	10660	10900	4700	0	−7180	−6080
58	4520	15980	9780	10740	3300	−3840	−10460	2080	7960	6740	760	−2760	−8780	−8100
60	4560	10100	11480	9460	6620	−4900	−5120	3020	8180	6020	−180	−4540	−9820	−6940

Max. profit = $19180 @ 8–14 days

**Wheat: First Ten Years: Total Futures Profits from
Trading Various Moving Average Combinations, Long and Short**

No. of days	Faster Moving Average													
	2	4	6	8	10	12	14	16	18	20	22	24	26	28
2	−113070													
4	−96940	−53780												
6	−51880	−43560	−54410											
8	−41720	−21230	−32200	−46550										
10	−37355	−27810	−23160	−11800	−17210									
12	−27740	−10200	−11800	110	1890	−8230								
14	−32800	340	5210	−1190	1540	−8490	−10570							
16	−25050	1820	11330	13670	23890	12330	5260	−2730						
18	−12390	3050	18140	21390	15280	17720	11080	6030	−2760					
20	−6060	13800	19660	33120	18730	21190	16040	9400	3020	−3223				
22	7900	21340	26850	29230	18580	22340	18690	13830	8530	−3280	13030			
24	8470	21610	32230	22120	25890	25780	26700	19820	7300	4890	1250	13860		
26	11110	17840	29510	27050	27590	28500	24650	19340	11130	16030	12140	16590	13080	
28	9310	24430	36450	32560	29810	37340	29760	18830	19260	9640	16800	19330	8330	11240
30	14490	19720	34670	33900	39060	40030	33690	26250	23470	11070	18270	23200	24530	6350
32	18430	27120	40150	33500	39970	38530	38150	34530	25530	15860	19510	24850	18730	14860
34	17050	34680	32250	30190	35230	39630	41470	36130	22470	17700	19500	27960	25550	20980
36	16960	33060	30230	34910	37900	43500	45470	28400	20860	17800	19070	23600	21270	19830
38	15680	28960	34230	38640	41850	44540	36150	26230	15950	16650	21430	23020	8500	10730
40	14480	27020	26010	33660	35070	38430	28790	20050	18360	18940	14080	14360	5490	−1690
42	16300	22760	28440	28580	31560	38340	21110	21020	19280	12490	13430	4800	−980	−5210
44	12110	27720	32790	29160	40300	32230	28190	22040	12460	11990	9850	3370	−650	−5610
46	20200	26450	32100	33650	36620	38400	30500	18110	14640	7780	6090	1950	−10270	−9110
48	25740	27830	36370	34120	37260	36440	26540	25610	13660	8090	4810	−400	−7700	−1020
50	26540	30970	36230	33590	36490	28650	27100	21280	15170	10390	10780	−3020	−8210	1570
52	28450	33810	27740	32760	34390	31870	20260	20420	25590	19530	10960	1440	2200	−3450
54	31830	34020	30040	31970	26940	23530	19660	16880	22160	15480	5890	560	6440	6210
56	35050	34310	30440	30080	28390	20510	20640	23550	17450	12990	7630	4920	11930	7390
58	32790	35610	33730	28130	27350	20580	16440	16540	16520	13360	7280	6760	5400	2310
60	25260	35860	33030	27420	28360	18050	12330	15140	12280	9960	9780	6980	4010	−30

Max. profit = $45470 @14–36 days

**Wheat: Second Five Years: Total Futures Profits from
Trading Various Moving Average Combinations, Long and Short**

No. of days	Faster Moving Average													
	2	4	6	8	10	12	14	16	18	20	22	24	26	28
2	−52770													
4	−44420	−32500												
6	−32360	−30580	−25560											
8	−28800	−26560	−24740	−14750										
10	−19080	−19420	−19860	−20000	−8800									
12	−13440	−9160	−9880	−12720	−16120	−5400								
14	−5560	−6420	−1200	−5780	−6560	−2600	−5950							
16	−9840	−2600	−3200	−5040	−1760	−1780	−9260	−2140						
18	−12200	−1820	−4400	−6780	−3680	100	−7840	−17080	−3760					
20	−6480	−3400	−5060	−6360	−1060	−3300	−7800	−12960	−11340	−8220				
22	−5520	−1280	−5800	−6000	−4020	−4240	−10820	−13860	−16920	−18500	−11840			
24	−3980	−3520	−3840	−6100	−4400	−9160	−10320	−17360	−14560	−11480	−16280	−11220		
26	−1240	−1200	−5620	−5280	−5840	−6100	−8820	−10200	−12800	−9420	−5660	−9480	−11580	
28	−1680	−2520	−4600	−5120	−4580	−7200	−10020	−11920	−12220	−9720	−12200	−14000	−15340	−6580
30	−3900	140	−4900	−7700	−4480	−7900	−6640	−10260	−8040	−7040	−13640	−13660	−9060	−9660
32	−4080	20	−5500	−10360	−6700	−9040	−8060	−9880	−10780	−8500	−11400	−10820	−7620	−12080
34	−1660	−80	−6860	−8520	−6900	−7560	−6120	−8740	−9500	−9780	−10000	−6560	−5780	−9720
36	−320	−3680	−8540	−7460	−5660	−10240	−8280	−8240	−6580	−7160	−9580	−5040	−10360	−13940
38	−1480	−4440	−6820	−7360	−7900	−8560	−7960	−6700	−7080	−8340	−12320	−8120	−10660	−11880
40	−3220	−7420	−8340	−5900	−9660	−9940	−7320	−7340	−5500	−9760	−8820	−9280	−12240	−13000
42	−7860	−9200	−7040	−8640	−11000	−13180	−9500	−8620	−6860	−10620	−7800	−9880	−11500	−11960
44	−9120	−8260	−9460	−14340	−12380	−10580	−9540	−7820	−10620	−10620	−10220	−10420	−12540	−15140
46	−9920	−10100	−7960	−9200	−14720	−14460	−9800	−7980	−10160	−10340	−10320	−12880	−14420	−14780
48	−10260	−12740	−7780	−9620	−14640	−13140	−11300	−10600	−10500	−11620	−11800	−12160	−14700	−14280
50	−11520	−13080	−8920	−9080	−15080	−11980	−10620	−12420	−11680	−11320	−11880	−11840	−13240	−10280
52	−10720	−12180	−8700	−8380	−13660	−10580	−10580	−13880	−12520	−8360	−11140	−9960	−11940	−11180
54	−10640	−11520	−7720	−8860	−13280	−14060	−13480	−9720	−9760	−8300	−10020	−7860	−10677	−8600
56	−11040	−9240	−6020	−10720	−10920	−13000	−12820	−8720	−10720	−6900	−6240	−7600	−10940	−7880
58	−9040	−9640	−56801	−11240	−9780	−11360	−9620	−6120	−7480	−3860	−7120	−8300	−8680	−8120
60	−9040	−8480	−2440	−10000	−12940	−11020	−7900	−7637	−6420	−5780	−5560	−7140	−7960	−7040

Max. profit = $140 @ 4–30 days

Hogs: First Ten Years: Total Futures Profits from Trading Various Moving Average Combinations, Long and Short

No. of days (Slower Moving Ave)	2	4	6	8	10	12	14	16	18	20	22	24	26	28
							Faster Moving Average							
2	−119519													
4	−96934	−42204												
6	−50248	−31014	−37334											
8	−21830	−16822	−30228	4656										
10	−4858	2086	9242	1816	10902									
12	11298	20860	21952	14074	1182	−4508								
14	12500	22780	18886	8856	188	−21972	−11494							
16	16736	17118	11754	6374	−1808	−6734	−13358	−3394						
18	16534	17394	6502	4606	−1838	−3712	−15484	−25386	9320					
20	14042	15714	8322	2806	6642	3322	−2086	−15618	−24278	5358				
22	12968	13584	10764	6686	9698	4532	−2514	−9532	−17668	−14806	−10758			
24	11548	17610	12764	10530	9380	−2228	1310	−4508	−5142	−15806	−16242	136		
26	9840	17732	11670	12600	11942	−3217	−4090	−4780	−8694	−9052	−5938	−11990	−6835	
28	11356	12756	8776	8534	6410	−2860	−296	−6430	−7316	−2604	−2040	−2172	−14494	774
30	12780	12048	11910	5448	868	−3604	−2882	−7338	−4508	1146	−1774	−3196	−13602	−21214
32	11990	15438	12206	6856	2238	−3006	−5766	−8382	−1374	1450	−1202	−5800	−11664	−11428
34	13600	19196	12216	5824	4822	−1182	−3420	−2460	−2734	1034	−4936	−5356	−5854	−6352
36	16202	14878	8072	9106	6546	1182	−2330	−750	−622	−6102	−5378	−2412	−6666	−4780
38	13112	13000	10260	9882	9954	5812	−1032	1058	408	2020	−2712	−5674	−5244	−6708
40	15004	17664	12372	13948	13144	4402	−1908	−530	−2188	6034	−440	4022	−1950	−1630
42	16084	19466	16254	14374	8694	4076	−186	708	2814	9260	4064	1656	22	1198
44	14054	17832	15230	15566	7792	4258	3016	8950	8008	11464	6894	1554	2858	8178
46	14074	21538	18204	15002	11532	8462	7318	10088	8804	15536	15838	7358	11114	13922
48	12434	22822	19718	14778	9168	4282	6670	10580	13744	16502	14486	12778	16608	12926
50	12150	21010	18460	15706	4492	7382	7854	13942	18136	17900	17642	14362	14842	13630
52	13172	20692	13060	13196	9068	8126	10554	12158	20252	19372	18836	12546	14132	14740
54	14722	21402	16682	10546	9078	8284	11132	14940	20264	20934	17028	10916	14032	13218
56	17305	17493	15319	10627	10855	14171	13685	14555	19975	15283	16463	14185	17207	12477
58	18400	15610	21502	13704	11266	13254	16080	12158	20146	13860	12660	16582	18662	14020
60	19805	16249	16407	15213	14427	15353	12737	12657	19969	13997	12285	15437	16209	11889

Max. profit = $22822 @4–48 days

Hogs: Second Five Years: Total Futures Profits from Trading Various Moving Average Combinations, Long and Short

No. of days (Slower Moving Ave)	2	4	6	8	10	12	14	16	18	20	22	24	26	28
							Faster Moving Average							
2	−67412													
4	−51022	−27416												
6	−33548	−22192	−25516											
8	−22039	−13048	−4810	−13162										
10	−13362	−6810	1636	−10118	−12538									
12	−8466	−6914	−4708	−8506	−8664	−7752								
14	−6664	−7284	−5166	2176	6858	1024	−934							
16	−3528	230	1600	5568	3374	2322	−1772	−118						
18	−452	3894	2918	7156	3024	3778	3140	−6306	1868					
20	570	3760	7822	11040	8008	7356	6642	566	−2426	788				
22	−1396	2122	7662	9874	7992	8600	79786	3880	2632	510	9268			
24	−4216	3202	8616	7204	8216	7202	10736	3776	3652	5116	−1092	9948		
26	−2614	8638	8968	6908	7964	7340	10966	9750	4638	6012	1778	2012	230	
28	−1172	8234	11956	10512	9256	11758	10954	8930	8316	8086	6734	3668	−934	4752
30	7188	8586	12200	12366	10164	13474	12766	11084	11020	39718	7384	5472	2290	422
32	2864	5482	11622	10100	10242	11694	12784	13160	11098	8390	6164	3994	4784	3264
34	4432	6388	10114	10286	9076	14592	11158	11942	11702	7488	4846	5910	4956	1950
36	3726	8704	8406	10682	12308	13598	11866	12219	8312	4162	1210	4118	3562	262
38	5128	8560	8158	5534	9428	13146	13708	11326	5180	3904	4564	3412	3154	260
40	5576	9528	10846	6618	7342	10458	13094	7416	4908	2518	2416	3246	2908	2810
42	5844	12570	9890	7212	10630	10632	10020	5454	3276	40	1870	2190	3112	3608
44	6996	12858	8022	5630	8370	9178	7590	2884	1944	−456	1240	4366	5982	5870
46	6198	9604	7410	6082	4680	7558	4302	2192	1392	−54	2946	2458	2186	2494
48	6340	8702	6750	4916	5402	7990	4130	356	1360	−172	2186	972	−1650	−1600
50	3656	7560	5748	5140	6306	5868	2858	−808	758	−1094	498	1086	−284	−854
52	48	6566	4114	3744	6116	5328	64	−3164	−658	−890	−566	664	730	42
54	844	4874	3240	3818	6152	2226	−1796	−4970	−1646	310	1750	610		−1484
56	−2470	1748	2742	3636	3094	1460	−1790	−3762	−2100	−1292	−404	2452	154	−1916
58	−3668	−44	752	2604	90	2418	−4316	−3300	−2412	−548	586	2038	−1942	−2698
60	−3252	−1240	350	2430	−2264	−182	−3232	−3354	−2346	−1568	700	2938	−1756	−510

Max. profit = $79786 @ 14–32 days

Live Cattle: First Ten Years: Total Futures Profits from Trading Various Moving Average Combinations, Long and Short

No. of days	Faster Moving Average													
	2	4	6	8	10	12	14	16	18	20	22	24	26	28
2	−66140													
4	−41304	16996												
6	6660	15224	3752											
8	4230	24948	−11892	16564										
10	13180	17348	2192	18112	17712									
12	19960	10032	15820	12644	6600	26844								
14	22480	8500	11312	19516	19304	−1276	15020							
16	22564	12796	17444	17340	8472	−1036	−8264	9112						
18	28448	20640	20308	22140	12860	7548	−1140	−15172	8268					
20	23784	15060	14136	16351	3288	5348	−312	−644	−6096	18436				
22	26480	22940	18832	12676	6300	8952	1412	4004	3920	−2168	14512			
24	25940	25272	14394	7624	1976	2764	−296	2488	844	5772	−2388	14376		
26	25044	26984	14268	19040	14304	948	−4224	264	1688	−604	−5032	−18048	432	
28	29484	23492	13144	15616	6912	−1476	−1816	−1384	5124	1260	−4844	−17952	−24440	−12136
30	29312	20672	14668	18000	8003	2768	5348	−144	−3976	−4248	−12196	−20908	−24248	−34184
32	25792	21008	11172	18792	7328	528	−444	−436	−1196	−10396	−12284	−18212	−24588	−24848
34	23068	20676	10268	9876	6232	−968	−6580	−7452	−8132	−8440	−14988	−22036	−31796	−21292
36	27780	22204	15424	10696	1032	−4372	−6412	−7156	−9816	−9480	−14576	−16528	−16584	−17044
38	30708	21492	18384	9228	−80	−3488	−6504	−9536	−6684	−10484	−15336	−18444	−14224	−9448
40	28536	22956	16412	9936	8840	−5820	−8284	−6804	−5556	−5904	−18996	−9112	−8964	−11600
42	26584	23120	10424	10096	9984	−4924	−2216	−3984	−8012	−11164	−5656	−8708	−9476	−9864
44	24788	18880	12416	9848	7344	332	4208	616	−4468	−10012	−3980	−9072	−9640	−12560
46	21796	18376	11092	10212	3636	2512	4068	−3024	−520	−8548	−5876	−8108	−10564	−19644
48	19568	14392	10208	11356	2652	2888	3404	−1020	−3192	−3352	−1636	−9276	−11500	−11180
50	15536	10972	9076	12336	8760	6592	1980	2512	−6824	−1892	−5360	−10612	−12568	−11960
52	14692	9680	9700	9192	9412	11844	4508	−612	−556	−6264	−6576	−8896	−9700	−14992
54	10116	9420	8056	9368	8960	5528	812	−484	644	−4236	−8568	−8904	−13360	−24779
56	9656	7236	2312	6088	6084	−460	−4968	2064	−1708	−3108	−10112	−11104	−19816	−22484
58	11696	2224	551	4464	−812	−4276	−6548	−1832	−3024	−2088	−8376	−16352	−20720	−20572
60	11348	6788	−4924	−924	−3984	−6052	−60	−2288	−5092	−5412	−5612	−19204	−22132	−24024

Max. profit = $30708 @ 2–38 days

Live Cattle: Second Five Years: Total Futures Profits from Trading Various Moving Average Combinations, Long and Short

No. of days	Faster Moving Average													
	2	4	6	8	10	12	14	16	18	20	22	24	26	28
2	−51936													
4	−30344	−28660												
6	−19604	−22204	−20156											
8	−9308	−15776	−25600	−12424										
10	−17868	−16924	−18732	−22040	−17584									
12	−15032	−16524	−18260	−25896	−31140	−11180								
14	−8028	−10260	−20224	−18440	−20700	−24648	−18680							
16	−8736	−12184	−19532	−15664	−19240	−9116	−8768	−12892						
18	−4456	−8568	−8660	−11268	−10336	−8272	−6492	−1408	−13160					
20	−4544	−10536	−7976	−7888	−6424	−9176	−9156	−160	−136	−3408				
22	−1000	−10736	−7296	−5876	−6628	−1848	−3328	−1028	7316	6524	−1184			
24	−1872	−13040	−10604	−5600	−992	−32	416	3136	3308	5836	−1480	−13444		
26	−5872	−12364	−15340	−6844	−596	−1256	3872	2844	3368	−1264	−5312	−13612	−12004	
28	−5984	−9184	−12544	−5020	−2336	−2440	2552	−1220	−3436	−7436	−8032	−10748	−14832	−16920
30	−3128	−11416	−12488	−7864	−4396	−108	−1000	−6504	−976	−6540	−6782	−11612	−16680	−14388
32	−4864	−8144	−8236	−9300	−3192	−4176	−8004	−3820	−6988	−6788	−4896	−8524	−9408	−8544
34	−5840	−3024	−4324	−6444	−6092	−9388	−8212	−2996	−8276	−2944	−9456	−13084	−9256	−6868
36	−2928	−2616	−452	−6276	−7176	−9172	−8204	−7648	−5300	−7880	−11200	−8540	−8160	−4488
38	−3456	−264	584	−5508	−8296	−8876	−6740	−8264	−6476	−8596	−9848	−9916	−7472	−2808
40	−104	384	2560	−5616	−4320	−7104	−4004	−5776	−6288	−6448	−6192	−6004	−8128	−7132
42	3828	1280	8	−3096	−4152	−5248	−6168	−3984	−6112	−3404	−6556	−10540	−8860	−6808
44	416	816	456	−1620	−4856	−3940	−6416	−5136	−5856	−1164	−4636	−7156	−9328	−8452
46	−464	3008	−424	−1204	−4432	−4596	−6456	−3680	−3808	−260	−3912	−6112	−9804	−10480
48	−5392	2336	−1036	−1092	−4012	−2820	−6380	−5336	−48	−860	−2660	−4904	−5600	−10604
50	−6688	1248	−632	420	−4240	−2124	−7004	−3440	104	−168	−4524	−8156	−6628	−6248
52	−9352	368	180	716	−2160	−2900	−5916	−4696	−2044	640	−8112	−5324	−9460	−11424
54	−10296	−2084	−1696	2848	144	−3300	−6764	−5892	−3408	−2352	−7440	−6384	−8700	−6264
56	−8432	−164	−2704	2792	1880	−1908	−2884	−4456	−892	−6376	−5728	−5332	−7208	−7576
58	−8384	−1644	−5060	−348	180	−4080	−6040	−316	−1860	−5096	−6680	−7920	−7056	−6660
60	−8408	−3960	−5868	−1928	2636	1036	−1904	−844	−3164	−2088	−3968	−8800	−8136	−6456

Max. profit = $7316 @ 18–22 days

Cotton: First Ten Years: Total Futures Profits from Trading Various Moving Average Combinations, Long and Short

No. of days (Slower Moving Ave)	Faster Moving Average: 2	4	6	8	10	12	14	16	18	20	22	24	26	28
2	–86240													
4	–105225	–34855												
6	–45225	–33015	–16335											
8	–29535	–14235	–59105	–70645										
10	–32715	–46855	–57065	–69455	–48010									
12	–19315	–45805	–57045	–48175	–15835	–2125								
14	–11635	–28355	–35275	–18245	–5675	–1615	12375							
16	–6925	–13465	–15835	7175	32585	25055	27135	32545						
18	4715	24475	12785	25645	33465	29335	49915	32805	30135					
20	20735	33175	25925	34835	31165	38765	62675	59665	57955	30415				
22	31975	26835	37665	38385	38205	44185	64735	63715	58715	27565	22345			
24	34665	36815	38125	39675	37335	57125	58025	55965	44565	14485	12915	3035		
26	32825	38155	38385	34685	43315	53365	48275	61205	31405	22465	31915	6025	17935	
28	22335	31945	19585	30605	37565	45805	47675	45705	34855	31605	35695	20165	4245	4815
30	22445	27055	19905	40615	42755	33875	53595	53975	44535	22205	27605	24865	17965	24115
32	18235	19995	14585	41955	42045	46255	46285	54635	42295	24125	28705	30035	22845	21405
34	18105	16965	26455	35775	41895	40865	55145	65445	36275	29445	36375	35165	23855	20165
36	13995	26585	37145	39925	49395	51555	55705	54185	37575	27925	28375	40615	18515	24815
38	27605	23345	36075	41165	45435	57795	61055	55595	39055	28395	29665	39605	20845	33015
40	19365	25165	27145	37525	37965	50985	60655	55905	34285	29995	34805	32845	17395	27005
42	34145	28985	26565	37855	37595	51455	52795	44675	34305	35305	35355	34155	28915	38385
44	26255	36425	33226	32775	26805	45195	45235	38525	36375	40155	44035	38385	31905	43885
46	25095	32445	39365	28355	26325	36085	42365	37485	39015	41435	43685	42645	32835	48165
48	28455	23825	36735	24555	26765	40585	38965	42905	38665	45355	44075	45115	42865	48645
50	25705	22125	29095	22185	32785	37005	37905	42755	50865	54425	51145	57875	57045	59495
52	1735	29965	28935	39305	34145	35925	39895	43275	49915	55695	57155	54235	55885	61815
54	20545	37815	32785	41415	35335	42145	43855	47745	51225	48035	52736	51385	56235	55555
56	19395	40015	37065	43045	40475	44335	39985	45635	49275	51775	52845	46385	58935	42615
58	16635	34175	42265	39735	43765	44285	43595	39665	45655	50575	46355	49685	44325	40805
60	15725	37065	42015	37785	48335	47365	47705	42125	42205	49025	37925	44685	46325	44615

Max. profit = $65445 @ 16–34 days

Cotton: Second Five Years: Total Futures Profits from Trading Various Moving Average Combinations, Long and Short

No. of days (Slower Moving Ave)	Faster Moving Average: 2	4	6	8	10	12	14	16	18	20	22	24	26	28
2	–50430													
4	–43830	–47840												
6	–39950	–45060	–34700											
8	–25380	–21790	–31990	–21490										
10	–12320	–17770	–28100	–17770	–7570									
12	–4410	–11320	–25080	–19260	–27110	–2780								
14	–2740	–13720	–13800	–6450	–24960	–18060	–2420							
16	–3600	–4400	–440	5580	1160	–2170	3820	21260						
18	5000	5130	8330	14970	14870	9600	18100	12460	10220					
20	13440	9640	11590	13910	17620	11640	12690	9610	9740	1250				
22	12780	13380	8900	18050	12170	7660	14350	5070	–840	–7110	–940			
24	13440	11580	12720	19600	17020	11060	10150	4290	320	2510	–3360	730		
26	8450	10340	6600	14200	9760	7340	10580	–1690	1730	–4240	340	2450	–90	
28	7370	9380	7150	7640	6180	9490	9440	2380	300	–3990	2220	910	–5460	–3240
30	6380	7960	9520	6330	3030	9730	9140	–2330	3430	–5120	3410	4110	–5920	–5860
32	4800	5870	10780	4890	4140	6150	1040	1770	3850	1370	–1190	320	–11190	–3710
34	1260	9520	15150	2250	4260	5270	–130	3620	3550	–870	–290	–1740	–7500	–3090
36	–1210	4490	7920	5720	2750	2340	2320	–600	2530	–760	3350	–2430	–1730	1470
38	2510	9850	6760	6640	6070	1100	620	–1600	1730	–1330	–140	1440	90	1690
40	4970	5260	6440	5400	3950	1540	–2250	–4870	–1760	–1000	2870	–340	–970	–140
42	3520	6730	6740	4150	6430	320	–1620	–5610	–950	–1580	–2140	–1870	–3070	3500
44	3350	7030	5640	5550	2250	–250	–3060	–4960	–3780	–3010	–530	–2260	–3860	2240
46	4160	4170	4650	5980	2030	2480	–3220	–3290	–2830	–2080	1330	–1960	–2720	8390
48	5980	3630	3870	6240	4410	4270	–2840	–590	–510	–2020	590	–7330	3180	9740
50	3770	210	3100	6500	7270	3620	–990	3650	–670	–3250	240	–5280	4750	5000
52	1000	1590	4550	4600	8540	6430	–980	3550	690	–5170	–1410	330	5330	–210
54	1300	–2620	5450	8720	9420	8130	–1720	3360	100	–4160	–1050	2230	810	1890
56	5800	–3050	–1450	8760	7710	10230	1950	2580	–1080	–1010	1000	820	680	4830
58	8680	–1560	–2890	8930	3750	7710	260	170	–1880	2750	3740	–1370	2560	6210
60	8250	–520	–2690	4820	3960	2610	2020	2240	2440	6260	3300	470	8000	11510

Max. profit = $21260 @ 16 day single move. ave.

Soybean Meal: First Ten Years: Total Futures Profits from Trading Various Moving Average Combinations, Long Only

No. of days	Faster Moving Average													
	2	4	6	8	10	12	14	16	18	20	22	24	26	28
2	−49190													
4	−35620	300												
6	−11280	1510	−2070											
8	−9230	14560	16050	8360										
10	−7390	17360	14510	13470	3170									
12	−800	32010	18160	8190	9950	−4830								
14	790	27260	21180	11340	15700	14700	16020							
16	2410	31680	24230	22610	21730	24170	28430	17520						
18	6270	26280	21550	32240	28630	27210	28400	11830	18230					
20	18080	33050	28630	34010	24630	29630	34280	21020	23970	19910				
22	24500	36110	30010	29530	25590	32180	30900	18540	22040	22610	28780			
24	24240	35330	37170	33750	28150	30200	26280	18160	30500	25990	18600	24690		
26	31120	39730	33490	27990	28770	28300	21570	22880	28860	23530	26760	15620	19490	
28	35570	38490	32610	26970	24440	27440	24960	22650	21960	22740	21140	15000	1640	20580
30	32520	37760	31920	31530	25800	23740	28650	21910	22510	15840	16190	15600	5830	10310
32	32020	36000	29940	26520	23970	21540	23390	25100	20670	19710	21120	17660	14210	11310
34	25970	32560	28380	24280	24040	25230	25530	26080	20490	22920	20050	21390	16520	19320
36	19220	29530	27990	25140	25270	26220	27320	23970	24000	24210	25160	18590	22810	21080
38	14810	26850	24160	23210	26610	30030	29910	23180	27400	25530	22430	25810	28600	23990
40	14990	26700	25610	25050	26150	30200	26810	24280	26140	22330	23380	26450	25120	25540
42	18380	22530	27890	25910	27100	30250	27040	23780	18050	20310	21350	23440	26810	24380
44	15210	24190	24950	26660	29880	29380	21850	19390	21730	21890	25790	21830	20680	24600
46	19480	24830	24630	25710	26980	24660	27550	21750	19020	22570	22760	21200	20530	23260
48	18520	23290	23480	23700	24800	25450	24860	19590	21930	24030	24060	22060	22530	23270
50	19760	21050	24960	24140	26980	24690	25150	20350	24120	27150	21770	21350	22650	24940
52	19860	22800	22830	24350	26580	26510	26630	21810	22550	24990	21820	22220	26470	23690
54	20350	25670	22670	29000	28630	27840	24830	22850	24480	23130	24260	24820	23200	22540
56	24670	26760	22440	28070	28700	26780	22800	22550	23510	24610	25510	23740	23040	20920
58	26310	28740	21020	23200	26830	26760	24240	21020	22890	21740	21250	24070	24210	20920
60	30990	26730	21750	24880	26230	21550	24370	25200	21410	24220	21210	21550	22720	22710

Max. profit = $39730 @ 4–26 days

Soybean Meal: Second Five Years: Total Futures Profits from Trading Various Moving Average Combinations, Long Only

No. of days	Faster Moving Average													
	2	4	6	8	10	12	14	16	18	20	22	24	26	28
2	−28880													
4	−27880	−20370												
6	−20530	−23340	−20560											
8	−18350	−18270	−20450	−14460										
10	−15550	−17890	−14440	−10130	−7090									
12	−14930	−10970	−6400	−500	−5210	−11580								
14	−15940	−10360	−7980	−2350	−3600	−8630	−8740							
16	−14870	−9000	−5600	−4780	−4080	−8680	−11150	−11070						
18	−9320	−5330	−4500	−3280	−3910	−4060	−6490	−7670	−5810					
20	−7870	−4570	−4420	−3710	−5310	−4680	−5230	−5800	−9290	−1890				
22	−8630	−5420	−4030	−2460	−3240	−5450	−5720	−5510	−7300	−6100	−1640			
24	−5850	−5020	−2880	−2500	−3190	−3340	−5250	−4330	−4370	−7320	−6720	−4300		
26	−4650	−4250	−2030	−3780	−4960	−4650	−5370	−5560	−3470	−4850	−4200	−3630	−3710	
28	−4560	−3730	−3200	−4210	−5430	−4030	−5810	−4880	−3200	−2580	−3140	−1810	−4060	−3920
30	−7720	−3330	−3520	−3900	−5430	−6810	−5910	−3380	−2260	−1750	−2560	−1240	−4310	−5940
32	−6730	−3180	−2490	−2140	−5830	−7210	−5530	−3490	−3460	−2430	−3690	−4660	−4800	−7700
34	−7940	−3410	−4570	−3540	−4070	−4820	−5140	−3120	−4660	−3670	−4650	−5590	−4660	−5280
36	−5150	−4620	−1380	−3900	−4560	−4780	−4170	−3500	−3840	−5120	−4490	−5700	−5750	−6010
38	−4580	−5210	420	−3260	−5540	−5880	−5780	−3420	−2980	−4770	−4480	−7530	−7630	−6780
40	−3640	−4500	−420	−3130	−6530	−6510	−5820	−3500	−4270	−3680	−5360	−7310	−7570	−7650
42	−5500	−3400	−110	−5320	−5960	−6300	−5930	−4190	−5110	−5470	−5910	−6810	−7230	−6750
44	−5280	−3420	−1850	−5450	−6320	−6090	−4090	−4630	−4650	−5850	−4840	−4860	−5680	−4030
46	−4530	−3460	−690	−4630	−5500	−5290	−2900	−2590	−3590	−6130	−5460	−5350	−5790	−4010
48	−2460	−990	−940	−3440	−5660	−3200	−1310	−1560	−2850	−3490	−1050	−3400	−3320	−2380
50	−4860	−370	−50	−4300	−5670	−3530	−1810	−1300	−2370	−1920	−390	−1500	−470	−2060
52	−980	−960	−820	−3900	−4840	−1590	−1490	−2590	−3100	−700	−720	−490	−1380	−1250
54	−510	370	−1280	−3290	−3940	−1430	−540	−2560	−560	150	−460	−1180	−1190	−240
56	560	590	−430	−2060	−170	−1280	260	−1310	−950	−80	−380	−1130	−130	−340
58	−1550	1160	−40	−1830	−1300	−610	−1600	−910	350	270	−430	20	180	420
60	−820	340	80	−1420	−1460	−1180	−1530	800	740	860	800	540	590	1480

Max. profit = $1480 @ 28–60 days

All Ag. Commodities: First Ten Years: Total Futures Profits from Trading Various Moving Average Combinations, Long and Short

No. of days (Slower Moving Ave)	Faster Moving Average													
	2	4	6	8	10	12	14	16	18	20	22	24	26	28
2	−609484													
4	−515183	−146728												
6	−200313	−122755	−124612											
8	−129755	3991	−140665	−92015										
10	−99588	6649	−49271	−41787	−26266									
12	−18777	44627	1697	−5697	2477	37796								
14	13555	80745	41983	43858	51517	19637	57636							
16	64925	100429	118603	157429	113399	120745	101363	98263						
18	93337	173719	142305	189561	158557	185581	153161	21497	119923					
20	136841	197009	176633	208532	174275	173135	164427	104083	68971	124701				
22	158113	215799	210161	198477	186123	196549	178323	150227	122337	77341	125879			
24	165083	226857	234543	219109	187781	199231	176289	154705	127357	74341	33805	116127		
26	191629	238961	219853	223885	196531	177496	150651	151609	111209	102009	99355	58707	98592	
28	194475	222583	197405	202845	170917	163589	151193	132861	122833	105071	114821	83891	−6359	80728
30	206707	203515	202933	203623	187816	161479	179911	144033	107051	84923	99615	102941	63975	21577
32	189737	201951	190113	217103	192621	174227	174925	161387	146955	125439	111899	96513	60993	54469
34	190353	205117	185499	191405	185869	184805	187055	182483	144269	144949	137571	123153	93245	85951
36	171627	205677	198001	205497	201993	203105	198293	180989	161097	145253	135021	143035	112495	101921
38	178055	199847	218069	219595	213609	230009	213549	181087	171859	160801	156697	161187	104567	105769
40	167475	209805	199299	213759	216609	212367	196483	187291	168471	167895	151029	145575	88721	92075
42	192413	207501	193983	215755	212053	213437	183633	223039	159917	17881	155933	132243	102931	112389
44	162397	218127	204342	209779	204461	209685	194829	186741	176195	168907	167469	126447	114843	116093
46	166775	213589	210551	198329	193573	207979	205761	184809	182199	169553	155887	138065	107625	123293
48	181687	195859	209001	190829	199315	205205	197299	200485	176287	182565	157225	139517	132773	140361
50	188841	187647	202701	194937	202107	201539	197369	195269	183127	199353	170777	150485	153099	156365
52	185859	213707	194695	223023	220485	203295	198457	198561	214921	196553	187845	156205	171417	154003
54	185123	227837	218553	241659	216793	213397	206849	188601	210693	183323	152146	155207	162147	154524
56	210266	236414	209216	236670	228854	211426	198222	196154	191882	179750	166196	162126	177486	143808
58	218031	219899	227558	219743	207809	206073	198807	179001	180007	178197	164089	176015	151737	140733
60	207658	227492	213148	210624	216978	204396	194322	185504	180712	180600	168228	173748	165052	144560

Max. profit = $241659 @ 8–54 days

All Ag. Commodities: Second Five Years: Total Futures Profits from Trading Various Moving Average Combinations, Long and Short

No. of days (Slower Moving Ave)	Faster Moving Average													
	2	4	6	8	10	12	14	16	18	20	22	24	26	28
2	−375698													
4	−288766	−204056												
6	−202479	−202126	−204712											
8	−156707	−161994	−167480	−120566										
10	−116770	−121844	−111336	−94228	−67992									
12	−87248	−85438	−81098	−60232	−91214	−48602								
14	−67002	−69034	−50920	−13574	−40552	−58384	−57474							
16	−57364	−48824	−36922	−16666	−18556	−42214	−55200	−21970						
18	−30775	−19504	−11402	8508	−14402	−24224	−17812	−55314	−24712					
20	−6374	−11916	−1314	8102	9944	−15900	−14824	−18454	−25322	−15480				
22	−11116	−12384	4446	16228	9244	−1308	66238	−8958	−13062	−32126	−26806			
24	−9368	−19088	9262	17434	22544	1710	−4838	−3738	920	−7888	−46882	−27696		
26	−11256	−9126	−6972	12914	8098	−384	−282	−5646	2696	−12952	−14364	−43110	−40984	
28	−12176	−6210	7552	15112	3920	−5872	806	−7460	−790	−15010	−19268	−34030	−64796	−26478
30	−3230	−1990	14622	8682	−7302	−2188	10106	−15640	1664	13078	−18278	−29980	−46610	−60616
32	−18320	−9142	20326	−2540	−13250	−6112	−12960	110	−12490	−21288	−19222	−30880	−44804	−50120
34	−17038	5244	21120	−4578	−11996	−8576	−5794	−5364	−19274	−15826	−23760	−26154	−27830	−31798
36	−10932	188	12224	−144	−3388	−11644	−9438	−14439	−13668	−29748	−26020	−14882	−23928	−34796
38	−6428	8706	14552	−5644	−6448	−14000	−3922	−22868	−29396	−35622	−23194	−17544	−31028	−25028
40	1792	2562	17616	−3538	−15468	−20766	−11370	−26600	−33840	−25780	−13316	−15558	−33090	−31642
42	−3258	8190	12378	−6084	−9182	−20226	−24328	−30840	−22746	−28424	−23086	−31040	−34438	−22300
44	1552	17814	98	−13980	−15726	−22612	−36206	−31172	−30112	−21090	−11736	−24180	−26176	−29042
46	−3546	7512	7556	−4762	−20812	−24358	−34804	−27358	−20006	−16574	−10006	−21214	−35898	−26396
48	−1942	7968	15974	−1706	−19670	−16610	−30510	−28140	−11538	−12572	−6204	−24572	−31280	−31474
50	68	6418	16956	5010	−13904	−17656	−31156	−21068	−6628	−9902	−11706	−28240	−20822	−23392
52	−11274	5574	14294	290	−3454	−11942	−25972	−22970	−8382	−2070	−15638	−10770	−17050	−26932
54	−8332	5520	15964	12126	−14	−13764	−37030	−22232	362	−4758	−12210	−5994	−18437	−18548
56	−4412	7494	9328	14678	3684	−10208	−29114	−12478	−2432	−1208	−2582	−6940	−22154	−17872
58	−9872	6402	−50149	10506	−2390	−11672	−31206	−4646	−2272	3986	−4774	−14582	−22328	−16338
60	−7240	−1410	4842	6292	−1738	−12666	−16176	−1145	3880	8074	−1098	−14462	−17912	−7046

Max. profit = $66238 @ 14–22 days

Chapter 7

Basis: The Economics of When and Where to Sell

Basis: The Economics of When and Where to Sell

Each commodity traded on a futures market has a parallel market for trading the physical commodity, or actuals. We refer to the latter as the cash market, or spot market. However, it is not a single place, like the futures market. Instead, cash markets are scattered across the landscape at the nearest elevator, auction barn, or buying station down the road from every producer.

The two markets have related, but distinctly different, prices arising from their respective price-making forces. These price differences and their relationship to each other constitute the study of **BASIS**. An understanding of the relationship and factors affecting it is one key to profitable marketing. Basis signals tell a commodity producer **when** and **where** to sell or buy the physical commodities on the cash market.

Place and Time Price Differences[1]

Price differences cause commodities to move from areas of surplus production to areas of deficit production, in both the location and time dimensions. The southeastern poultry production region depends upon midwest grain production as a feed source. Export markets in Toledo, New Orleans, and Portland, Oregon, depend upon interior production areas to supply commodities for export shipment. Price differentials between production and consumption or export locations must be sufficiently wide to pay the costs of transportation and handling. Otherwise, there is no incentive to move commodities between locations.

In a time context, concentrated harvest periods create short-term surplus supplies relative to immediate consumption needs. During later periods, consumption exceeds production (there is no production until the next harvest). Price differentials over time encourage storage until the supply is needed for consumption; these differentials ration supply over time.

On a given day, location and time factors cause many prices to exist for a particular commodity. For example, on a November day, some prices for No. 2 yellow corn were:

By Location $/bu.		By Time $/bu.	
Gulf	$2.15	Central Illinois	$1.84
Chicago	1.95	Dec futures	1.95
Cincinnati	1.90	Mar futures	2.07
Decatur	1.86	July futures	2.20
Des Moines	1.66	Sept futures	2.26
Omaha	1.72		
Iowa farm	1.50		

These individual corn prices are functionally related to each other and actually existed on that particular day. As such, they should be viewed as one price. Differences in price reflect differences in location and time of delivery.

Basis is the name of the spreads between prices. It is defined as "the number of cents per bushel that, on a particular day, a local commodity cash price is **above or below** the price of some futures or cash price at another market location."[2] On this November day, corn prices were: December futures = $1.95; Chicago cash (spot) = $1.95; central Iowa farm price = $1.50; and Gulf = $2.15. In basis terms, Chicago cash was **even** with the December futures, Iowa farm price was 45 cents **under** December and the Gulf price was 20 cents **over** the nearby December futures. Additional basis definitions are given in Table 7–1.

Futures markets exist for both storable and non-storable commodities. Significant differences in basis concepts exist for the two. Price and basis relationships for non-storable commodities will be discussed in Chapter 9.

TABLE 7–1: Basis Definitions

Basis: The difference between a futures contract price and a local cash price.

> Futures Price $1.95 December Corn CBOT
> − Local Cash Price $1.66 Des Moines November Cash Bid
> (Spot or Flat Price)
> = Basis −$0.29 *Under* Nearby December

Negative Basis: Futures contract price is higher than cash price. Quoted as: "cash is *under*."

Positive Basis: Futures contract price is lower than cash price. Quoted as: "cash is *over*."

Strong: Price difference is relatively small (cash is 10 cents *under* Dec. Futures).

Weak: Price difference is relatively large (80 cents *under*).

Option or Option Month: The specified futures contract from which the basis is calculated.

Nearby: The next option month during which delivery on a contract could be made. On Feb. 10, March is nearby corn contract month.

Distant: Some option month later in the year. On Feb. 10, July is a distant corn contract month.

Old Crop: Futures contracts on the crop that has most recently been harvested.

New Crop: First futures trading month of crop not yet harvested: Dec. corn, Nov. soybeans, Dec. cotton, July wheat.

The Basis Thermometer

It is the move in the basis, to strengthen or weaken, that generates profits or losses in the basis market. Strengthening of the basis means that the cash price increases **relative** to the futures price. Similarly, a weakening of the basis means that the cash price decreases **relative** to the futures price. A few readers experience difficulty in putting the basis concept into focus. Their major problem seems to be in losing sight of the fact that it is the **relative** prices (the cash versus the futures price) that is

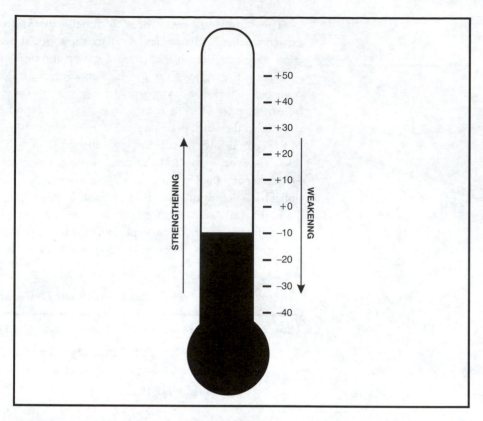

FIGURE 7–1: The Basis Thermometer. Source: Schap, Keith, **Commodity Marketing,** *A Lender & Producers Guide to Better Risk Management,* **American Bankers Association and the Chicago Board of Trade, Chicago, IL, 1993, page 46. See also pages 49 to 51.**

important and **not the absolute** price level that is the current issue. To assist in clarifying these misconceptions, we offer the "Basis Thermometer" in Figure 7–1.

Like a thermometer that measures air temperature, the basis thermometer measures the temperature of the market. In this case, it measures how "hotly" the market participants are willing to contest the ownership of agricultural commodities, here and now, versus some more distant time and/or place. One major difference is that, particularly out in the country, the basis thermometer has a negative reading, –20 or 20 **under** the March future because Ames, Iowa is spatially removed that far from the Chicago market. On **rare** occasions, when the market desperately wants Ames corn, here and now, it will bid a positive premium: +7 cents **above** the March futures. The further out in the country, the higher the cost of moving grain to a central terminal, the lower (more negative) the basis thermometer. Twenty below may be cold, but 30 below is colder and while 10 below is still cold it is warmer than 20 below—it is all relative and conditioned by what one is accustomed to. When the basis moves from –20 to –10 the basis is strengthening. When the thermometer falls from –10 to –20 the basis is weakening. Eighty below is not only cold, but it is a weak basis. Even though 80 below is cold, it is 10 better than the –90 basis norm in that "brutal" part of the world.

While the cash is usually below (under) the futures, there are places where the basis is normally positive. The Gulf (New Orleans) cash corn basis is usually +25 (above) the futures. It sometimes improves to +30 and often deteriorates to less than

+5. The same is true for Norfolk, VA and Portland, OR. Why? Norfolk and Portland are 40 cents and half a continent closer to European and Asian markets.

Components of Basis

An explanation of the difference between a futures contract price and a local cash price of storable commodities requires understanding several different components. The first component is the amount that Central Illinois cash price is below the nearby futures contract. This difference arises from handling costs in loading and unloading, the handler's profit margin, and the costs of storage until grain can be delivered against a futures contract.

The second component is the amount by which cash price in the country is below the cash price in Central Illinois because of transportation cost. If transportation were free, or if the cost of transportation were not related to distance, then the price of a commodity would be the same in all locations. Transportation is not free, and it is related to distance. Thus, commodity prices on a particular day vary considerably from location to location.

The third component of basis arises from the carrying charges incurred in holding commodities from harvest until they are used on some future date. Storage is a market function that must be performed by someone who is willing to incur the costs of providing that service. The returns to storage are offered in the market by basis spreads between cash price and different monthly future contracts. As time of delivery on a futures contract approaches, the reduction in carrying costs causes futures and cash price to converge. The time relationship of prices is an important part of basis.

The effects of these various components are summarized in Figure 7–2 as they contribute to price differences that existed on any particular day. The value of corn is not one price, but several, as different components are bid into the market to account for the place and time differences.

July Futures Price	**$2.20**
Basis Due to Storage and Carry	−0.13
March Futures Price	**$2.07**
Basis Due to Storage and Carry	−0.23
Central Illinois Cash Price	**$1.84**
Basis Due to Location	−0.18
Des Moines Cash Price	**$1.66**
Basis Due to Location	−0.16
Iowa Farm Price	**$1.50**

FIGURE 7–2: Components of the Corn Market Basis on a Particular Day in November.

Relationship of Cash and Futures Prices

Two features characterize the relationship between cash and futures prices over time, as illustrated in Figure 7–3.

1. Movement up and down together, called "parallel fluctuation," and
2. Convergence of cash and future prices.

Each feature has its reason for occurring, and each is important in understanding hedging as a business tool.

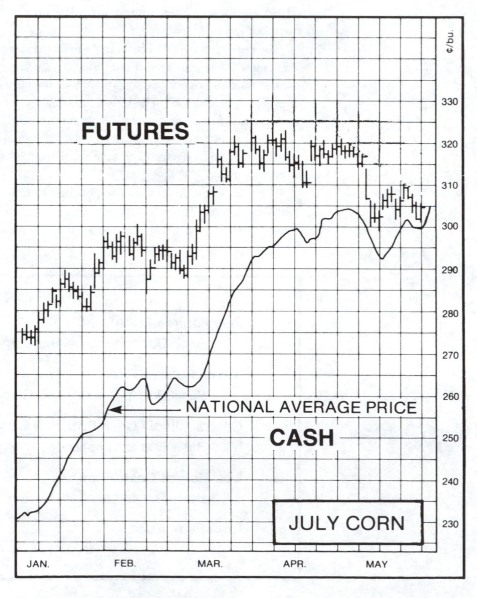

FIGURE 7–3: Illustration of Fluctuation and Convergence of Futures and Cash Commodity Prices.

Parallel Fluctuation of Cash and Futures Prices

Cash and futures prices tend to move in unison because they are affected by the same fundamental forces of supply and demand. These forces are usually first registered in the futures market. The trading pits offer convenient facilities for trading and are an arena for price discovery. Dealers in the actual commodities, such as local elevator managers, cotton merchants, livestock buyers, or exporters, look to the futures market for guidance with respect to the cash price they bid or ask for a commodity. Stronger futures prices encourage them to raise cash bids; weaker futures often lower cash bids. Even if a local grain elevator manager does not hedge wheat purchases in the futures market, but instead calculates the offering price to farmers on cash price in a terminal market, that local price is still linked to the futures market, since the terminal's bid is based on the futures price. Participants in a trade typically use a futures price as reference in negotiating the cash price. But the ultimate determinant of value is what buyers are willing to pay and sellers are willing to take in a cash transaction in the spot market. Reluctant sellers can withhold supply from the market and force cash prices higher, or they can overload the market at harvest and force prices lower.

Invariably, the question arises: Which price actually controls which? The answer is neither. The cash price influences the price of futures, and the futures price influences cash. On occasion, the cash leads the futures, but most often, futures price leads the cash.

The foregoing does not mean that cash and futures prices are the same in dollars and cents. It only means that the law of one price operates so that most price differences can be attributed to transportation costs between locations and carrying charges that reflect holding costs over time. If price differences between locations exceed transportation costs, or bids among different months exceed carrying charges, a speculator, seeking a profit, will buy cheap in one market and sell dear in another. Arbitrage trading moves cash and futures prices into the "proper" relationship to each other, as determined by fundamental economic factors.

Once relative values are established, the tides of rising or falling price level affect both prices. It is the parallel fluctuation of futures and cash price that makes it possible to hedge for price protection.

Convergence of Cash and Futures Prices

Cash commodities can be delivered to satisfy the obligation inherent in a short futures contract. Consequently, during the period of delivery against a maturing futures contract, the cash price of a commodity and its futures price are forced to be equal, save minor transaction costs and some uncertainty, at the point of delivery. The basis spread between the two prices approaches zero, no matter how wide the difference between the two prices might have been on some earlier date. Market forces squeeze cash and futures prices together on this date and at this location.

Consider the situation of a terminal elevator manager in Chicago. The firm has representatives on the floor of the Chicago Board of Trade and a storage facility half-filled (Figure 7–4). The elevator can receive grain into storage or make deliveries from it. To maximize profit, the manager wants to buy grain "below the market" or "sell it above the market." If the futures price is higher than cash, the manger will "sell" a contract to establish a delivery obligation at a price higher than the value of the grain in the warehouse. Upon maturity of the futures contract, the manager only needs to deliver a shipping certificate for grain and collect the higher price. On the other hand, if the futures price is less than cash, the manager will "buy" a futures contract and hold it in order to accept grain delivered against the obligation. The manager will have bought grain below the market price. Thus, the terminal operator's practice of arbitrage between the cash and futures market forces the two prices to be essentially equal in months when futures contracts mature. The opportunity to make or take delivery against futures contracts is the key element that forces prices to converge.

The convergence relationship creates the opportunity to capture a storage profit from basis change, as we shall soon observe.

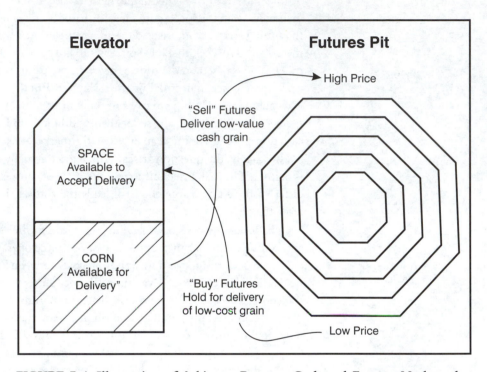

FIGURE 7–4: Illustration of Arbitrage Between Cash and Futures Markets that Causes Price Convergence.

Basis and Hedging

There are two reasons for hedging. One is for price protection; the other is to capture a storage return. Each use respectively depends on parallel fluctuation or on convergence of cash and futures prices. The first can be illustrated with a pre-harvest hedge, the second, by the storage hedge.

The Pre-harvest Hedge: Protecting Price Level

A farmer who is concerned that price level might decline during the growing season can protect against that risk by selling a new crop futures contract, that is, by hedging. Suppose on July 1, December corn futures are trading at $3.70 per bushel. If this price level is satisfactory, a producer could hedge away effects of price change. The approximate price of $3.70 is protected by the equal, but opposite, positions in two markets—a long position in the cash, offset by a short position in the futures. The producer will not be in a position to realize the full $3.70 price, since the Ohio farm is not located in Chicago or along the Illinois River and he or she would have shipping and handling fees in making delivery to satisfy the futures contract. The hedge will protect a farm gate price equal to $3.70 less delivery costs and handling fees incurred in delivering against the futures contract.

But few producers deliver against futures in Chicago or Illinois river terminals. Most sell cash grain on their local market and offset their futures obligation by buying back (offsetting) the futures contact. They enjoy price protection without the cost and hassle of distant delivery.

Knowledge of local basis becomes essential in order to translate a given futures price into a probable cash price for local delivery. The producer "localizes" the futures price to estimate the cash price he or she will receive by placing the hedge. The futures price, plus or minus basis in the local market in the month of delivery, is the price that the futures market is offering for grain delivered to the local elevator during a particular month. A historical basis record might show that cash corn price averaged 10 cents under December futures during mid-December for the producer's location. By subtracting the 10-cent average basis from the futures contract price, the anticipated "localized" price is $3.60.

Actual basis at time of delivery may be greater or less than the average 10 cents and affects the final results of the hedge. If the basis is 20 cents under at delivery, the local price will be $3.50. If the basis is zero, the protected price is $3.70. Thus, a futures hedge removes the effects of price level change but does not remove basis level risk. Since year-to-year basis fluctuation is usually less than price level fluctuation, hedging fixes an approximate price that can be fairly accurately projected.

The Storage Hedge: Earning a Storage Profit

Hedging can be used to earn a storage return. It is done to take advantage of expected convergence of cash and futures prices. In this use, a hedge is a profit-making tool, as contrasted to a risk management tool otherwise used to protect a price level.

It is common business practice for a warehouser to use a storage hedge to hold commodities over a period of time, when basis is expected to narrow. A storage hedge has two purposes: (1) to protect the value of inventory against adverse price level change, and (2) to capture a storage return during the storage period. The alternative is to store commodities unhedged and speculate on price. Few can afford that luxury. There is no guarantee that price level will rise over the storage period to pay the costs of storage. It may fall dramatically. The storage hedge protects the elevator operator from a price level change. It also serves to lock in a storage return. This tool can be used effectively by farmers who own or rent storage space.

A storage hedge is usually established at harvest. Relatively low-priced cash grain is deposited in storage facilities, and an equal amount of higher-priced futures contracts are sold. The futures position counteracts the effect of price level change while allowing basis change to occur. As long as the difference between cash and futures prices shrinks, or moves to a premium of cash over futures, there is a basis gain. Figure 7–5 shows five different combinations of price changes that occur to produce the expected convergence of cash and futures price. In each case, the narrowing of basis earns a storage return, irrespective of price level change.

For example, at harvest in October the cash soybean price is $6.40, and the March futures trading at $7.40. From a basis history, a southwest Ohio farmer determines that the Cincinnati basis in January averages 35 cents under the March future. By localizing the March contract, the farmer calculates that the futures market is offering

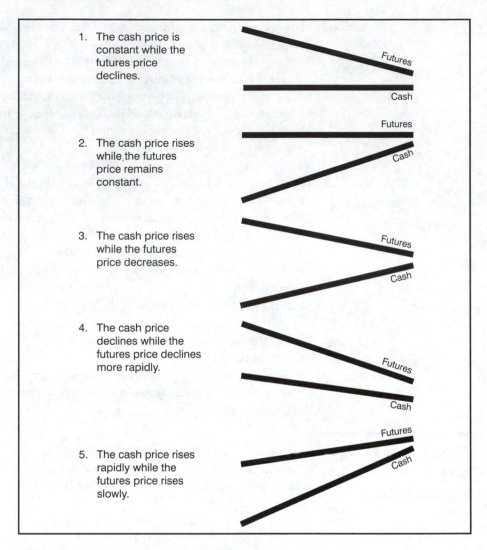

1. The cash price is constant while the futures price declines.

2. The cash price rises while the futures price remains constant.

3. The cash price rises while the futures price decreases.

4. The cash price declines while the futures price declines more rapidly.

5. The cash price rises rapidly while the futures price rises slowly.

FIGURE 7–5: Cash-Futures Relationships Causing Basis to Narrow.

about $7.05 for January delivery. Compared to the harvest bid of $6.40, the farmer expects to capture a 65-cent gross storage return by a storage hedge. The farmer stores and hedges. In January, basis is 35 cents under. The farmer sells cash soybeans and lifts the hedge. The results can be traced as follows:

Date	Cash Market	Futures Market	Basis
Oct. 15	Store SB worth $6.40 for later sale	Sell March SB contracts @$7.40	$1.00
Jan. 15	Sell Cash SB @ $6.65	Buy March SB contracts @ $7.00	0.35
Gain (loss)	+0.25	+0.40	+0.65

Results: Jan. 15 Cash Price	$6.65
Plus Futures Gain	0.40
Net Price	$7.05
Fall Value of SB	$6.40
Storage Return	$0.65

In this example, futures price level declined and cash price increased. Without the hedge, the farmer would have received $6.65 for the grain, or only a 25-cent storage return. The hedge protected against price decline in the futures market by capturing 40 cents of price protection. These numbers illustrate situation #3 in Figure 7–5. Each of the other four situations will produce the same results if the prices produce a 65-cent convergence in basis.* Situation #5, with a rising futures price level, is the only one to have negative results. Thus:

Date	Cash Market	Futures Market	Basis
Oct. 15	Store SB worth $6.40 for later sale	Sell March contracts @ $7.40	$1.00
Jan. 15	Sell Cash SB @ $7.20	Buy March SB contracts @ $7.55	0.35
Gain (loss)	+ 0.80	(– 0.15)	+0.65

Results:		
Jan. 15 Cash Price		$7.20
Less Futures Loss		–0.15
Net Price		$7.05
Fall Value of Beans		$6.40
Storage Return		$0.65

The storage return of 65 cents is the same as the other examples because of narrowing basis. However, the cash price is reduced by 15 cents due to the futures loss; the farmer would have netted $7.20 instead of $7.05 had he or she not placed the storage hedge. This example emphasizes the fact that a hedge eliminates price level improvement as well as providing protection against falling price level, while capturing the storage return. If a warehouser were certain that futures price level would rise, he or she should delay setting a storage hedge and speculate on both price level and basis. Only when the price level started to fall would the warehouser set the hedge. In actual practice, storage managers cannot afford the risk of not hedging. They cannot predict when price will fall, nor how fast. Three limit down moves in price on 1 million bushels of stored grain generates a financial disaster of $900,000 loss.

Profitable storage hedging is based upon the expectation of cash and futures price convergence. A problem of storage hedging is that basis levels cannot be predicted exactly. The farmer or warehouser will know the basis level on the day of storage. However, he or she does not know what the basis level will be at the time of the cash sale, at a later date. While historical averages provide a guide as to what to expect, basis fluctuation is a totally different market situation from price level fluctuation. A storage hedge instituted in October with the expectation of a "normal" basis of 35 cents under March during January may actually produce a larger or smaller return than anticipated. If basis is weaker than 35 cents when the cash soybeans are sold and the futures contract lifted, the storage return will be less. On the other hand, a stronger

---

* The 65 cents is a storage return—not storage profit. The cost of facilities, risk, management, and money tied up in the grain must be subtracted from the 65 cents to determine net profit from storing.

basis than 35 cents will yield a larger storage return. The storage hedge removes price level speculation but does not eliminate basis speculation. Variations in basis notwithstanding, basis is normally much more predictable than the general price level because of required convergence. This fact enables hedgers to substitute basis risk for price level risk through use of futures contracts. Skill must be developed in forecasting basis change to be successful with storage hedging, however.

Cash and Futures Prices Over Time

Cash and futures prices are related over time by carrying charges. This relationship depends upon four conditions:

1. Storable commodities are harvested at one time of year and used at fairly even rates throughout the year. Hence, inventories must be carried forward.

2. Costs are incurred in the process of storing, financing, and maintaining quality of the commodity.

3. There is virtually no cost in holding futures contracts.

4. Cash and futures prices will be essentially equal in the month of maturity at delivery points.

A soybean processor in Decatur, Illinois has a constant supply requirement each month to keep the plant operating at capacity. If, at harvest, the processor decides to cover the processing needs for next July, he or she must have storage space to hold the inventory. That space has fixed investment, depreciation, and maintenance costs. Additionally, the processor must operate the space and invest capital equal to full value of the grain. On the other hand, July futures contracts provide the right to receive soybeans needed to keep the plant operating in July. The only cost of holding futures contracts is the interest on the margin money deposited and brokerage fees for the transaction. Whether harvest price level is high or low, the cost of holding soybeans in storage is about 2 cents per week, or about 80 cents per bushel for 9 months. If the difference between the October cash and the July futures is more than 80 cents, the processor can hold long futures contracts at lower cost and wait for delivery of soybeans. If the basis spread equaled carrying charge, the soybean processor would be indifferent as to whether the firm owned the physicals or futures. The effective cost of soybeans as a production input for July would be the same.

Figure 7–6 depicts a widening price spread, or divergence, between futures and cash as we move back in time from the end of the crop year in July to harvest the previous November. Theoretically, cash price at harvest differs from the end of a crop year futures price by the exact cost of holding the cash commodity over time. Furthermore, prices should diverge an equal amount each day. If actual cash price change followed its theoretical path, carrying charges would exactly equal the amount of convergence.

The theoretical relationship of carrying charge between cash price and futures options forms the benchmark for basis analysis.

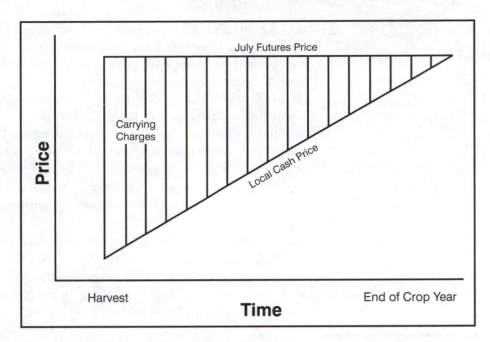

FIGURE 7–6: Theoretical Convergence of Cash and Futures Prices Over Time from Harvest to End of Crop Year.

Basis Level Negotiation in a Market Economy

In the real world, basis spreads never exactly fit the expected theoretical relationship. The deviations from the expected offer profit opportunities and require analysis of the basis market. Understanding what causes deviation prepares commodity industry participants to take profitable advantage of them.

Rather than conforming to what should exist as carrying charges, actual basis level is a negotiated price. The negotiation process is predicated on the fact that somebody has to pay the costs of holding grain from harvest until it is used. The decision to convert grain into cash (by the producer) or cash into grain (by the user) depends upon the calculation of costs involved and returns expected. If the spread between cash price and a futures contract is greater than holding costs (excess carry), there is an incentive to buy cash grain from a motivated seller (farmer with no on-farm storage), hedge away the effects of price level change, and hold the commodity until it can be sold or is needed for processing. On the other hand, if the spread will not cover holding costs (less than full carry), a commodity processor can protect against the possibility of price level change and at the same time guarantee delivery by buying a futures contract and holding it for delivery. In this case the processor is guaranteed receipt of the grain when needed but does not incur the cost of holding grain.

The relative offerings of cash commodities versus futures contracts determine basis. Heavy cash sales relative to futures widen basis. Light cash sales relative to futures narrow basis.

The carrying charge component of basis becomes the **market price for storage space.** When the available supply of a commodity is large relative to space available, and grain is flowing to the market rapidly, the cash price is weak relative to the futures so that the basis spread is wide. The market is signaling that the commodity is not needed for use at the current time and should be held in storage by the producer until later. The demand for storage is greatest at harvest, and therefore, the price for storage is typically the highest, or basis weakest at that time. Once harvest is complete and commodities are in storage, basis narrows quickly. Thereafter, basis level depends upon reluctance or eagerness of sellers to move grain out of storage and into use. When offers to sell are limited and the commodity is flowing slowly to the market, but demand is strong, basis is strong.

The demand for a flow of cash grain into the market, relative to reluctance of farmers and warehouser to sell, sets the tone of basis negotiation. If rivers are frozen, and roads and railways clogged, a New Orleans exporter will offer a premium cash price to avoid demurrage fees on waiting ships. That premium is expressed in the market as a strong basis level. It ripples through the market to entice sellers to deliver grain.

Negotiations to determine basis take place in two different price relationships. The first concerns the relative values of cash grain and the nearby futures contract. The second concerns price spreads between various futures contract months. Both are related to what should theoretically exist, but both are modified by current and anticipated conditions.

Cash Versus Nearby Futures

On any given day, the cash price in Chicago is determined by trading between large buyers and sellers of grain. Terminal elevator operators who own grain, and exporters, processors, and merchandisers who need grain, are constantly bidding in cents per bushel under or over the nearby futures prices for grain for immediate delivery. If demand for cash grain is strong and/or the current available supply is small, buyers become eager bidders, and elevator operators holding stocks become reluctant sellers. Without any change in worldwide supply and demand conditions to affect the futures market, local demand for cash trading produces strength in the cash market. The cash market rises relative to futures price and strengthens the basis. Conversely, the demand for cash grain for immediate delivery may weaken because of a decline in processing or export requirements. In this case, cash price in Chicago may fall relative to the futures price, producing a weaker basis.

Farmer Psychology and Basis

An important determinant of basis results from farmer reactions to price level. Farmers are willing sellers at high prices and reluctant sellers at low prices. At what they judge to be low prices, they are willing to defer selling decisions because they feel downside price moves are less likely than improving prices. They are willing to continue to speculate on cash price. At high price levels, they sell more readily because they gauge that downside risk exceeds the probability of higher prices.

Since farmers are typically attuned to local cash prices more than to futures prices, their market actions significantly affect basis levels. When futures prices are high, spot cash prices are likewise strong. Farmers are willing cash sellers; there is no need for buyers to bid a premium to entice farmer selling. On the other hand, when futures prices are low, cash prices are low and farmers are reluctant sellers. Buyers are required to bid a basis premium to entice sellers. Basis is, therefore, wide at high price levels and narrow at low price levels. Farmer reaction to price level and the resulting basis levels are summarized in Figure 7–7.

Elevator operators are well aware that they earn larger profits in high-priced markets than they do in low markets. The opportunity to buy cash grain at wide basis discounts and subsequently lay off hedges to capture potential basis gain is improved

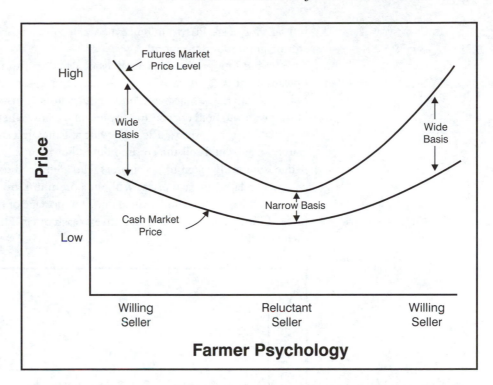

FIGURE 7–7: Illustration of Farmer-Seller Reaction to Price Level and Resulting Basis Levels.

when farmers are willing sellers. At low prices, narrow basis does not offer significant storage returns.

Contract-to-Contract Futures Spreads

The second basis negotiation determines the spread between futures prices in different contract months. Prices on various contract options are related to each other by carrying costs over time, theoretically calculated to cover holding costs. In actuality, they are negotiated between buyers and sellers.

Futures contracts trade for delivery in selected months rather than daily. Hence, the convergence is reflected in contract option prices that normally stairstep upward. For example, if carrying costs to hold soybeans were calculated to be 10 cents per bushel per month, various futures option prices should be related to harvest price and to each other, as shown in Figure 7–8. An October 1 harvest-time price of $6.00 per bushel would be exactly equivalent to the following August 1 futures price of $7.00. The $1.00 basis spread between the October cash and August futures would equal "full carry," the $1.00 cost of holding soybeans for ten months. All futures contracts would be related to the October cash price and to each other by the cost of carrying grain from futures month to futures month.

In practice, the spreads between futures prices for different delivery months do not always equal full cost of carrying inventory. Price differences between contract months can be more than storage costs, producing a situation called "excess carry." The spread in prices is greater than the cost of carrying grain into the future. This situation seldom exists for very long because commodity traders are quick to capitalize on market differences that exceed costs of storage. On the other hand, month-to-month differences between successive futures contracts can be less than full carrying costs. Price differences between contract months are less than the cost of storing grain month to month.

The bidding process that establishes whether futures contract prices stairstep upward to reflect a full carry, less than carry, or excess carry takes place between warehousers (who may be farmer-producers, elevators, or users) and speculators. A warehouser wants to hold the commodity in storage only when he or she can sell a futures contract at a price spread sufficiently wide to generate a storage profit. Otherwise, the warehouser would sell the commodity rather than store it. The opposite side of a hedge is typically taken by speculators who are forced to pay a premium for futures contracts relative to cash price. Without a premium, no hedgers would offer to sell futures; all would be selling cash grain. The shortage of contract offerings causes the price of futures contracts to rise relative to cash price. Thus, the stepwise price rise is established.

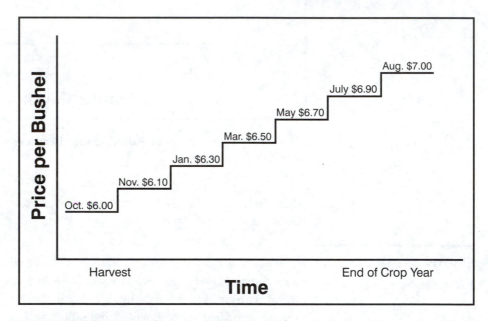

FIGURE 7–8: Stairsteps in Soybean Futures Contract Prices to Reflect Cost of Full Carry, From Contract Month to Contract Month.

In holding actuals and selling futures, the warehouser benefits when the spread between cash and futures in various months is as wide as possible. The warehouser is locking in a differential between what the cash commodity costs and what he or she can guarantee by hedging the price level for future delivery. The warehouser chooses the contract month that offers the widest spread relative to holding costs. Heavy offerings of contracts in one month depress price relative to other months, emphasizing the contract-to-contract steps in price.

Speculators who practice spread trading add emphasis to contract price spreads. On one hand, spread trading narrows contract-to-contract basis as traders bid excess returns out of the market. On the other, spreading widens basis, if there is no "carry" bid into the relative prices. Speculators can remove the effects of price level change from their trading plan and limit their speculation to basis level change through spread trading. If the price spread between nearby and distant futures is small, a spread trader would establish offsetting futures positions in the market by selling the nearby and buying the distant months. He or she knows that if warehousers are motivated to sell cash commodities early in the season, a shortage of supplies later will lead to higher price. Cash markets will pull the distant futures price upward. Less than full carry between contract months produces a market bias that should produce a profit from the spread position, as calculated in Table 7–2. The trader profits from the widening basis as market forces react to excess supply in the present relative to shortages later. Spread trading adds to the stairstep price effect by increasing selling pressure on nearby contracts and buying pressure on distant ones.

**TABLE 7–2: Illustration of Profitable Spread Trading
of Corn Futures Contracts**

Set Spread		Lift Spread		Results	
January 20		February 20			
Sell March	276	Buy March	296	Loss	20 cts
Buy May	280	Sell May	316	Profit	36 cts
Basis	–4	Basis	-20	Net	16 cts

There are economic reasons for the nearby futures option to be higher than cash price, and deferred options higher one to the other over time. But each step is a price negotiated in the market place rather than a fixed amount. The cash-to-nearby option and the contract-to-contract spreads in futures develop as a bidding contest between hedgers who wish to widen the spread and speculators who wish to narrow it. The bidding process establishes a "normal" relationship between different futures contract months. Warehousers want to earn returns based on costs calculated from new cost of facilities, full depreciation, and prime interest rates on capital invested in inventory. However, if they have older facilities with lower costs, or if they are willing to accept lower rates of return, they will calculate costs significantly below those of a newly built facility. They then offer contracts at less than the full annual carry. They will not offer zero carry, as the speculators want, because their alternative is to sell cash commodities and not offer contracts.

In a competitive market environment, "normal" carry is bid into the market place by the offerings of cash or contracts by hedgers and speculators. The market may offer full carry through wide basis, or less than full carry due to narrow basis, depending upon who prevails.

Local Basis Year to Year

Country basis changes do not simply mimic Chicago basis charges. Relatively low country prices, resulting from a widening of basis, may be triggered by higher transportation charges, by a weakening of local demand for grain, by a local supply larger than the storage space available, or by a shortage of railroad cars. Conversely, a short local supply, strong local demand for grain for use or inventory accumulation, and a favorable change in freight rates could result in narrowing basis and raising local prices relative to the futures price.

Just as basis can vary week to week or month to month in any year, it can vary from one year to the next. The reasons may indicate a permanent shift in the relationship or a temporary aberration that will be quickly corrected. A permanent change may result from a change in local demand, as livestock feeding increases or decreases in the area. Likewise, opening or closing of a grain-processing plant in the community, abandoning or restricting use of railroads, and changing barge freight rates will affect local basis positively or adversely.

Local basis is most affected by size of the crop and available storage space. In years of large crops and storage shortages, basis tends to be weaker, especially during harvest months. The amount of farmer-owned storage relative to size of crop has a significant effect upon harvest-time basis. Long lines of trucks waiting to unload at elevators encourage wide basis. Short lines cause more aggressive bidding and strong basis, as elevators try to fill their now empty bins.

Basis Records and Charts

Since basis varies from year to year and place to place within a year, commodity producers and users need to know their individual local historical basis patterns. Basis patterns provide the information for timing cash sales or purchases and for deciding whether or not to store a crop, what contract month to use for hedging, and when to turn an unusual basis situation into a profitable opportunity. In addition, knowledge of basis can identify the most profitable contract month for hedging price level.

Each producer faces a different basis climate depending upon the cash market alternatives available. It is necessary for individuals to maintain their own basis records; a national average is not satisfactory. A producer should at least maintain a weekly record of cash and futures prices. Select a day of the week to record local cash bids and the futures prices. Friday's prices give the producer a longer time span to reflect on the information until the market opens on Monday. But all too often, Friday's trading is distorted by speculators avoiding positions over the weekend or elevators taking protection against possible price changes before trading resumes on Monday. A midweek day (such as Wednesday) is preferable. The extra work of maintaining a daily record during historically strong selling months can better identify the most profitable day to sell.

One handy method is to clip the market report from a newspaper, such as *The Wall Street Journal,* and paste it in a notebook (Figure 7–9). Record relevant local prices on the same page. This method offers the advantage of recording all contract months. Another way is to create a Table with vertical columns to show date, nearby futures, local cash price, and basis (Figure 7–10). A sheet must be maintained for each commodity sold or bought. If the producer has two or three alternative cash markets, it is best to record all cash bids. Also, one may need to record several contract months. The clipping is a handier method than the Table for all these details.

Plotting Basis Charts   The daily or weekly prices can be converted into a basis chart useful in visualizing basis patterns. To make a basis chart, set some base price equal to zero and plot basis deviations around it. Several types of basis charts can be drawn. They differ only in which price is chosen as the base. The illustrations in Figure 7–11 show the appearance of each of the methods. One method is to set the local cash price equal to zero and plot futures prices relative to it. This method has the advantage of showing sev-

FUTURES PRICES

Monday, January 3, 2000

Open Interest Reflects Previous Trading Day.

GRAINS AND OILSEEDS

CORN (CBT) 5,000 bu.; cents per bu.

	Open	High	Low	Settle	Change	Lifetime High	Lifetime Low	Open Interest
Mar	204½	205¾	200½	200¾	– 3¾	270	195¼	211,737
May	211¼	212¼	207¾	208	– 3¼	261	202½	61,464
July	217¾	219	214½	214¾	– 3¼	278½	209	58,065
Sept	225	225¼	221¾	222	– 2	257	215¾	16,869
Nov	229	229	228½	228¾	– 2	245	222½	559
Dec	233½	234½	231	231½	– 2¼	279½	225¼	33,477
Mr01	242¾	242¾	239¾	239¾	– 1¾	246½	233¾	869
Dec	253	253½	251½	252	– 1½	263	246½	1,114

Est vol 72,000; vol Fri 0; open int 384,452, +306.

OATS (CBT) 5,000 bu.; cents per bu.

	Open	High	Low	Settle	Change	Lifetime High	Lifetime Low	Open Interest
Mar	110	110	107	108¼	– 1	135	107	7,020
May	114¾	114¾	113½	113¾	– 1	133	112¾	2,759
July	112¾	113	111¾	111¾	– ¾	124¼	110½	1,466
Sept	117¼	117½	116¾	116¾	– ¼	130	115¾	649
Dec	123¾	123¾	123¼	123¼	– ¾	135	123	788

Est vol 1,200; vol Fri 0; open int 12,682, –15.

SOYBEANS (CBT) 5,000 bu.; cents per bu.

	Open	High	Low	Settle	Change	Lifetime High	Lifetime Low	Open Interest
Jan	461	466	455¾	456½	– 5¼	632	415	13,916
Mar	472½	474¾	463¼	464½	– 5¼	598	423½	58,472
May	478½	482	471	471¾	– 6¼	554	432	24,780
July	485	489	477½	478½	– 6½	647	440	23,882
Aug	488	488	478½	478½	– 6½	541½	441	2,434
Sept	489½	489½	480	480½	– 6¼	541	450	680
Nov	495	497	485½	486½	– 6	631	453	10,227

Est vol 42,000; vol Fri 0; open int 134,425, –508.

SOYBEAN MEAL (CBT) 100 tons; $ per ton.

	Open	High	Low	Settle	Change	Lifetime High	Lifetime Low	Open Interest
Jan	148.00	148.00	145.50	146.40	– .30	164.00	123.30	10,410
Mar	146.80	148.70	145.70	146.70	– .10	162.00	126.30	31,427
May	148.50	148.50	146.00	146.30	– .90	162.60	127.30	20,929
July	150.00	150.00	147.20	147.50	– .80	164.20	130.00	16,676
Aug	148.10	148.10	147.20	147.20	– .80	164.00	131.00	5,439
Sept	150.00	150.00	147.20	147.20	– 1.10	164.00	132.00	3,120
Dec	152.50	153.00	150.70	150.80	– 1.00	169.00	135.50	6,748

Est vol 13,000; vol Fri 0; open int 96,632, –198.

SOYBEAN OIL (CBT) 60,000 lbs.; cents per lb.

	Open	High	Low	Settle	Change	Lifetime High	Lifetime Low	Open Interest
Jan	15.80	15.90	15.43	15.44	– .31	24.90	15.38	10,615
Mar	16.02	16.11	15.67	15.70	– .32	23.95	15.67	53,022
May	16.38	16.38	16.01	16.02	– .30	23.50	16.01	23,579
July	16.70	16.70	16.32	16.32	– .31	22.30	16.32	19,837
Aug	16.80	16.80	16.45	16.45	– .31	21.00	16.45	4,590
Sept	16.60	16.60	16.60	16.60	– .31	21.70	16.60	2,847
Oct	16.80	16.80	16.75	16.75	– .31	22.25	16.75	3,174
Dec	17.35	17.35	17.02	17.04	– .35	20.62	17.02	13,494

Est vol 15,000; vol Fri 0; open int 132,060, –776.

WHEAT (CBT) 5,000 bu.; cents per bu.

	Open	High	Low	Settle	Change	Lifetime High	Lifetime Low	Open Interest
Mar	249	251½	245¼	247½	– 1	340	236½	85,073
May	260¾	263	256	258	...	322	246¾	10,091
July	273	273	265¾	267¾	– ¾	347	256¾	21,326
Sept	281	281	276	277½	– 1½	335	266½	1,395
Dec	293½	293½	290	292¼	– 1	345	280½	2,983

Est vol 19,000; vol Fri 0; open int 120,912, –236.

WHEAT (KC) 5,000 bu.; cents per bu.

	Open	High	Low	Settle	Change	Lifetime High	Lifetime Low	Open Interest
Mar	277	277½	272½	274¼	– 2	361½	262½	40,982
May	288	288	283½	284½	– 1¾	340½	272¾	7,922
July	298	298¼	293½	294¾	– 2¼	366	282	11,966
Sept	306¼	306¼	302½	304	– 1	346	291	690
Dec	314	316	314	315½	– 1½	354	302	638

Est vol 4,872; vol Fri 0; open int 62,198, +1.

WHEAT (MPLS) 5,000 bu.; cents per bu.

	Open	High	Low	Settle	Change	Lifetime High	Lifetime Low	Open Interest
Mar	321	321	315	316½	– 1½	386	312½	12,319
May	327½	329	323½	325	– ½	389	320	4,952
July	336½	336½	331¼	332½	– ¾	380	327¼	2,296
Sept	...	344	339¼	339¼	– 5¼	385	334½	1,018
Dec	...	...	...	348	+ 1	390	344	210

Est vol 2,470; vol Fri 0; open int 20,832, +480.

CANOLA (WPG) 20 metric tons; Can. $ per ton

	Open	High	Low	Settle	Change	Lifetime High	Lifetime Low	Open Interest
Jan	255.20	255.50	251.40	251.70	– 2.30	353.50	250.00	4,159
Mar	260.20	260.50	256.50	256.90	– 2.10	335.00	255.40	33,216
May	263.90	263.90	260.50	260.70	– 2.30	306.50	260.00	8,269
July	266.70	267.00	264.70	265.00	– 2.00	309.00	260.00	8,584
Aug	...	...	...	267.90	– 2.10	302.50	264.00	450
Sept	...	...	...	270.90	– 0.60	295.90	271.00	305
Nov	272.00	272.00	271.50	271.50	– 0.50	315.90	269.0	854

Est vol 4,900; vol Thu 4,271; open int 55,837, –2,019.

WHEAT (WPG) 20 metric tons; Can. $ per ton

	Open	High	Low	Settle	Change	Lifetime High	Lifetime Low	Open Interest
Mar	129.00	129.00	128.30	128.30	– 0.20	151.50	126.50	6,248
May	...	...	...	131.70	– 0.20	141.00	129.50	626
July	134.10	134.10	134.10	134.10	– ...	146.30	132.80	315
Oct	...	...	...	138.10	– ...	141.50	137.00	531

Est vol 133; vol Thu 61; open int 7,720, –133.

BARLEY-WESTERN (WPG) 20 metric tons; Can. $ per ton

	Open	High	Low	Settle	Change	Lifetime High	Lifetime Low	Open Interest
Mar	113.50	113.50	112.40	112.40	– 1.00	129.00	112.40	6,337
May	116.00	116.00	115.50	115.50	– 0.90	125.40	115.10	3,525
July	...	...	...	117.30	– 0.70	126.60	116.50	1,248
Oct	...	...	...	120.50	– 0.70	127.90	118.50	4,339
Dec	122.50	122.50	122.50	122.50	– 1.00	123.50	121.10	344

Est vol 156; vol Thu 539; open int 15,793, +15.

FOOD AND FIBER

COCOA (NYBOT)-10 metric tons; $ per ton.

	Open	High	Low	Settle	Change	Lifetime High	Lifetime Low	Open Interest
Mar	840	846	820	830	– 7	1,910	800	40,478
May	869	870	846	856	– 7	1,697	827	17,484
July	893	896	873	883	– 6	1,478	855	8,340
Sept	921	921	900	910	– 8	1,494	886	9,446
Dec	957	957	940	946	– 8	1,336	925	5,862
Mr01	...	...	...	984	– 8	1,362	965	4,834
May	...	...	...	1,008	– 8	1,222	1,002	2,932
July	...	...	...	1,034	– 8	1,245	1,060	2,704
Sept	1,080	1,080	1,080	1,062	– 8	1,246	1,080	993

Est vol 3,120; vol Thurs 3,217; open int 93,073, +724.

COFFEE (NYBOT)-37,500 lbs.; cents per lb.

	Open	High	Low	Settle	Change	Lifetime High	Lifetime Low	Open Interest
Mar	122.25	124.00	116.10	116.50	– 9.40	149.00	83.50	29,132
May	125.00	125.50	119.00	119.50	– 8.90	150.00	85.85	8,243
July	127.50	127.50	121.50	122.00	– 8.75	149.00	87.50	3,229
Sept	128.50	128.50	125.00	124.10	– 8.55	148.50	89.75	2,714
Dec	...	...	...	125.50	– 8.75	150.50	92.00	2,073
Mr01	128.00	128.00	127.00	127.00	– 8.75	153.85	94.25	431

Est vol 8,132; vol Thurs 5,695; open int 45,822, +758.

SUGAR-WORLD (NYBOT)-112,000 lbs.; cents per lb.

	Open	High	Low	Settle	Change	Lifetime High	Lifetime Low	Open Interest
Mar	6.15	6.18	6.05	6.10	– .02	10.00	5.06	81,338
May	6.30	6.31	6.21	6.24	– .07	9.05	5.26	47,581
July	6.42	6.43	6.30	6.34	– .10	9.15	5.40	28,371
Oct	6.53	6.55	6.42	6.48	– .10	9.00	5.68	9,631
Mr01	6.48	6.50	6.48	6.48	– .10	7.48	6.00	7,585
May	...	...	...	6.49	– .10	7.38	6.25	1,045

Est vol 12,427; vol Thurs 9,518; open int 175,618, –580.

SUGAR-DOMESTIC (NYBOT)-112,000 lbs.; cents per lb.

	Open	High	Low	Settle	Change	Lifetime High	Lifetime Low	Open Interest
Mar	18.20	18.20	18.20	18.20	– .04	22.30	17.50	2,215
May	18.47	18.50	18.47	18.50	+ .01	22.33	18.40	2,178
July	19.18	19.18	19.18	19.18	+ .18	22.39	18.80	2,292
Sept	...	...	...	19.75	– .10	22.38	19.16	1,516
Nov	...	...	...	19.17	+ .26	22.30	18.75	819
Ja01	...	...	...	19.05	+ .15	21.75	18.75	1,036
Mar	...	...	...	19.23	+ .33	12.15	18.80	367
May	...	...	...	19.48	+ .39	19.65	18.85	711

Est vol 117; vol Thurs 50; open int 11,134, .

COTTON (NYBOT)-50,000 lbs.; cents per lb.

	Open	High	Low	Settle	Change	Lifetime High	Lifetime Low	Open Interest
Mar	50.45	51.10	50.35	51.07	+ .33	67.90	48.60	32,867
May	51.96	52.50	51.85	52.36	+ .20	75.40	50.10	13,381
July	53.45	53.75	53.35	53.71	+ .10	73.25	51.50	8,016
Oct	55.40	55.40	55.20	55.10	– .40	68.50	53.45	425
Dec	55.90	55.90	55.55	55.63	– .44	64.50	53.75	3,837
Mr01	...	...	...	56.83	– .45	60.45	55.25	338
May	...	...	...	57.43	– .45	59.70	56.50	105

Est vol 9,500; vol Thurs 11,062; open int 59,058, –1,940.

ORANGE JUICE (NYBOT)-15,000 lbs.; cents per lb.

	Open	High	Low	Settle	Change	Lifetime High	Lifetime Low	Open Interest
Jan	85.50	88.60	85.50	88.25	– .70	133.00	76.00	735
Mar	83.75	85.20	83.75	84.05	– 1.15	131.00	78.75	13,341
May	83.75	84.60	83.75	83.80	– .70	101.60	81.10	3,479
July	84.50	84.50	83.90	83.55	– 1.20	102.50	82.80	1,602
Sept	84.00	84.25	84.00	83.50	– .75	102.95	84.00	653
Nov	84.50	84.50	84.00	84.20	– 1.05	103.95	84.00	148

Est vol 2,500; vol Thurs 5,526; open int 20,099, –1,954.

	Corn	Soyb.	Wheat
Cash Price	178¼	439¾	224
Basis	22¼	16¾	23½

FIGURE 7–9: Illustration of Method to Record Cash and Futures Prices for Calculation of Basis.

Commodity: CORN

Date	Local Cash Price	July Futures	Basis	Date	Local Cash Price	July Futures	Basis	Date	Local Cash Price	July Futures	Basis
Oct. 1	$2.98	$3.66	−68¢	Oct. 7	$2.48	$3.29	−81¢	Oct. 5	$1.96	$2.53	−57¢
Oct. 8	2.98	3.65	−67	Oct. 14	2.47	3.36	−89	Oct. 13	1.90	2.58	−68
Oct. 15	2.85	3.74	−89	Oct. 21	2.38	3.27	−89	Oct. 20	1.88	2.55	−67
Oct. 22	2.97	3.85	−88	Oct. 28	2.36	3.27	−91	Oct. 27	1.89	2.45	−56
Oct. 29	3.18	3.87	−69	Nov. 4	2.39	3.25	−86	Nov. 4	2.03	2.46	−43
Nov. 5	3.16	3.93	−77	Nov. 11	2.33	3.16	−83	Nov. 10	2.15	2.54	−39
Nov. 12	3.18	3.95	−77	Nov. 18	2.35	3.10	−75	Nov. 17	2.19	2.58	−39

FIGURE 7–10: Illustration of Method to Maintain a Basis Record.

eral different futures delivery months on the same chart. It focuses attention on the value of the cash commodity relative to different futures contracts. It is continuous through time and does not come into existence, mature, and expire like a futures contract. Its use might satisfy a hedger who is seeking to determine which futures contract month offers the greatest basis spread relative to cash.

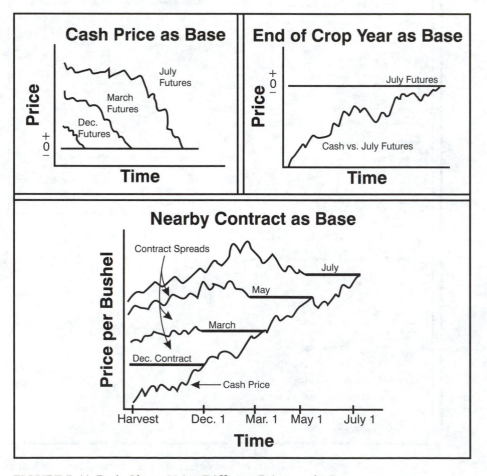

FIGURE 7–11: Basis Charts Using Different Prices as the Base.

An alternative method of basis charting is to set the last maturing contract of a crop year equal to zero and plot cash prices relative to it. This method shows how cash price converges on futures as expected. It emphasizes deviations from the expected theoretical relationship and flags profit opportunities. The disadvantage of this type of chart is that it focuses on the distant futures contract, while trading is usually conducted on the nearby. Furthermore, it typically spans only 9 or 10 months from harvest to the end of old crop year, not a longer marketing year. This method is the simplest and most straightforward method for farmer-warehousers.

The third method is to set the nearby futures contract equal to zero and plot daily cash quotes and the other futures options relative to it. This method illustrates the cash gaining on futures, and it focuses on the nearby futures contract commonly used in actual trading. As the first day of the delivery month arrives, the base is switched to the next maturing contract. This method spans an entire crop-marketing year and is continuous. The other futures contract months, plotted as deviations above the nearby option, help identify which contract offers the greatest potential basis gain. This last type is most applicable to decision making by commercial warehousers who daily trade basis, since it incorporates the effects of both the first two types.

Historical Charts and What They Show

Historical basis charts can be developed from information available from local elevators. Five years of information should provide enough time span to reflect the local pattern. To develop the history, the producer should secure cash and futures prices for the period and calculate the difference between the two prices for one day of each week during each year, as shown in Table 7–3. The next step is to average the weekly basis figures into a long-term average. Finally, a plot of the annual and average for the period of years will produce a long-term historical picture, showing when basis has been wide or narrow. The average is of little value in timing cash sales in the current year, because no two years have the same basis-making factors. But it often presents a generalized pattern of usually profitable selling months.

TABLE 7–3: Historical Corn Basis (Wednesday Prices), Cincinnati Cash Versus July Futures

Week		Year 1	Year 2	Year 3	Year 4	4-Year Average
Oct.	1	.04	−.11	−.21	−.36	−.16
	2	.10	−.08	−.25	−.11	−.16
	3	.00	−.12	−.30	−.39	−.20
	4	.07	−.15	−.25	−.45	−.20
	5	.13	−.18	−.19	−.30	−.14
Nov.	1	.05	−.24	−.11	−.27	−.14
	2	.08	−.17	−.03	−.18	−.08
	3	.02	−.17	.08	−.14	−.05
	4	.06	−.21	.14	−.11	−.05
Dec.	1	.11	−.17	.09	−.11	−.02
	2	.04	−.15	.05	−.12	−.05
	3	.15	−.08	.07	−.15	.00
	4	.14	−.03	.05	−.16	.00

(Continued)

TABLE 7–3 (Continued)

Week	Year 1	Year 2	Year 3	Year 4	4-Year Average
Jan. 1	.08	−.01	.05	−.18	−.02
2	.13	−.02	.02	−.14	.00
3	.13	.05	.04	−.09	.03
4	.14	.07	.11	−.05	.07
Feb. 1	.09	.07	.09	−.09	.04
2	.07	.02	.13	−.07	.04
3	.03	.04	.11	−.09	.02
4	.08	.03	.10	−.06	.04
Mar. 1	.12	.11	.08	−.02	.07
2	.13	.13	.08	−.05	.07
3	.14	.08	.12	−.03	.08
4	.14	.09	.15	−.04	.09
Apr. 1	.16	.05	.17	−.05	.08
2	.11	.05	.23	−.01	.10
3	.13	.08	.23	−.01	.11
4	.12	.09	.24	−.01	.11
May 1	.11	.12	.19	.02	.11
2	.14	.09	.19	.07	.12
3	.17	.07	.20	.00	.11
4	.15	.06	.20	.02	.11
June 1	.16	.05	.20	.00	.10
2	.20	.04	.32	.01	.14
3	.22	.02	.18		.14
4	.20	−.01	.34		.17
Jul. 1	.13	.00	.04		.06
2	.07	.01	.05		.04
3	.00	.00	.00		.00

The average basis for three years of data for soybeans in the Cincinnati and Wilmington, Ohio, markets is presented in Figure 7–12. Several interesting insights can be gleaned from the picture. The average illustrates the narrowing that is expected to occur, but it is not the straight line that theory suggests. In interpreting the picture with likely events, late December showed a narrow basis because farmer-sellers may have been in Florida or Acapulco, or they were delaying income into the next tax year, but the trade still needed a supply of cash grain. In February, basis widened when farmers sold crops to pay taxes. In April and May, tillage and planting took precedence over shipping grain, causing buyers to bid a premium (stronger basis) to keep grain flowing. In June, as fertilizer and seed bills came due, farmers were motivated sellers; the grain trade "rewarded" this motivation by weakening basis.

The Cincinnati basis reflects a large market area, sensitive to many market forces. The Wilmington market essentially feeds into Cincinnati with few other alternatives and consequently represents transportation costs and handling charges. Thus, the Cincinnati-Wilmington basis is fairly constant.

In this market, the historical average identifies January and April-May as profitable basis selling months. February and June are months to avoid. Each location will have its own characteristic pattern. A historical basis chart will reveal interesting conditions that exist at each location.

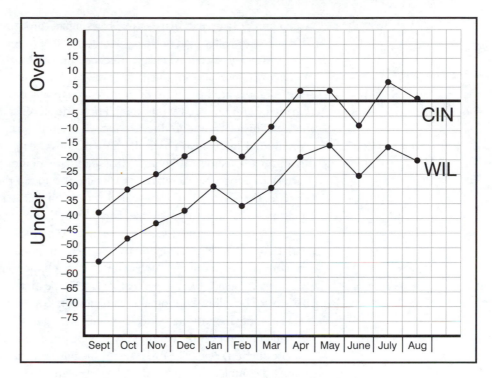

FIGURE 7–12: Soybean Average Basis for Three Years, Monthly Cash Price Versus August Futures, Cincinnati and Wilmington, Ohio.

A historical average of basis provides a standard for evaluating basis sales opportunities. On a blizzard day in January, Cincinnati basis was 10 cents over the March corn contract for four hours. Compared to the historical average, a cash bid 10 cents over March futures was an *exceptionally strong* basis. The Great Lakes and Mississippi River above St. Louis were frozen, foul weather had slowed rail traffic and stopped trucks, but exporters in New Orleans needed grain to load waiting ships. They reflected those needs with a very strong bids, until their needs were met by sellers around Cincinnati. The point is that commodity producers should know the typical basis patterns for their market and be ready to respond when the market signals unusually profitable opportunities. During periods of "abnormal" basis, a daily record flags the situation to help capture extra profit. Better still, a close working relationship with buyers, so they will call when an exceptional situation occurs, can mean extra dollars in the bank for a farmer-grain merchandiser. A historical basis chart is the first tool to identify those opportunities.

Basis Signals for Cash Sales and Purchases

A strong or positive basis is an indication that, for some reason(s), buyers are willing to pay a higher than usual price (relative to futures prices) for a commodity for immediate delivery. These conditions may be confined to a limited area or prevail nationwide. They may last for a few hours or several days. An abnormally strong basis is a market signal for producers to deliver their production and make cash sales. Correspondingly, a weaker than normal basis signals producers to hold grain. It is saying that there is greater than normal discount relative to futures prices.

Basis signals dictate when a producer should make cash sales. But seldom do both a high price level and advantageous basis level exist simultaneously. In order to capture both, the producer must be prepared to employ the marketing tools described in Chapter 2 and reviewed at this point. High price level and weak basis are the signal to hedge. This permits a producer to lock in the high price with futures contracts but continue to speculate on basis. Because of convergence, there is a greater chance of significant basis improvement rather than basis deterioration. On the other hand, narrow basis occurs when farmers are reluctant to sell their production. They judge price level to be too low; they expect it to go higher, so they delay sales. The grain trade bids a basis premium to maintain flow of the commodity. Low price and strong basis call for executing a basis contract, or a cash sale and "buying the Board" to replace the cash speculative position. Both methods result in immediate delivery of the commodity in response to basis signals, but they delay fixing the price level until later, when the producer expects price level to be higher. They both capture a strong basis price but allow continued speculation on price level.

A combination of high price level and strong basis, as is usual in drought years, dictates a cash sale if the commodity is on hand, or forward contracting if it is still in the production stage. The cash sale results in immediate delivery of the commodity. Forward contracts for delivery at a later date anticipate a lack of supply availability at the later time and indicate that users are lining up supplies. However, it is seldom that elevator managers will bid away their basis profit months ahead of expected grain delivery.

Knowledge of normal basis patterns allows a producer to evaluate whether a forward contract offer is generous or not. Prior to harvest, a local elevator may offer $3.20 for corn delivered in March. Noting that March futures price is $3.25, (only—5 cent basis) and knowing that the usual March basis is 15 cents under, the producer can gain 10 cents above normal basis. If the price level is satisfactory, a producer can agree to the contract, knowing the decision has captured a narrow basis and a good price.

The general rule for farmer-sellers is to move their production with basis contract or cash sales when nobody else is selling and basis is strong. They should hold their grain when others are selling and basis is weak. Commodity buyers, like cattle feeders, should take the opposite stance. They prefer an exceptionally weak basis. They should buy physical inputs when basis is weak. With basis discounts larger than normal, they can earn storage returns for their feeding operations. Those who do not have storage facilities can forward contract for futures delivery at a fixed price. Cash contracts allow buyers to fix the price and basis but delay receipt of the grain. The cattle feeder, pork, or poultry producer can reduce feed costs by locking in large discounts offered by a wide basis. Correctly reading the basis market will signal when the situation exists.

Contract Month to Place Storage or Production Hedge

Commodities can be priced for future delivery in any one of several alternative contract months. Price differences between various contracts are bids to grain owners for storage payments. The differences between contract prices reflect the extent of carrying return being bid in the market. As such, they can help the commodity producer decide which month is best to use for hedging his or her production. The fact that the futures price is higher for a July contract than a March contract does not necessarily mean July is the most profitable month in which to hedge. The calculation should be based upon the **returns to carry** offered by futures prices compared to the grain marketers holding costs.

Whether the market is offering a return equal to costs of carry, excess of carry, or less than carry can be evaluated from relative futures prices. As a practical matter, futures prices for different delivery months seldom offer excess carry opportunities. Commodity traders capitalize on spreads that offer returns in excess of carrying costs and bid them out of the market. At the other extreme, futures prices can reflect a negative carry. A particularly strong demand for cash grain for immediate delivery can push the price of nearby contracts above distant months. It produces an inverted market, as in Figure 7–13. The market is penalizing owners of grain for storing until later; it is signaling that the market wants the grain now rather than later. Immediate strong demand or limited current supplies produce inverted markets. The degree of inversion into futures months depends on how desperately buyers want grain and how reluctantly owners are willing to part with it. An inverted market is most likely to occur in a year of short supply, that is, a small crop in relation to overall demand. Some users are willing to pay a higher price for cash grain they do not yet need—storage costs notwithstanding—in order to assure themselves an adequate supply and to avoid what they fear may be higher prices later in the season. But as cash prices rise significantly, other users find input prices too high to allow them to make a profit. They begin reducing the size of their poultry flocks, cattle on feed, or sow herds. High price is performing its function by rationing limited supply. Users typically overreact by cutting use too severely. In four or eight months after harvest, price level is lower than harvest because of the decline in demand. An inverted market tells commodity producers to sell immediately. It tells commodity users to wait until later to make purchases. All too often, both parties ignore these market signals, to their eventual regret.

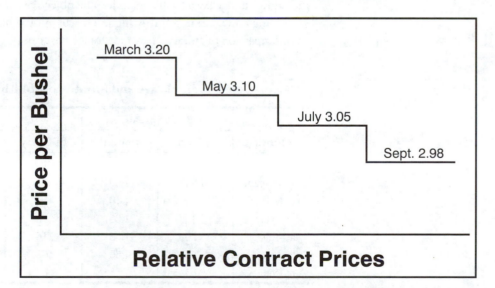

FIGURE 7–13: Inverted Futures Market, Prices Demonstrate Negative Carry.

It is easy to select the month for hedging in excess carry markets–set hedges as far out as possible. By contrast, an inverted market says *"Sell Now"* or use nearby contracts to hedge. The decision process for selecting the option to use for hedging is not so easy between these extremes. The decision depends on which contract month offers the greatest net return relative to costs of carrying stored grain.

Suppose that during fall soybean harvest, the cash price offered for soybeans is $5.50. The January futures is $6.50, the following September option is $7.05, and the other contract months look like this:

October Soybean Prices

Current Delivery $5.50	Futures					
	Jan. $6.50	Mar. $6.60	May $6.75	July $6.95	Aug. $7.00	Sept. $7.05
Spreads						
Oct.-Jan.	1.00					
Jan.-Mar.		0.10				
Mar.-May			0.15			
May-July				0.20		
July-Aug.					0.05	
Aug.-Sept.						0.05
Cumulative						
Returns Being Offered:	$1.00	$1.10	$1.25	$1.45	$1.50	$1.55

For simplicity, assume that the producer's local market usually has a zero basis in month of delivery.* The market is offering $7.05 ($5.50+$1.55) for September (11 months from now) delivery. Should the producer store and use **next** September's contract to hedge in the $1.55 storage return? Not necessarily, as further analysis, including storage costs, will conclude. The objective is to maximize return in excess of holding costs, not necessarily highest price; but the highest **net** price.

Comparison of storage return to storage cost can be calculated in this fashion:

Calculation of Cumulative and Month-to-Month Net Return to Storage

Market Months	Basis Spreads		Added Cost of Carry	Net Return to Storage	
	Added	Cumulative		Added	Cumulative
	$/bu.	$/bu.	$/bu.	$/bu.	$/bu.
Harvest-Jan.	1.00	1.00	.44	.56	.56
Jan.-Mar.	.10	1.10	.16	(.06)	.50
Mar.-May	.15	1.25	.16	(.01)	.49
May-July	.20	1.45	.16	.04	.53
July-Aug.	.05	1.50	.08	(.03)	.50
Aug.-Sept.	.05	1.55	.08	(.03)	.47

---

* This assumption is invalid for most farmer markets. However, it does not invalidate the analysis. The important decision criterion is the change in basis, not its absolute value.

In the calculation, the producer allocates all his or her fixed on-farm storage costs to the first month of storage, because placing the soybeans in the bins eliminates other alternative uses, and the producer insists on covering all fixed costs immediately. Monthly charges thereafter reflect 12 percent interest charges on capital tied up in inventory and 1.5 cents per month shrinkage and quality risk. Storing to January would yield 56 cents net storage profit, but holding across the next spread to March would reduce storage profit by 6 cents. The May-July spread offers profit but does not make up for the January-May loss. The analysis clearly defines January as the hedging month with intentions of moving the grain by then. Later returns do not cover monthly storage charges. Interest returns from money in the bank exceed interest returns on inventory, without the danger of spoilage loss.

Rolling the Hedge

Taking the analysis one step further, assume that in January, basis did not converge to zero as the producer expected. Instead, it is 20 cents under January. If this occurs, the hedger has two options: (1) sell in the cash market and buy back the futures contract, accepting the weak basis, which would produce a 36-cent return instead of the expected 56 cents, or (2) roll the hedge forward to another futures month. To roll a hedge forward, the hedger buys back the January contract and simultaneously sells another month to reestablish the position in the deferred futures. To determine whether to roll and what month to roll into, the hedger must again evaluate the spreads that exist. On January 10, cash and futures prices for soybeans look like this:

Prices on January 10

| | Cash 5.45 | | Futures | | | | | |
		Jan. 5.65	Mar. 5.75	May 6.00	July 6.20	Aug. 6.25	Sept. 6.30
Spreads							
Jan.-Mar.			0.10				
Mar.-May				0.25			
May-July					0.20		
July-Aug.						0.05	
Aug.-Sept.							0.05

Analysis of Spread Return Versus Cost of Carry

Months	Amount of Spread	Added Cost of Carry	Net Return to Storage	
			Added	Cumulative
Jan.-Mar.	.10	.16	(.06)	(.06)
Mar.-May	.25	.16	.09	.03
May-July	.20	.16	.04	.07
July-Aug.	.05	.08	(.03)	.04
Aug.-Sept.	.05	.08	(.03)	.01

The analysis shows that it is *potentially* profitable to roll into the July contract and continue storing until early July (betting on a 7 cent gain).

Notice that by January 10, the futures price level had declined to $5.65 compared to $6.50 when the farmer established the initial hedge. The decline does not affect the farmer-warehouser. First, the original storage hedge protected the inventory against value decline; and second, the decision to sell cash grain or continue storage is made by analyzing contract spreads rather than price level or speculative urges.

The final results are evaluated in July. Suppose July cash and futures are trading at $6.00. Basis did reach zero as anticipated. An evaluation of spreads into the following months of August-September does not promise additional profit. The producer would sell the cash and lift the hedge with the following results:

Sell Soybeans for cash	$6.00
Futures gain on Jan. Contract (6.50 –5.65)	0.85
Futures gain on July contract (6.20 – 6.00)	0.20
Gross Price	$7.05
Cash value in October	– 5.50
Storage Return	$1.55
Storage costs (–0.44 –0.16 –0.16 –0.16)	– 0.92
Profit from Storage	$0.63

Relationship of Futures Price Between Marketing Years

Each harvest begins a new marketing year with its own calculation of demand and supply arithmetic. Price levels may be higher or lower than the previous year. Since futures prices reflect both old crop and new crop conditions, the futures price in the last delivery month of one marketing year may differ sharply from futures prices for the first delivery month of the next marketing year. In theory, if the same or similar supply-demand conditions were to exist in both years, December futures for a new crop year should trade at the same level as the previous December, not the end-of-year contract. For example, if the remaining supply of old crop corn is expected to be adequate for the needs of feeders, processors, and exporters during the summer and early fall months, but a bumper new crop is in prospect, stairstepping corn futures prices might resemble Figure 7–14. The new crop would be priced at a significant discount to the old crop.

The large down step in price does not say that old crop and new crop prices are totally unrelated. If there is an overabundance of old crop supplies as the marketing year draws to a close, and if a big new crop is in the offing, the large impending harvest will almost certainly depress end-of-year old crop prices. Conversely, if old crop supplies are barely adequate for the remainder of the year, and a small new crop is expected, the possible shortage will strengthen both old and new crop prices, and the new crop might trade at a premium to old crop.

Significant disparities between crop years can present profitable hedging opportunities for producers. Because speculators are accustomed to watching month-to-month spreads in old crop years, they sometimes forget that the new crop contract has its own set of price-making forces. Even though the old crop year might have been a short crop and produced high prices, next year is another planting season. When the most distant old crop contract and the first new crop contract trade at a pre-

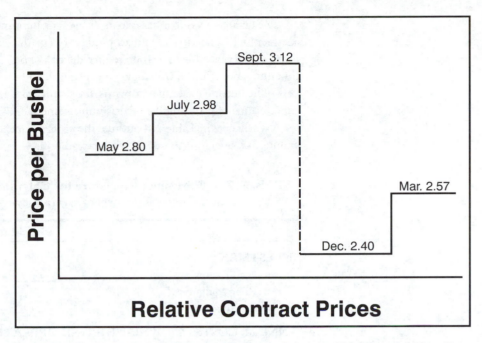

FIGURE 7–14: Relationship of Futures Contract Prices in Old Crop and New Crop Years.

mium to the nearby futures, farmers can use the new crop contracts to establish profitable hedges on growing crops or planned production. Since producers typically react to high early season prices by boosting production in the growing year, new crop prices decline as the season progresses.

To Own or Not to Own Grain Storage

Commodity producers are constantly faced with decisions about whether to build storage facilities, or if facilities exist, how long to store grain. Storage is one of the market functions that must be performed by someone–farmers, elevator operators, processors, or consumers. Whoever performs the function incurs the costs and receives the return from it. A storage hedge can capture a storage return, but it does not guarantee a storage profit. There are costs of holding commodities in storage. They arise from owning the facilities and investing funds in the commodity. A storage profit is generated when basis gain (storage return) exceeds these additional holding costs.

Huge demand for space during harvest typically produces price discounts in excess of full carrying costs. When such discounts exceed annual costs of storage, farmers have an incentive to store their production at home or in commercial facilities. If elevators charge more for space than farmers would incur, then farmers can profit by building storage. On the other hand, if surplus storage space prevails, relative to the size of the locally produced crop, elevators will likely charge less than what on-farm storage costs would be. Elevators are filled and emptied several times each year, and therefore their per-unit storage cost is lower than most farmer-owned storage facilities.

The farmer-producer has two incentives for building on-farm handling and storage facilities. The first is tied to production in that convenient on-farm facilities improve harvest efficiencies and reduce the risk of field loss from harvest delays. The timeliness benefit of farm storage means not having to wait in elevator lines at harvest. The second incentive converts the farmer-producer into a warehouser with expectations of profiting from basis improvement.

A worksheet in Table 7–4 outlines the process for calculating annual storage ownership and operating costs. A farmer-warehouser must recognize that costs exist

TABLE 7–4: Worksheet Procedure for Calculating Annual Ownership and Operating Costs of Grain Drying and Storage

	Building	Equipment	Drying
INVESTMENT			
1. Storage structure	$_____		
2. Handling equipment		$_____	
3. Drying equipment			$_____
ANNUAL COSTS OF OWNERSHIP AND OPERATION			
FIXED COSTS			
4. Depreciation—investment divided by years of life	$_____	$_____	$_____
5. Insurance—mid-value times rate (.004)	_____	_____	_____
6. Interest—mid-value times rate (_____ %)	_____	_____	_____
7. Taxes—assessed value times tax rate	_____	_____	_____
8. Total annual costs	_____	_____	_____
9. Annual fixed cost per bushel (item 8 divided by bushel capacity)	_____a	_____b	_____c
VARIABLE COSTS PER BUSHEL			
Variable Drying Costs:			
10. Drying fuel: LP gas _____ ¢/gal. × .0165 × _____ pts. moisture removed			$_____
11. Electricity for drying _____ ¢/kwh × .01 × _____ pts. moisture removed			_____
12. Total variable cost			_____
13. Total drying cost (items 9c & 12)			_____
Variable Storage Operating Costs:			
14. Electricity for aeration _____ ¢/kwh × (.05 to .1)			_____
15. Loss from excess drying (3¢/bu. for each 1% under 15.5%)			_____
16. Dry matter shrink .5% × _____ price of corn			_____
17. Insurance on grain .00475 × _____ price of corn			_____
18. Rodent & insect control .002/bu.			_____
19. Property taxes (if applicable)			_____
20. Risk of grade loss (.01 to .10/bu.)			_____
21. Management costs (5% of value)			_____
22. Extra handling 0–5¢/bu.			_____
23. Total variable storage operating costs			_____
24. Total storage cost (items 9a, 9b, and 23)			_____
25. Interest on inventory per month: _____ % interest rate divided by 12 × _____ price of grain times months stored			_____
26. **TOTAL CARRYING COST**			_____

beyond owning the facilities. He or she must incur additional costs of extra handling, shrinkage, quality control, and risks of loss with on-farm storage. The interest cost of holding inventory must be recognized. The farmer-warehouser could be earning interest on money invested in a savings instrument if he or she cashed in inventory rather than holding grain in storage.

The drying function should be considered part of the production process. Though closely associated with storage, it should be evaluated on its own profitability as compared to commercial rates. Nevertheless, efficient use of storage typically requires access to a grain dryer.

The first task is to estimate annual fixed costs of ownership. These depend upon total investment in facilities. However, costs must be annualized. Investment should be broken down into drying and storage components. Annual fixed costs arise from the DIRTI-five: depreciation, interest on investment in facilities, repairs and mainte-nance, property taxes, and insurance on structures and equipment. Depreciation is calculated by dividing investment less salvage value by years of life. Insurance and interest charges are calculated on midlife value of the investment. Taxes are charged according to assessed value times the local tax rate.

In addition to the procedure outlined in Table 7–4, a shortcut method to estimat-ing annual costs is to multiply investment by an appropriate percentage rate, as shown in Table 7–5. The 14.1 and 21.4 percent rates reflect the DIRTI-five annual fixed costs. By multiplying the original investment by these rates, annual ownership costs can be estimated. Storage facilities requiring an investment of 50 cents per bushel would have annual fixed costs of about 7 cents per bushel (50 cents × .14); those costing $1.00 per bushel would cost 14.1 cents annually. This can be compared to the storage charge levied by commercial elevators in comparing alternatives for least cost.

TABLE 7–5: Annual Cost of 20-Year Storage Structures and 10-Year Equipment, Calculated as a Percentage of Original Investment

Item	Structures (20-Yr. Life)	Equipment (10-Yr. Life)
	(% of Original Investment)	
Depreciation	5.0	10.0
Interest (half-rate)	7.0	8.0
Repairs and Maintenance	1.0	3.0
Taxes	0.7	—
Insurance	0.4	0.4
Total	14.1%	21.4%

The next step is to estimate variable costs incurred by placing grain in storage. The first item is the variable fuel and electrical costs for drying grain to a grade stan-dard of 15.5 percent and adding annual fixed drying costs to determine total drying costs. Drying should be estimated as a cost separate from storage. Though drying is necessary for safe storage, it could be done commercially. The on-farm drying cost can be compared with commercial rates to determine which is most economical.

To calculate variable storage costs, the farmer must add the costs of aeration to maintain quality, and extra drying to reduce moisture to safe levels for summer stor-age. Lowering moisture to 13 percent incurs fuel costs and causes weight loss; both must be included. In addition, grain handlers experience dry matter shrink during storage. The farmer-warehouser should include the costs of insurance on stored grain, costs of rodent and insect control, taxes on inventory where they are levied, and a charge to reflect the risk of maintaining quality. Extra management cost is also required for storage. On-farm storage requires charges to cover the costs of extra handling and shrinkage due to handling. Suggested coefficients for calculating these costs are included on the worksheet. The sum of variable costs are then added to per-bushel fixed costs to arrive at total storage cost.

Most of the above costs are incurred with the initial decision to store, and except for extra drying to lower moisture levels for safe summer storage, these costs do not change much during the storage period. The final item, interest on stored grain, is a function of time and increases day by day. Farmers have the option of having their money invested in the bank, earning interest, rather than having their investment in grain. And grain can be converted to cash any business day. Thus, an interest charge should be assessed on grain inventory value. Adding interest charges to storage cost produces total carrying cost.

Examples of cost estimates from Illinois researchers for storing and carrying corn and soybeans in storage are presented in Tables 7-6 and 7-7.[3] Total on-farm storage costs for corn range from 24.6 cents for one month of storage to 58.4 cents for nine months. The cost of soybeans ranges from 28.9 cents for one month to 85.3 cents for nine months. As long as basis gain exceeds these additional (marginal) costs, storage will produce extra profit.

TABLE 7–6: Farmer Costs of Storing and Carrying $2.80 Shelled Corn for Varying Periods of Storage

Item	Months Stored			
	1	3	6	9
	(¢/bu.)			
On-Farm Storage				
Storing				
Annual Bin Costs	9.8	9.8	9.8	9.8
Handling & Shrink	4.0	4.0	4.0	4.0
Other	8.0	8.0	8.0	8.0
Total Storing	21.8	21.8	21.8	21.8
Carrying Charges				
Interest (12%)	2.8	8.4	16.8	25.2
Extra Shrink (15.5%–13%)	—	—	9.4	9.4
Extra Drying	—	—	2.0	2.0
Total Carrying	2.8	8.4	28.2	36.6
Total Storing & Carrying	24.6	30.2	50.0	58.4
Rental Commercial Storage				
Minimum Charge (3 mos.)	15.0	15.0	15.0	15.0
Monthly Charge (1.5¢)	—	—	6.0	12.0
Shrink (15.5% to 13%)	9.4	9.4	9.4	9.4
Interest (12%)	2.8	8.4	16.8	25.2
Total Storing & Carrying	27.2	32.8	47.2	61.6

Source: R. B. Schwart and L. D. Hill, "Costs of Drying and Storing Shelled Corn on Illinois Farms."

TABLE 7–7: Farmer Costs of Storing and Carrying $7.00 Soybeans for Varying Periods of Storage

Item	Months Stored			
	1	3	6	9
	(¢/bu.)			
On-Farm Storage				
Storing				
Annual Bin Costs	9.8	9.8	9.8	9.8
Handling & Shrink	4.0	4.0	4.0	4.0
Misc. Items	8.0	8.0	8.0	8.0
Total Storing	21.8	21.8	21.8	21.8
Carrying Charges				
Interest (12%)	7.0	21.0	42.0	63.0
Insurance & Conditioning	.1	.3	.4	.5
Loss & Damage	?	?	?	?
Total Carrying	7.1	21.3	42.4	63.5
Total Storing & Carrying	28.9	43.1	64.2	85.3
Rental Commercial Storage				
Minimum Charge (3 mos.)	20.0	20.0	20.0	20.0
Monthly Charge (2¢/mo.)	—	—	12.0	18.0
Interest (12%)	7.0	21.0	42.0	63.0
Total Storing & Carrying	27.0	41.0	74.0	101.0

Source: R. B. Schwart and L. D. Hill, "Costs of Drying and Storing Shelled Corn on Illinois Farms."

The tendency for basis to narrow as the delivery month approaches provides farmer-warehousers with payments for storage. Figure 7–15 compares the theoretical basis pattern, reflecting carrying charges, to an actual historical average basis. It is

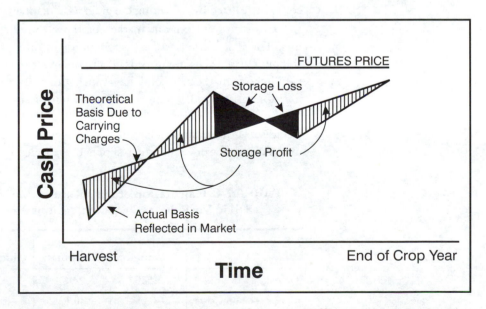

FIGURE 7–15: Illustration of Storage Profit Due to Differences Between Carrying Charges and Basis Change.

the difference between basis gain, producing payments for storage, compared to carrying charge costs that yield storage profit to warehousers. As long as basis gain exceeds storage cost (slope of basis line greater than slope of cost line), storage is profitable. When basis gain is less than storage cost, storage is not profitable (slope of basis line less than slope of cost line).

A farmer-warehouser makes the decision to build or not to build storage depending upon whether or not anticipated basis gains over the life of the bins will be greater than the total life cycle cost of the bins and interest cost of holding grain. In **any one-year after** construction, however, fixed bin costs are irrelevant. The decision to store depends upon the variable cost of storage for that year compared to the expected basis gain. If a farmer erects a set of storage bins and basis patterns change drastically in two years, the decision of whether or not to dispose of the bins depends on only the variable cost of carrying grain and the annual opportunity cost calculated on the salvage value of the bins as compared to expected future basis gains.

Location–Where to Sell

Price variations between different locations cause a commodity to flow from areas of surplus production to deficit consumption areas. The price difference is related to transportation cost, but not perfectly. Deciding **where** to sell depends on which location will yield the highest net return to the producer at the farm gate. Different markets have different prices and are worth shopping for. But the farmer must know what the transportation costs will be to move commodities to each market. On the day identified by basis signals to make a cash sale, the seller should deduct transportation charges from the various price offers at different markets. Highest net price at the farm gate is the decision criterion. An example calculation of net returns on corn for a farmer located in southwest Ohio is shown in Table 7–8. The highest net price could be realized by delivering the grain to Cincinnati.

Basis variation from one market location to another is explained by several factors. The first is the actual transportation cost difference. Barges offer lower-cost shipping than trucks or railroads. Therefore, river markets usually bid higher prices than interior markets. Second, how much gross margin is the elevator or buyer taking? One may handle grain for 5 cents while another wants 10 cents per bushel. Finally, location differences depend upon the elevator operator's astuteness as a merchandiser. If he or she is adept in finding high sale price alternatives, that manager may bid a higher price to farmer-suppliers than a less active merchandiser. The rea-

TABLE 7–8: Calculation of Farm Gate Value to Determine Where to Sell in Various Markets with Various Transportation Costs

	Wilmington	Cincinnati	Chicago	New Orleans	Norfolk
Delivered Price/bu.	$2.25	$2.43	$2.57	$2.53	$2.95
Transportation Costs	–0.05	–0.12	–0.35	–0.45	–0.66
Net to Farmer	$2.20	**$2.31**	$2.22	$2.08	$2.29

sons behind the differences are not important to a seller; the farmer's concern is finding the buyer who offers the highest net farm gate price.

Summary

This chapter emphasizes the difference between a price level sale and a basis level sale; they are two separate, but related, markets. Price level reflects worldwide demand and supply conditions; basis level reflects local demand for, and supply of, storage space. Market analysis of basis signals will tell the producer when and where to make cash sales.

We emphasized two features of the relationship between cash and futures prices–parallel movement and convergence. We reviewed their effects on types of hedges.

Our analysis looked at storage returns offered by futures contracts and costs of holding commodities in storage. As long as **added returns <u>exceed</u> added costs**, there is economic reason to store. Basis records and charts for individual producers guide their decisions.

Finally, we explained how to read basis market signals. Abnormally weak basis is the signal to withhold commodities from the market, delaying cash sales. Abnormally strong basis signals an immediate sale as the proper response. By separating basis decisions from price level decisions with the appropriate marketing tools, a farmer-producer is in a better position to capture advantageous prices in both markets.

Questions for Study and Discussion:

1. What two elements of marketing determine basis price levels?
2. How does the Law of One Price explain different prices that exist on a particular day for a commodity?
3. List and explain the two features that characterize the relationship between cash and futures price.
4. Where and when are cash and futures prices the same? What assures this relationship?
5. How are cash and futures prices related over time?
6. Why do grain futures contract prices normally stairstep upward?
7. What is the foundation of basis level negotiation in a market economy?
8. How does farmer reaction to price level affect basis?
9. What factors cause local basis to differ from Chicago basis?
10. Explain the two reasons for hedging price level.
11. Give examples of the five situations that produce narrowing of basis.
12. How does a warehouser with storage space earn a storage profit?
13. Why is it important that each producer-storer maintain his or her own basis record?
14. What do different basis charts show?

15. What are the basis signals for cash sales and purchases?

16. Explain the procedure for calculating which contract month to place hedges.

17. What is "rolling the hedge"? How is it done?

18. Why is the relationship of contract prices between marketing years important?

19. Explain the procedure for determining where to sell.

Basis Exercise:

Maintaining a basis record is an interesting exercise to help you learn more about the subject. Use a form such as the one suggested in this chapter to maintain a weekly record. Secure futures prices and local bids from newspaper or radio reports or from a local elevator for commodities you are producing. Local basis is computed by subtracting cash price from a selected futures option. The basis numbers can be plotted on a chart to visually illustrate its pattern over time. Three to five years of historical information can be used in identifying both average and extremes in a local market.

The end-of-year contract (July for fall harvested crops and May for wheat) focuses on convergence of cash and futures over time. It is important in deciding whether to store or not and in estimating the amount of storage return to expect. It can also signal when to sell cash grain. Since most farmers are long-term storers, rather than daily basis traders, the end-of-year option serves them best in its simplicity.

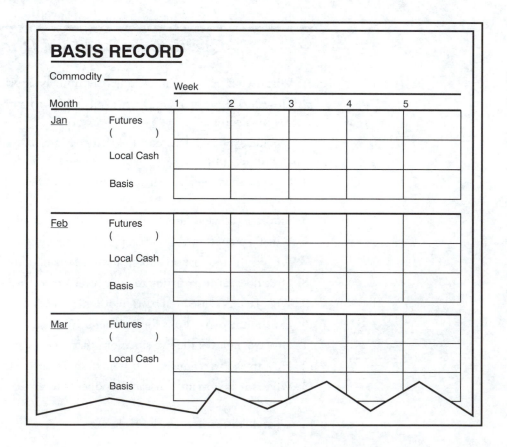

Knowledge of basis and a feel for why it fluctuates will hone a producer's ability to make profitable basis transactions.

Homework Problems:

1. On October 25, you are harvesting corn and the elevator is bidding $2.10 for cash corn, and December futures are $2.40. Will a hedge in the December option return you a storage profit? Outline the procedure for making this analysis of profitability.

2. On October 10, you are harvesting soybeans and you have empty storage bins that you could use, or you could sell the beans to an elevator for $6.00 per bu. November futures are trading for $6.60. What facts do you need to determine whether you should sell at harvest or sell in January when you think January soybeans futures will reach $7.50?

3. You were through harvesting corn on October 25 and have your 20,000 bu. stored in excellent condition in a new 20,000-bu. bin. On December 1, you note that cash corn is selling for $2.20 and March futures option is $2.30. What should you do if you think the price of March futures will reach $2.45 by March 1, when your elevator normally bids cash price 12 under March? How much more per bushel would you receive by your decision, assuming your guess about price level is correct?

4. At planting time (April 15), you note the elevator is bidding $2.60 for new crop corn delivered at harvest while December contract is $3.00. You have on-farm storage that would cost 2 cent per bu. per month (variable cost) plus 5 cents per bu. for the extra handling (in and out of storage). You have a loan on which you are paying 12 percent interest that could be reduced at harvest if you wanted, but which could be extended to May 1. May futures are $3.20 and your records show the elevator normally pays 10 under May during the last weeks of April. What should you do to maximize the storage income from your crop by April 15?

5. It is mid-August. The summer drought scare is in full swing. Your crop looks good, and futures are: corn $3.95 Dec., $4.10 March; soybean $8.75 Nov., $9.50 March. You have 100,000 bushels storage and expect 100,000 bushels of corn production from your 800 acres. In addition, you anticipate harvesting 40,000 bushels of soybeans. Fall harvest bid on corn is $3.80. Historical corn basis is: harvest –70, Dec. –20, March –10. Interest cost is 6 cents per bushel per month. You determine price level is at its peak so you decide to price your crop. Normal soybean basis is: harvest –1.00, Nov. –40, and March –10. How can you merchandise your bin space profitably? Even though price level may change, normal basis is expected to exist both Oct. and March. How much profit do you expect to earn on your storage space? What tool did you use to lock in the profit? Explain.

Endnotes for Chapter 7:

1. _____, *Improving Margins Using Basis,* Grain Merchandiser Series (Chicago Board of Trade, Chicago, IL, n.d.

2. Bailey, *Understanding Basis—The Economics of WHERE and WHEN* (Chicago Board of Trade Clearing Corporation, Chicago, IL).

3. For further information on procedures and coefficients for calculating costs, see R. B. Schwart and L. D. Hall, *Costs of Drying and Storing Shelled Corn on Illinois Farms,* Circular 1141 (Champaign: Cooperative Extension Service, University of Illinois, Aug., 1977).

Chapter 8

Financing a Hedging Plan

Financing a Hedging Plan

Importance of a Marketing Plan

Agricultural lenders who finance farm production join hands with farmers in their financial success or failure. By its nature, farming is a speculative process subject to serious risk and uncertainty in both production and pricing. Consider, for example, an 800-acre corn farm producing an average yield of 130 bushels (Table 8–1). Yield fluctuations of 20 bushels per acre change gross receipts by $41,600. In comparison, market price swings of $1.20 per bushel change gross receipts $125,000. It takes yield variability of 60 bushels per acre to cause the same gross receipt variability as a $1.20 price swing. The farmer and lender must decide which is more likely to occur, a 60-bushel yield variation or a $1.20 price swing. A production plan, without a marketing plan, covers only part of the risk that lenders share with farmer-borrowers.

TABLE 8–1: Change in Gross Farm Receipts Due to Change in Yield and Corn Price, Example Farm

Situation: 800 acres of corn; 130 bushels per acre base yield, $2.60 base price			
Production		**Marketing**	
base price $2.60		base yield 130 bu.	
Yield	**Gross Value**	**Price**	**Gross Value**
120	$249,600	$3.20	$332,800
130	$270,400	$2.60	$270,400
140	$291,200	$2.00	$208,000
Change high to low		Change high to low	
$41,600		$124,800	
17% change		60% change	

Lenders are accustomed to financing production plans without requiring marketing plans, thereby aiding and abetting cash market speculation. A Purdue survey of 37 agricultural banks showed that only 17 financed hedging accounts, with the average number of accounts per bank being six.[1] A lack of demand from borrowers was cited as the major reason by most of the 20 banks not financing hedging accounts. A second, but less cited, reason was lack of knowledgeable bank personnel available to supervise the accounts.

Farmers and ranchers are typically cash market speculators. They often double-up on their price speculation with unpriced inventory stored in bins, *in addition to* an unpriced growing crop. Thus, it is common for farmers to speculate on the price of **two** crops simultaneously.

Lenders and farmers need to form a team to improve farmer marketing. All too often a farmer's marketing strategy is dictated by "a payment is due" or "the banker says to sell." There is a better way. A planned strategy is better than a forced strategy. Knowing when and how to control cash market speculation can be a key to financial success for farmers and financial security for lenders.

Various marketing tools permit a borrower to protect against price and basis risk. The lender and borrower must mutually agree to use the appropriate marketing tool at the appropriate time. The degree of price speculation according to the marketing tool used to make pricing decisions is summarized in Table 8–2. A cash sale or forward contract eliminates all price speculation. It also eliminates the possibility for capturing more profitable opportunities if prices rise during the year. Basis contracting or hedging with futures removes respective parts of cash market speculation. The highest degree of speculation is practiced with unpriced inventory.

TABLE 8–2: Summary of Degree of Price Speculation and Marketing Tools That Hedge Various Speculations Away

Degree of Speculation	Type of Price Speculation	Type of Contract (Tools)
Highest	Price Level and Basis	Unpriced Inventory (grain in bins & crop in ground)
	Price	Delayed Price or Basis Contract
	Basis	Futures Contract
	None	Forward Contract
Lowest	None	Cash Sale

Before farmers and lenders begin using some marketing tools to reduce cash market speculation, they must recognize that adequate financing is a prerequisite to a sound marketing program. In particular, hedging in the futures market requires adequate liquidity. Without adequate financing, hedging increases financial risk. The producer who needs $25,000 to meet potential margin calls due to a price move against hedge positions, but only has $20,000, will be liquidated when margin calls are not answered. Invariably the market will turn the other way and move past the original hedged price. Thus, there is a double loss. Production will be sold at the lower price, and the futures loss will have to be deducted from that lower price. A hedging plan, without a financing plan, increases risk rather than reducing it.

Funds are required in a hedging account for initial margin and additional (maintenance) margin to cover short-run setbacks. Initial margins are "good faith money" deposits to insure the financial integrity of the futures position. Table 8–3 outlines the margin requirements for agricultural commodities as established on several exchanges by one firm. They are illustrative and subject to change. With a $600 initial margin per 5,000-bushel contract, a 100,000-bushel corn grower would require $12,000 as initial good faith money to hedge that level of production. The deposit must be made with the broker before a futures trade can be initiated.

Initial margin remains a cash asset for the farmer. Most brokerage houses allow the deposit to be held in Treasury Bills. The initial unit is $10,000, and subsequent $5,000 units can be purchased. With the initial margin invested in Treasuries, the farmer receives an interest return on the investment. Otherwise, an idle balance held by the brokerage house does not earn interest for the farmer.

Since it is highly unlikely that any hedger will place a position at the extreme of a price move, additional liquidity for maintenance margin is required to hold a futures position once it has been established. Maintenance margin is the minimum balance that must be maintained in the brokerage account. Should the balance fall below maintenance levels, the broker will make margin calls for additional deposits to bring the balance back to minimum level. Due to volatile futures price fluctuations, margin calls can be large and required on short notice. A hedger of 40,000 bushels of soybeans who sets hedges $1.25 per bushel below the top of a price run up, or a hedger of 100,000 bushels of corn who sets hedges 50 cents too low, would each need $50,000 ready cash to meet margin calls.

TABLE 8–3: Hedger's Initial and Maintenance Margin Requirements

Commodity	Initial Hedge	Maintenance (Min. Bal.)
Chicago Board of Trade		
Wheat	$ 700	$ 700
Corn	600	600
Oats	500	500
Soybeans	1,250	1,250
Chicago Mercantile Exchange		
Live Cattle	1,000	750
Lean Hogs	1,000	800
Feeder Cattle	1,000	750
Mid-America Commodity Exchange		
Wheat	140	140
Corn	100	100
Soybeans	250	250
Live Cattle	375	375
Lean Hogs	675	500
New York Exchanges		
Cotton	1,500	1,500
Frozen Orange Juice	1,000	750
Sugar	1,000	750
Other Markets		
Minneapolis Wheat	500	400
Kansas City Wheat	500	400

The amount of liquidity required for hedging depends upon the price volatility of the commodity being hedged and the type of pricing strategy that is practiced. Our experience with a particular system suggests that lenders should consider a **minimum** 60 cents per bushel of corn, $1.00 per bushel of soybeans, 75 cents per bushel for wheat, 20 cents per pound of cotton, and $10 per hundredweight for livestock. Different pricing strategies will have their own liquidity requirements. These numbers mean that Midwest grain farmers may need $50 to $100 per acre liquidity commitments for a hedging program. This is in addition to the $150 per acre for seed, pesticides, fertilizer, fuel, and other out-of-pocket production costs. A total loan package represents a $250-per-acre liquidity requirement. How much of that liquidity is supplied by the farmer and how much by the lender is open to negotiation between the two.

The suggested liquidity figures assume that the pricing strategy will establish hedges near market extremes but still provide a margin of error. The amount required to meet margin calls is the difference between the hedge price and the extreme of any subsequent price move, if the farmer practices fixed hedging to hold the position until maturity. The soybean market that went from $4.00 to $13.00 per bushel in 1973 would have required huge margin deposits had hedges been established before the top was reached. Likewise, corn that might have been hedged at $4.00 in 1996 would require an additional margin of $1.50 per bushel when the market topped at $5.50 in July. It is the inability to project total liquidity that might be required in a hedge that frightens many lenders. Unfortunately, bankers seem to lose little sleep over crop sales of $1.25 for corn or for $3.50 soybeans—crops that often could have been sold at three times those prices with a sound marketing program.

It should be remembered, however, that margin money outstanding against a hedge position does not constitute a loss for a hedger.★ The outflow of cash for maintenance margin when price moves against a position will be returned to a hedger when the cash sale is made. Because of parallel movement of cash and futures prices, the increased inventory value will be about as much as the deficit in the future position.★★ But during the time period that maintenance margin is required, the change in inventory value produces a paper profit, while margin calls are cash transactions. A liquidity problem arises until the cash commodity is sold. It is this liquidity need that requires financing.

The true costs of hedging are limited to the interest charges on margin money while it is outstanding and to the commissions paid to the broker. The margin deposits are not losses. They are simply lost pricing opportunities that all too often cause regret—regret that one did not wait for a better price. On the other hand, when a hedge position produces a surplus in the futures account, the surplus is not a profit. The inventory value of the cash commodity will have declined approximately equal to the gain in the futures position. The futures account accumulates cash while the inventory loss is still on paper, until the commodity is sold at a lower price. The lower cash price plus the futures gain raises the realized price to the level protected by the hedge. Lenders, as well as farmers, need to remember the offsetting gains and losses in both markets when hedges are set for price protection. Instead of bemoaning $5.00 corn pricing opportunities, farmers and their lenders should be asking: "how many *additional* years of production can be sold at these bonanza prices?"

★Review the parallelism discussion in Chapter 2 to understand offsetting gains and losses.

★★ This assumes the producer has full production to sell at the higher cash price. Often, drought-reduced crops, leading to the price rise, leave the farmer with less than total production. The futures deficit thus exceeds increased inventory value.

The Lender's Role[2]

Lender Policies for Hedge Loans

Lenders who decide to finance marketing plans must recognize the importance of profitable pricing decisions and must consider the borrower's income and ability to repay. Many lenders have seen good producers suffer from poor marketing. They see improved pricing as a way to protect the quality of their loans. A sound hedging program should enhance a customer's borrowing capacity because it protects earning power. The lender who has a secured position in both the commodity and the hedging account should have a stronger loan.

Lending institutions will need to develop policies and practices to cover lines of credit for agricultural borrowers to use in futures market hedging. When the lender decides to finance hedging, the first step is to formulate a statement of policy that governs the administration and control of all hedge loans. The statement might say: "The lender agrees to grant loans for the purpose of commodity hedging and will extend funds for financing initial margin, maintenance margin, and broker commissions that arise out of trading contracts in the futures markets."[3] In addition to the formal statement of policy, a lending institution should make clear its position on the basic elements of a hedge loan and develop criteria for loan officers to follow.

Efforts to educate loan officers on hedging practices should be an early step toward a sound and effective loan program. Loan officers must become familiar with alternative marketing tools and the mechanics of hedging to properly evaluate loan requests and monitor accounts.

The institution should define what percentage of initial and maintenance margins will be lent to qualified borrowers. Some lenders will advance total financing of margin requirements; others will advance a percentage of each margin. Another variation is to require borrowers to provide an initial margin while the lender agrees to provide liquidity for maintenance requirements. Some lenders may finance the total margin requirement on commodities in storage, but only a portion on anticipated production since production risks still exist.

Once the hedge loan is granted, a lending institution must always be able to provide all the funds required to follow through with the plan. If the lender cannot provide all the liquidity the market requires, the borrower's position in the futures market will be liquidated, and funds deposited to meet prior margin calls will be lost.

It is impossible to accurately predict the maximum commitment needed to meet margin calls, because no one knows the extremes the market will attain once hedges are set. If the lender anticipates that legal loan limits may be approached, he or she should secure a commitment for an over-line loan from another bank to meet full requirements.

Lenders must keep in mind that loan quality is not damaged by margin deposits, regardless of large they may be, as long as the producer has commodities as collateral. A hedge that is liquidated prematurely because of inadequate liquidity can be disastrous for both borrower and lender.

In developing the mechanics for establishing hedging loans, lending institutions must:

1. Develop procedures to evaluate loan requests,
2. Structure operating agreements,

3. Define documentation required to secure hedge loans, and

4. Establish methods to monitor the hedge account.

Evaluating Hedge Loan Requests

Policy guidelines should define who is a qualified borrower for a hedge loan. Normal credit standards and criteria should be used. Furthermore, the lender must be certain the borrower understands different techniques and implications of using marketing tools to set pricing decisions. The lender should be convinced the borrower has a clearly defined marketing plan and pricing strategy and can manage the emotions of a hedge account. The lender should also determine that the borrower's production or production capacity meets the specifications of futures contracts with regard to quantity and quality.

A balance sheet statement and a profit and loss projection determine a borrower's financial condition. Analysis of the first shows the borrower's ability to withstand adverse production or pricing risk. The second spells out cash flow requirements and price levels required to produce profit for repayment.

The borrower's pricing strategy must be evaluated against the financial strength on the balance sheet in formulating a marketing plan consistent with risk-bearing ability. Weighing the degree of acceptable cash market speculation against risk-bearing ability of the borrower is the first step in evaluating the marketing plan that a lender agrees to finance. A high-equity operator can afford to be a cash market speculator much longer than a low-equity operator. A highly leveraged farmer might be held to the least speculative alternative with early cash contracting.

Selecting the marketing tool to confirm pricing decisions and determining how it is used, depends upon the experience and knowledge of the farmer-hedger. An experienced hedger is more likely to have the knowledge and discipline to move in and out of cash market speculation when it is financially advantageous to do so. The lender and borrower must decide if fixed hedging, selective hedging, or flexible hedging best fits the farmer's knowledge, experience, and financial resources. No single plan fits all situations, but a specific, pre-agreed plan and pricing strategy solve potential problems. It also provides a way to do a historical analysis on how well the plan would have performed in past market situations.

Structuring Operating Agreements

Due to the speculative element present in farmer use of futures markets, a special working relationship is required between borrower and lender. Lenders abhor the idea of financing speculation. (But they do so each time they finance a spring operating loan based on cash markets without a marketing plan.) Forward planning and stated procedures formalized into a written operating agreement prevent misunderstandings. The agreement provides the details of the understanding reached by both parties.

The borrower's production estimates, marketing plan, and pricing strategy should become parts of the agreement. They can be reviewed from time to time and used as a standard for evaluating progress. How much production will be committed to the plan? Will the producer set pricing decisions on a calendar schedule, determine price objectives from technical analysis, or employ a trend following strategy? It should specify whether true hedging or selective hedging will be followed. The plan should be specific about setting and lifting hedges and about what criteria will be used to make the decisions.

The agreement should describe how the flow of funds will occur. In some cases, funds are transferred directly from the bank to the brokerage house on a call from either the broker or borrower. In other cases, they move from the bank to the farmer, then to the broker. Wire transfers are quicker and save mailing details, though they are more costly. Excess brokerage account funds often go back to the bank directly to keep the outstanding balance as low as possible to reduce interest charges, or to earn interest on surplus balances. The borrower and lender should negotiate beforehand what interest rates will be paid on credit and debit balances in the hedge account.

Periodically, the hedging account should be settled with payments to the borrower or against the loan balance. The agreement should spell out time of settlement. The logical settlement time is when cash transactions are made, but it need not be so limited. For crops, settlement could occur at the end of the crop year. For livestock, settlement could be made as each contracted group of animals is sold. It might take a week's, a month's, or two month's production to equal the contract amount. Even though the producer sold $45 hogs in the cash market, that producer could take home a $56 effective hedged price as he or she settled up on the way past the bank. When cash price was higher than the futures contract, sale of cash animals would be the time to pay the deficit balance on the futures position to the lender.

The agreement should also state that the customer will not speculate in commodities that are not produced. Borrowers should agree to use a reputable broker who understands the farmer's objectives and marketing plan. The loan could be made callable to protect against any terms of the agreement being violated.

Lender Documentation of Hedge Loans

In order to secure a lien on funds used for trading futures, contracts, lenders need to have three documents: (1) the note, (2) the security agreement, and (3) the hedge agreement and assignment of hedge account, known as a tri-party agreement.

1.) The note should be written for the amount of the maximum borrowing anticipated by the hedging plan, the interest rate, and the maturity date. It can be either a time or demand note. Since agricultural products may become marketable at times different from original plans and expectations, a demand note may be more advisable. It may eliminate the need for drawing up a new note, whereas a time note might mature before the commodity becomes marketable.

2.) The security agreement provides the lender with a lien on the assets used as collateral on the loan. It further spells out the lender's interest in receipts from sale of the inventories used as collateral. Whatever collateral is required to secure a production loan, the amount should be increased to secure the hedging loan.

3.) The third document is a hedge agreement and assignment of hedge account. It is used by the lender to secure funds in the brokerage account as part of collateral. With this provision, the effects of lower prices on the farmer's products are protected by surplus in the brokerage account. Conversely, a deficit in the brokerage account is offset by higher inventory value of the farm products.

In addition to the borrower and lender, the tri-party agreement further involves the broker in a three-way business arrangement. The document authorizes the lender to expand the loan to meet margin calls. The broker also has authority to request and receive additional margin as required, directly from the lender. Operating in the other direction, it allows the lender to withdraw excess margin funds from the brokerage account to reduce the borrower's loan. The lender can make an "excess margin call" on the broker. This document further authorizes the brokerage firm to furnish copies of the hedge transactions to the lender. These make monitoring the hedge much easier for the lender.

Several provisions should be included in the tri-party agreement. These provisions should be listed after naming the parties concerned and after making the statement of intentions of the loan to finance a hedging account. The following provisions may be included:[4]

1. "The lender agrees to advance funds to the borrower for hedging. The borrower agrees to deposit the funds with a broker and to use the funds only for hedging as defined in the marketing plan.

2. "As security for the funds, the borrower grants a security interest in the hedge account, an assignment of the funds presently in the account, and those to be accumulated in the future.

3. "The broker is authorized and directed to pay funds from the hedge account to the lender upon request. Further, the borrower agrees not to withdraw funds from the account without the consent of the lender.

4. "The lender will advance funds for margin requirements directly to the broker.

5. "The borrower gives the lender power of attorney to operate the hedge account if necessary.

6. "The borrower may execute transactions in the account, unless prohibited by the lender.

7. "The lender is authorized to liquidate the account when the lender deems its necessary for his own protection.

8. "Any funds paid by the broker to the lender will be applied to the borrower's loan. The interest rate charged on outstanding balances will be _____%; the rate paid on surpluses will be _____%.

9. "The lender is authorized to receive from the broker, and the broker is authorized to deliver to the lender, copies of all futures transactions made in the account.

10. "The agreement will remain in full force until canceled in writing.

11. "The borrower warrants to the lender that the account has not been alienated or assigned.

12. "The agreement is signed by the borrower and the lender.

13. "The agreement acknowledgment is made by the broker or brokerage firm."

After the agreement has been completed and signed, the original should be filed and recorded in accordance with local, state, and federal laws. The lender and borrower each should retain an acknowledged copy.

Monitoring Hedge Loans

The tri-party hedge agreement between farmer, lender, and broker provides measures of control over activities in the account. For the most part, the borrower should manage his or her own sales decisions or employ a market consulting service. Lenders are in the business of making loans and providing advice as requested, but not trading commodities. Understandably, most lenders are reluctant to help manage a borrower's marketing decisions. However, in order to safeguard funds, a lender may insist on being informed of market positions and may reserve the right to intervene to prevent losses for the borrower and ultimately the bank.

The broker provides copies of all brokerage statements to both borrower and lender. These statements serve as a report to the lender of market activity and status of the account. The statements can be used as a check to determine whether a borrower is following the marketing plan.

Brokerage firms provide customers with several account statements, as shown in Figure 8–1.

1.) The "confirmation of sale" itemizes trades that are made. It reports the position taken, number of contracts, commodity traded, and price information. Confirmations of sale are issued within 24 hours of the transaction and mailed to appropriate parties.

2.) The "statement of purchase and sale" (Daily Activity Statement) reports each time a position in an account is offset and lists date of transactions, prices, profit or loss for each trade, commissions paid, exchange fees charged, and customer's ledger balance.

3.) The "statement of open trades" shows the status of an account at the close of trading day. This report notes the status of open trade equity for positions not yet closed out. Equity is the difference between initiation price and the settlement price on close of the report. It is included automatically with each monthly activity statement. Daily statements of open trades are available upon request.

4.) The "monthly activity statement" is similar to a bank statement, as it lists all credits or charges to an account for a 30-day period. It reports the current bottom-line balance.

Additional loan monitoring requires a ledger of cash flows and inventory records on the physical commodity. A separate ledger for the commodity account should record the amount of margin funds advanced or excess margins received and repaid on the loan. This ledger can be balanced with the month-end statement issued by the brokerage firm. The borrower may be required to furnish a listing of inventory on hedged commodities, especially when advances are requested. In addition, a collateral inspection should be made occasionally by a loan officer or third party to verify inventory and reconcile the listings provided by the borrower.

With proper understanding of operating procedures, documentation, and monitoring, loans for hedging can be adequately secured without undue time demands on lenders. Improved earnings for the producer, gained from executing an effective marketing plan, improved the quality of an operating loan. The alternative is for the lender to finance cash market speculation.

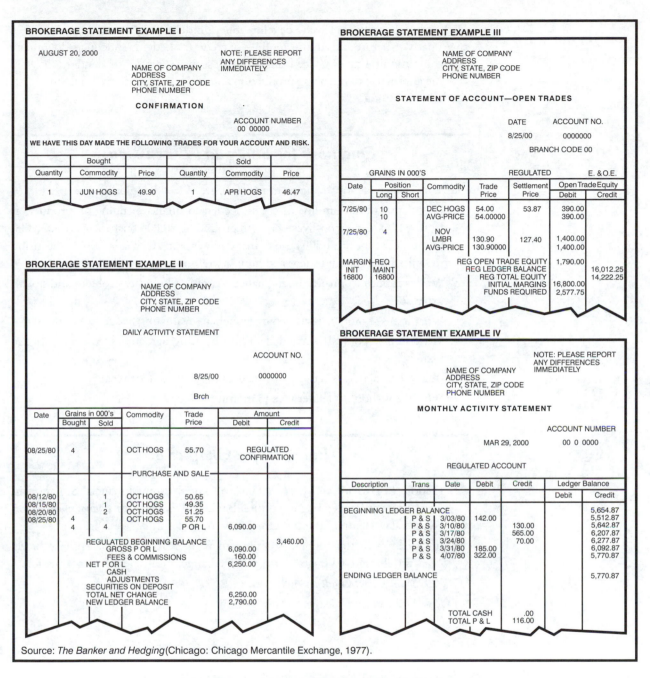

FIGURE 8–1: Examples of Brokerage Firm Account Statements.

Summary

Lenders usually are called upon to provide financing for hedging as well as production. A hedging plan must have adequate liquidity to carry it out. Without adequate liquidity, financial risk is increased rather than minimized by hedging. Earning power of loans can be enhanced by a sound marketing plan and pricing strategy.

Lending institutions have to develop policy guidelines and operating procedures to properly structure and administer lines of credit for hedging. Adequate documentation and monitoring of hedge loans properly secure them for the lender.

The alternative to financing production and hedged marketing is to finance cash market speculation.

Questions for Study and Discussion:

1. Why is adequate financing an important part of a hedging program?
2. Calculate a minimum line of credit needed to finance hedging on production from a 1,000-acre crop-livestock farm producing 80,000 bushels of corn, 30,000 bushels of soybeans, 1,500 head of market hogs, and 200 head of live cattle, using the initial margins and suggested liquidity requirements in the chapter.
3. Why is money to finance maintenance margins a liquidity problem and not an expense?
4. What items should be reviewed by lenders in evaluating hedge loan requests?
5. What items should be included in an operating agreement governing hedge loans?
6. What documents are required by lenders to secure a hedge loan?
7. What steps should a lender take to monitor a hedge loan?

Endnotes for Chapter 8:

1. Freddie L. Barnard and J. H. Atkinson, "The Current Financial Situation in Agriculture as Viewed by Agricultural Bankers," *Purdue Farm Management Report* (West Lafayette, Ind.: Purdue University, April 1982) 7.
2. Information in this section was gleaned from *Risk Management Guide for Ag Lenders* (Chicago: Chicago Mercantile Exchange, N.D.). Used by permission.
3. *Risk Management Guide for Ag Lenders* 6.
4. *Risk Management Guide for Ag Lenders* 16.

Chapter 9

Livestock Marketing

Livestock Marketing

Characteristics of Perishable Commodities

Perishable agricultural commodities, such as live cattle, hogs, eggs, broilers, and feeder cattle, have an entirely different relationship to futures markets than storable commodities. The differences change the role of futures pricing because they eliminate storage decisions from pricing and they generate price cycles that must become part of a production and pricing plan. Thus, pricing decisions must be made differently.

Corn, wheat, and cotton have relatively fixed planting, production, and harvest dates. Timing of the production cycle is fixed by environmental conditions. Production from a short harvest period must be rationed into use over a longer consumption period. Storable commodities can be held for several months, a year, two years, or longer, if properly stored. They can be drawn out of storage and into consumption by high current prices relative to lower price offers for future delivery.

Livestock products, in contrast, do not have the same storable characteristics or marketing flexibility. Production is virtually continuous. Once the breeding decision is made, production will be forthcoming in due time. When livestock are market-ready, they cannot be withheld from the market, nor can meat products be preserved and stored economically for long periods. Prior to being market-ready, production cannot be moved into consumption ahead of schedule. Producers do not have the flexibility of withholding production from the market in response to low prices, nor can they bring immature livestock to the market in response to high prices, without incurring substantial market penalties. Non-storable commodities conform to rigid production schedules that dictate that they move into consumption when, and only when, market-ready, give or take a few days or weeks. The price the product will bring depends upon what consumers will pay at that time. An unlaid egg can be neither poached nor fried, only done without; surplus eggs sink in value because of limited alternative uses.

Individual futures contract prices under such conditions are considered "true" forecasts of the equilibrium price that will prevail in the future at the time of maturity. The market participants' assessments of supply and demand conditions that will exist as livestock mature, determine the equilibrium price they bid into futures prices. Unlike grain, various livestock contracts are not tied, one to the other, by carrying charges.

However, in the real world, futures prices have not been accurate forecasts of cash prices when the future becomes the present. Based upon early data, Leuthold concluded that futures prices for live cattle and live hogs tend to be biased and inefficient forecasts of cash prices.[1,2] Garcia and Martin studied later data and introduced methods for including cyclical, seasonal, and feed-grain cost factors into their analysis. They concluded:

In a forecasting sense, cattle futures appear to add little information beyond that available in lagged cash prices. Live hog futures appear to have performed the forecasting function well, relative to both live cattle futures and lagged cash prices, except during the period when feed grain prices were highly variable. However, when forecasting was most difficult, hog futures failed to predict wide price swings.[3]

Even though futures prices on a particular day are thought to be the best estimates of value at the time, these studies concluded that futures prices are poor predictors of realized cash prices, because of supply and demand adjustments made in the interim.

What is the farmer or rancher to do if he or she cannot rely on futures as accurate price predictors? How can they decide whether or not to produce, or when to make pricing decisions? First, from a fundamental perspective, the farmer might try to develop skills as a price forecaster to gauge when he or she feels the futures market is wrong. If futures are higher than he or she expects the cash price to be, the farmer can hedge production; if prices are lower, the farmer can remain unhedged. Second, the farmer can calculate costs of production and make decisions to hedge whenever he or she is satisfied with profits the futures are offering. It is a matter of saying, "I don't know what the price will be, but I will be satisfied with the profit the market is offering in order to reduce my risks." Finally, the farmer can forget fundamentals and preferably, develop a viable and profitable trend-following technical trading strategy.

Fundamental Price Forecasting in Livestock [4]

Hog price forecasting, based upon fundamentals, is tied to measures of pork demand and supply projections. Supply projections can be based on quarterly pig crop reports. Upon issuance of each report, hog price forecasts for the next 12 months can be calculated from the supply projections. These forecasts should be compared to futures pricing opportunities and plans developed for the sale of each quarter's production. This procedure would involve modifying previous plans for the next three-quarters and new plans for only the farthest-out quarter. The producer would be making pricing plans at the same time that breeding plans are made.

The above decision pattern applies when no change in production is anticipated by the producer. However, increasing or decreasing production should always be considered an option. When feed supplies are small and higher input prices drive up the cost of production, or hog numbers are so large as to drive prices down, production will be unprofitable. The market is dictating reduced production. On the other hand, low feed costs or reduced pork supplies signal producers to increase production to capture higher profits. The forecasting reference to the pork producer for changing production level is the hog cycle.

The time to consider production level changes is late summer, when feed supplies are known and intended hog production information is available. At that time, producers can use the futures market to lock in the cost of inputs for 12 months. They can hedge-buy soybean meal and corn supplies and hedge-sell planned production if relative prices offer profits. This idea was introduced earlier when we discussed locking in profitable cost-price spreads before initiating production increases.

Supply estimates for beef production can be developed from semi-annual cattle inventories and quarterly cattle-on-feed reports. The demand for beef can be estimated from disposable personal income and consumers' willingness to spend money for beef. The other factor in predicting price is the supply of competing products. Other meats, such as pork, lamb, veal, broilers, and turkeys, are most important. Other foods, such as eggs, dairy products, and vegetables, can also affect the demand for beef, but only slightly.

The resulting price forecast, compared to cost of production, guides producers in deciding whether or not to fill feedlots. If a profit level is not bid into the market by relative production costs and sale price, the producer may decide to leave the lots empty. Whether or not to leave feeding facilities idle can be a month-to-month decision. Breeders do not have as much flexibility as feeders because breeding herds are difficult to liquidate and reestablish.

A producer who intends to use futures contracts to establish prices should start planning cattle sales about six months before cattle are placed on feed. During that time, futures price relationships between feed, feeder livestock, and fed cattle will hopefully adjust to equal or exceed a profit objective. The time to establish hedges to capture margins offered by futures price relationships depends upon price forecasts and the willingness to accept the risk of lesser profit in exchange for the opportunity for higher profit.

If the feeding enterprise is entered into without forward pricing finished cattle, prices must be constantly monitored in order to decide whether to cover the sale via hedging or not. The price should be fixed when the probability of further gain is less than the probability of loss from the existing price level, or when further loss from lower prices would endanger economic survival of the enterprise. It is better to hedge in a small loss rather than suffer large losses due to a further price decline.

While the above reasoning appears logical, an analysis of various cattle-feeding hedging strategies to implement that rational have not produced encouraging results.

Using Hedging to Fix Cattle Prices

A Kansas study reports the results on cattle feeding profits of several hedging strategies over a 10-year period.[5] The operation was patterned on a 20,000-head (one-time capacity) feedlot with approximately 400 head of feeder cattle scheduled in every week and 400 finished cattle sold every week. Costs of inputs and prices of finished cattle were based on current quotations throughout the period. Weekly, annual, and 10-year average feeding profits were calculated on a per-head basis. The feeding period was 147 days, so 505 lots of cattle were involved in the simulated analysis.

Seven program options were considered:

1. All lots fed out unhedged (the unhedged program was a benchmark or control for the other alternatives),

2. All lots routinely hedged,

3. Hedge only when the futures hedge price equaled or exceeded the calculated break-even live-cattle price,

TABLE 9–1: Average Profits, per Head, from Seven Alternative Cattle Hedging and Contracting Programs, May 1965 to December 1974.

Alternative	1965	1966	1967	1968	1969	1970	1971	1972	1973	1974	Average 10-Year Profits	Variance	Lots Hedged
								($)					
I. Unhedged	36.30	13.68	2.97	18.67	26.77	–1.46	21.52	29.85	8.55	–53.11	9.55	1,079.737	0
II. Routine Hedge	15.51	16.12	14.01	6.23	2.23	–6.98	–7.36	–1.54	–29.06	–2.67	0.18[1]	417.243[1]	505
III. Futures ≥ Break-even	19.05	17.22	14.01	16.71	18.98	–5.32	21.45	27.28	2.05	–11.31	11.81	980.095	218
IV. Futures ≥ Cash	29.89	23.48	14.01	19.36	26.10	–6.23	10.48	29.85	–10.89	–0.23	13.08[2]	732.439[1]	204
V. Futures ≥ Break-even & ≥ Cash	29.89	23.33	14.01	19.36	26.10	-4.76	21.32	29.85	2.05	–12.28	14.43[3]	1,060.335	145
VI. Seasonal Hedged (Fall)	31.65	17.22	6.66	14.39	28.29	0.96	10.61	24.24	20.80	–44.48	10.38	907.302[3]	174
VII. Contracted	19.69	17.43	–1.33	13.72	21.42	–0.97	2.06	21.43	–29.98	–34.38	2.41[1]	199.556	0

[1]Indicates that difference as compared to unhedged value is statistically significant at the 1 percent level.

[2]Indicates that difference as compared to unhedged value is statistically significant at the 10 percent level.

[3]Indicates that difference as compared to unhedged value is statistically significant at the 5 percent level.

4. Hedge only lots when the futures hedge price equaled or exceeded the cash price for finished cattle at the time feeder cattle went into the feedlot,

5. Hedge only lots when the futures hedge price equaled or exceeded *both* the break-even price and the current cash cattle price,

6. Hedge only lots finished during the September to December period, and

7. Sell finished cattle by cash forward contract the day feeder cattle went into the feedlot, at the contract price offered for finished cattle that day.

Option 1—Average profits on completely unhedged operations for the 10 years were $9.95 per head. The trend in profits was typified by instability. Under this system, 349 lots (69 percent) returned profits, while 156 lots experienced losses.

Option 2—Routine hedging produced average profits of 18 cents per head, significantly below Option 1. In fact, routine hedging was the worst alternative in 7 out of 10 years. Routine hedging of all lots on the day they entered the feedlot improved the profit on 183 lots, but reduced the profit on 332 lots.

The major reason routine hedging and other price-fixing techniques did not improve profits in this study was that cattle prices were in a general up trend for most of the period studied, as illustrated by Figure 9–1. Hedging nullified windfall profits that otherwise would have accrued during a general rise in price level.

Option 3—This is one of several selective hedging programs. Cattle were hedged only when the effective futures hedge price (futures adjusted for location basis) equaled or exceeded the break-even price. This strategy generated an average profit of $11.81 per head. This program assumed that when prices did not meet the hedge criterion, the cattle were fed unhedged. Fifty-seven percent (287 lots) did not meet the

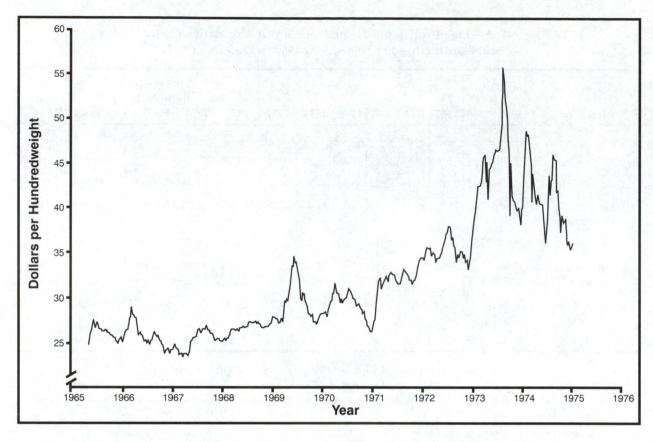

FIGURE 9–1: Prices of Choice 900 to 1,100 Pound Slaughter Steers, Kansas City, May 1965 to December 1974. Source: John H. McCoy, *Cattle Hedging Strategies*.

criterion for hedging. Profits increased on 112 of the 218 hedged lots, while profits decreased on 106 lots.

If cattle had not been placed on feed when the effective futures price was less than break-even, average profits would have equaled $10.36 per head. This variation of Option 3 would have left the lots empty part of the time. Profits on the animals actually fed were less than profits from unhedged cattle. Thus, there was less return to total costs during the period studied.

Option 4—Hedging was done when futures were equal to or greater than current cash price. Profits were $13.08 per head. Forty percent (204) of the 505 lots were hedged. Leaving the lots empty 301 times, when the criterion was not met, dropped profits to $3.21 on those actually fed out.

Option 5—This option showed the highest profit of all the strategies tested. The combination specified that cattle would be hedged only when the effective futures price equaled or exceeded both the break-even price and the current cash price. Profits were $14.43 per head. But only 29 percent of the lots were hedged. Leaving the lots empty when the two criteria were not met dropped average profits to $7.10 per head.

Option 6—A seasonal program of hedging produced average profits of $10.38 per head. Only cattle finished for market during September through December were hedged; all other lots were finished unhedged.

Option 7—The final option did not use the futures market but assumed that the day cattle went on feed they were contracted for delivery when finished. Profits were $2.41 per head, largely because that system precluded windfall profits during rising price trends.

Basis Problems and Opportunities

In the Kansas study, McCoy and Price noted particular problems in their analysis because of basis variation.[6] For satisfactory hedging, the basis relationship should be relatively constant, or at least predictable. The record shows substantial instability in the Kansas City basis for live cattle (Figure 9–2). Before 1973, the Kansas City basis averaged about minus $1.00, ranging from approximately minus $2.50 to plus $2.00. After 1973, however, the basis ranged from minus $7.40 to plus $4.40, with an average of approximately minus $2.00. The 10-year study used a basis of only minus $1.00 in deciding when to hedge. However, the realized basis was often much worse at maturity. A hedge sale that originated with an expected minus $1.00 basis, but ended with minus $3.00 basis, produced less than the expected price protection. Thus, several lots in the study that were hedged to cover break-even price ended with unanticipated losses.

Figure 9–3 illustrates the same general scenario. In this figure, the Omaha, Nebraska basis for live hogs is plotted against the nearby Chicago hog futures between 1975 and the end of 1997. A range of $5.00 above the futures to $5.00 below

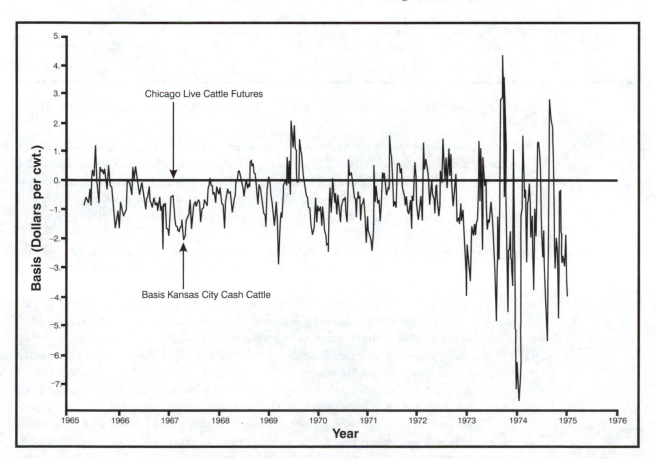

FIGURE 9–2: Kansas City-Chicago Live Cattle Basis, May 1965 to Dec. 1974. Source: John M. McCoy, *Cattle Hedging Strategies*.

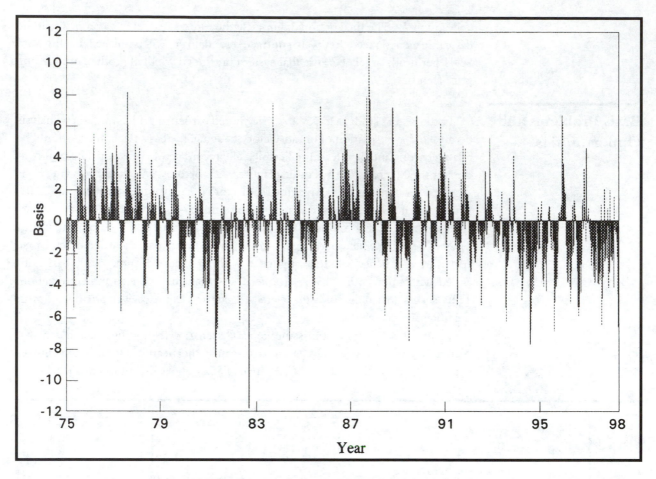

FIGURE 9–3: Omaha Basis: Live Hogs Vs. Nearby Chicago Futures, 1975 to 1997.

the futures is a common event. On rare occasions, the basis goes below –$10.00 and above +$10.00. Thus, a $10.00 basis gyration, representing $25.00 per head (250 pounds × $0.10), is not at all uncommon. Those swings represent real money and emphasize the real challenge of attempting to hedge away livestock price-level risk. The gains and losses in cash and futures markets do not offset each other when there are $25.00 basis swings.

Basis decisions for non-storable commodities are thus meaningless. Unlike storable commodities, there is no way to capture strong basis levels, nor avoid poor basis levels. Live cattle contracts are traded on a bimonthly schedule, but finished animals must move into slaughter and consumption when, and only when, ready. Cattle finished in mid-May cannot be delivered against an expiring April contract, nor profitably held until mid-June to deliver against the June contract. A weak basis during May cannot be avoided. Similarly, a favorable basis in April cannot be captured because the animals are not ready for market. Hedging non-storable commodities with futures contracts, therefore, does not provide total protection against price change because of basis variation. The basis variation is beyond the control of producers, and it is impossible to either capture strong basis or avoid weak basis at the time the cash sale is made. The cash sale must be made when animals are market-ready.

Only during delivery months are cash price and futures price related by the transportation and transaction costs of delivering against contracts. The only way a hedger can avoid a bad basis situation is to deliver production to satisfy the futures contract. Basis projections could be more meaningful and predictable if futures contracts on non-storable commodities were traded every month and delivery problems were minimized. Cash settlement of contracts, rather than physical delivery of livestock, could be beneficial in stabilizing basis. In point of fact, procedures are currently undergoing changes. Feeder cattle contracts (50,000 pounds) are now settled by cash rather than actual physical delivery. Lean hog carcasses, yielding 74 percent of live hogs are also settled in cash. Live cattle contracts, not offset in the futures market, still require actual physical delivery, but to more convenient slaughter plants or commission firms. Knowledgeable experts anticipate that all livestock futures contracts will be settled by a cash transaction in the foreseeable future. Until those pesky procedural changes come to final fruition, hedging livestock with futures contracts poses a serious hazard to a pricing plan because of basis variation.

Basis variation can eliminate hedge profit. The feed lot operator who buys feed and feeders at a basis premium to the hedged prices effectively pays more for the inputs than the hedged costs would suggest. Similarly, if the finished product is sold at basis discounts, gross receipts are reduced. The feeder may have a futures profit of $50.00 per head locked into a pen of cattle, but if he or she pays $14.00 per head more for feeders (2 cents above hedged price on 700 pounds) and $13.50 more for corn (54 bushels per head times 25-cent basis premium), and receives $22.50 less on finished cattle (2 cents below hedged price on 1,100-pound cattle), the hedged profit margin is eliminated (a total of $49.50 higher costs and reduced returns). Profitable hedging does not guarantee profitable production because of basis variation.

Viewed from the opposite perspective, basis variation on hedged feed purchases, hedged feeder cattle purchases, and hedged live cattle sales can be sources of additional profits. Futures transactions could lock in a generally acceptable profit level, but astute cash buying and selling to improve basis could yield significant additional returns. The feeder who can buy feed ingredients at basis discounts in excess of carrying costs will lower his or her cost of feed inputs below the hedged price. Likewise, if able to make cash purchases of feeder cattle below the hedged price, the buyer will have lowered the effective purchase price of feeders. On the product selling side, the feeder who can make cash sales of finished animals at higher than hedge prices will improve profits above the hedged selling price. Used as a marketing tool, with both a basis and price perspective, hedging removes price level speculation while focusing attention on basis level trading. Astute basis trading is as important to profitability in livestock feeding as fixing price level decisions.

Behavior of Non-Storable Futures Contract Prices

Hieronymus studied cash and live cattle futures over the period of 1973 to 1975.[7] His observations can be helpful in formulating a pricing strategy for hedging price level. The years of study covered a period of unusual price variability. The demand for meat was strong as income rose. During 1973, there were production cutbacks on meat and price ceilings were imposed by presidential decree. During 1974, poor

TABLE 9–2: Cash Cattle and Cattle Futures Prices, Chicago Mercantile Exchange, Monthly Averages of Daily Closing Prices, Dollars per Cwt.

Month	Choice Steers Omaha	Feb. 1973	Apr. 1973	June 1973	Aug. 1973	Oct. 1973	Dec. 1973	Feb. 1974	Apr. 1974	June 1974	Aug. 1974	Oct. 1974
1973												
Jan.	40.31	42.84	42.71	42.70	41.42	41.28						
Feb.	43.07	44.47	43.84	44.09	43.43	42.62	42.50	42.34				
Mar.	45.28		45.92	45.74	44.91	43.80	43.32	43.37	43.28			
Apr.	44.76		44.38	43.25	43.11	42.65	42.81	43.06	42.99			
May	45.71			45.91	45.52	45.15	45.25	45.35	45.03			
June	46.65			47.63	47.24	47.14	47.21	47.35	47.02	46.83		
July	48.11				50.93	51.75	52.41	52.59	51.87	51.68	50.94	
Aug.	53.39				56.72	55.64	56.50	57.07	56.67	56.72	55.92	
Sept.	44.79					43.48	45.40	48.11	48.31	48.42	47.91	47.42
Oct.	40.65					41.71	44.91	48.30	48.94	49.16	48.59	47.89
Nov.	37.47						40.56	45.28	46.27	46.38	45.71	45.02
Dec.	38.87						38.64	46.06	48.00	48.13	47.60	46.86

Month		Feb. 1974	Apr. 1974	June 1974	Aug. 1974	Oct. 1974	Dec. 1974	Feb. 1975	Apr. 1975	June 1975	Aug. 1975	Oct. 1975
1974												
Jan.	47.22	50.95	53.80	54.30	52.93	51.81	51.60					
Feb.	46.22	46.57	48.61	51.50	52.26	52.07	52.14					
Mar.	42.42		43.17	46.23	48.32	48.83	49.36					
Apr.	40.86		42.20	45.65	47.64	49.91	47.02	46.96				
May	40.04			41.04	41.59	39.69	39.26	39.13	37.06			
June	37.50			36.47	37.21	37.49	37.42	37.12	36.73			
July	43.60				46.02	46.30	45.73	45.51	45.41	45.65		
Aug.	46.52				48.46	46.70	45.19	44.81	45.38	46.10	46.33	
Sept.	41.24					39.41	38.11	41.21	41.77	43.02	43.30	
Oct.	39.72					41.09	42.96	44.06	44.03	45.29	45.02	
Nov.	37.77						38.53	40.57	41.28	42.46	42.24	
Dec.	37.13						38.12	41.14	41.90	42.60	42.68	42.43

Month		Feb. 1975	Apr. 1975	June 1975	Aug. 1975	Oct. 1975	Dec. 1975	Feb. 1976	Apr. 1976	June 1976	Aug. 1976	Oct. 1976
1975												
Jan.	36.17	37.10	38.00	38.70	38.82	38.69	38.83					
Feb.	34.68	35.08	36.72	37.54	37.22	36.81	36.93					
Mar.	36.14		37.96	38.59	37.65	36.77	36.90	37.02				
Apr.	42.68		42.50	43.29	41.08	39.27	38.92	38.74				
May	48.35			48.47	43.86	40.33	39.39	38.81				
June	52.21			53.42	48.63	43.51	41.74	40.67	40.17			
July	50.41				46.77	41.91	40.84	40.37	40.05	40.61		
Aug.	47.84				47.27	42.07	42.04	42.77	42.98	43.65		
Sept.	49.64					47.79	47.21	45.29	44.45	44.92	44.77	
Oct.	48.27					47.49	44.59	41.69	40.81	41.72	41.53	
Nov.	45.49						45.83	43.48	41.90	42.79	42.16	42.05
Dec.	45.62						47.03	43.70	41.71	42.81	42.69	42.16

Note: Read across the table to discern the relative cash and various futures options prices that existed in any month. Read down the columns to see how the price of cash or a particular option changed over time.

Source: Hieronymus 168.

crops led to herd reductions, forcing large meat supplies to market. Prices fell significantly in early 1975 but rebounded later as available supplies declined.

The futures price gyrations of that time (reported in Table 9–2) led Hieronymus to the following conclusions about futures pricing of live cattle:

"All prices rise and fall together.

"Variation in cash price and nearby deliveries is greater than with more distant deliveries.

"There is substantial independence of price change in the more distant deliveries and the cash.

"The dominant influence on cattle futures is the cash cattle price.

"The dominance of the cash price suggests that the market seems to think cash price is as it should be and will not change.

"Traders are very skeptical about the validity of recent price change. They view a change as an aberration which will not last, and they assume that prices will return to their earlier level.

"And finally, markets do look ahead, anticipate supplies to come to market at a later time, and adjust prices in line with expectations."

These generalities are important in understanding some of the differences between non-storable and storable futures. Non-storable commodities do not have a reason to reflect a market carrying charge. Therefore, the price in each contract month should be the same as every other month, save for changes caused by anticipated supply or demand changes in each respective contract month. This explains the dominance of cash price in setting nearby futures price level and the unresponsiveness of distant contract months to fluctuations in nearby contracts. Since real changes in future demand or supply cannot be accurately predicted, the present situation dominates the price-making forces. Unpredictable events that occur over time cause futures prices to fail miserably as an accurate forecast of cash prices. As with grain contracts, futures prices are a daily pricing offer to accept or reject; they are not accurate price forecasts because price-making forces change.

Making Livestock Hedging Work: Bear Spread Hedging

Textbook fixed hedging in distant months when production is to be market ready seldom works to capture a high price level in livestock. To gain pricing rewards from the futures markets, livestock producers need to employ a different strategy.

Livestock hedgers are frustrated trying to set hedges with futures contracts because of the time lag between nearby high prices and lower prices on distant contracts that coincide with when they will have market-ready animals. A reduced supply of immediately available market-ready animals causes strength in cash prices as packers bid for animals. Strong cash price leads the nearby futures contract price higher. But more distant contract months typically do not rise as strongly as nearby months, if at all. Breeders and feeders cannot capitalize on price protection for expected production when current cash prices are high, and distant futures offer significantly lower pricing opportunities.

To illustrate the point, suppose the cash hog market and futures rise to $60 per hundredweight and then decline to $40 over a 12-month period at a steady rate. A producer selling equal numbers each week realizes the annual average price of $50.

A fixed hedging strategy would sell contracts in each contract month, hold those contracts until the live hogs are sold, and liquidate the contracts upon sale of live animals. Since not all calendar months have future contracts, it is necessary to place two contracts in the next month to cover non-traded months. The prices in Table 9–3 are futures quotes for respective contracts that traded on November 1, 1999. A farmer producing 2,200 head of hogs annually would market 180 head each month, or about one contract. The price locked in by executing hedges on November 1, 1999 averaged $52.43. After subtracting a $2.00 basis (applicable for Ohio) and 20 cents commission, the average net hedged price is $50.23. Hedging results are almost exactly the same as doing nothing and speculating in the cash market!

TABLE 9–3: Lean Hog Futures Prices of Various Contract Months, as Traded November 1, 1999

Contract	Price per Cwt.
December	$45.20
Jan. sales on Feb. contract	45.20
February	48.95
Mar. sales on Apr. contract	50.80
April	50.80
May sales on June contract	58.70
June	58.70
July	57.60
average	52.43
less Ohio basis	– 2.00
	50.43
less commissions	– 0.20
	50.23

However, suppose that cash hog price rose to $70.00 before starting the 12-month decline to $40.00. In this instance, the cash market speculator would average $55.00 per cwt. The fixed hedger, who locked in $50.23, would be worse off. With contract-to-contract price discounts, such as on November 1, 1999, fixed hedging will match the results of not hedging only when hedgers execute their positions at the extreme top of futures market prices. When distant options are discounted less, hedgers might gain. But if distant contracts are discounted more, hedgers lock in lower returns than cash marketers.

The secret to capturing returns in livestock hedging is to execute futures contracts equal to a full year's worth of production in the nearby contract (trading at a premium to other months), hold the position until the futures market breaks, and then lift all positions at once, when the bottom of the trend is signaled. When the break occurs, the nearby option typically falls more sharply than the deferred contracts. Since the break may not occur while holding the nearby option, the contracts must be rolled forward into the next nearby month, waiting for the break. Three or four rolls may be required.

A recent hog price situation serves to illustrate the procedure. The weekly plot of the various contract months in Figure 9–4 shows the strength in nearby contracts when strong cash markets led the futures contracts to profitable highs. A pork producer looking at cash prices in August and planning to expand production would not have extra animals market-ready for at least ten months. The October contract traded up to $64.00. At the same time, the June contract was trading at $54.00 per hundredweight. This meant that June lean hog futures were offering $22.00 ($10.00 per cwt. × 2.20 cwt.) less per head. Farmers usually expect current prices to hold at a constant level into the future. Each farmer thinks he or she is the only person who sees strong price as a market signal to expand, and does not believe other producers will increase production. By discounting distant contracts significantly, commodity traders disagree.

It is psychologically difficult for farmers to hedge when distant contract prices lag substantially below current cash prices. Producers remember being stung by hedging too soon in the market rise and they do not want to risk another drubbing. Figure 9–4 shows that prices did fall in response to increased production as the June contract dropped to $46.00. A June contract sold at $54.00, when the October contract peaked, would have offered some protection. A June hedge set at the contract high in January of $56.00 would have provided more price protection, but picking tops is virtually impossible. More important, June hogs hedged at $54.00 in October smell much better than those finished hogs sold for $47.00 in June.

Instead of traditional fixed hedging, concentrate annual marketings in the nearby month. In August, hedge the coming year's production with the October contract at $63.00. Liquidate that contract at $57.00 in September and roll **all** hogs into the February contract at $57.00. Liquidate at $57.00 in January to roll one year's production into the June contract at $55.00. Finally, liquidate the June contract at $48.00 in June. In summary: ($63.00 – $57.00 = +$6.00) + ($57.00 –$57.00 = $0.00) + ($55.00 –$48.00 = +$7.00). Thirteen dollars on a full year's production is far better than June 2000 hogs sold in August 1999 at $54.00 and lifted at $48.00 in June 2000. In our arithmetic, $13.00 is a larger number than $6.00 at one-fourth, despite two extra round-turn commissions.

The example illustrated above is a solution to the livestock hedging problem and it involves *bear spread hedging*. It permits capturing the strong nearby futures price level and relies on anticipated producer response to present high prices as they expand production. It also depends upon futures prices eventually flattening out month-to-month in characteristic fashion, because they have no carrying charges to reflect.

Bear spread hedging is initiated by using some method (preferably trend following) to decide that the futures price of the nearby contract has reached its high, or is sufficiently high to be acceptable. At this time, hedges on 12 months of production are executed by using the nearby contract—selling the October contract in early September near the top depicted in Figure 9–4. There is no way that the producer can speed up production to have animals available for delivery against the nearby contract, as a true hedge would dictate. But as the nearby contract reaches the time of maturity, the price protection is rolled into a more distant contract month, such as February. This is done by simultaneously lifting the nearby (with a buy) and resetting the hedge in another contract month (with a sell). As the price level declines, the second contract continues to provide the price protection sought. When the second contract approaches maturity, the protection is again rolled into another contract month, such as June, by the same process described above.

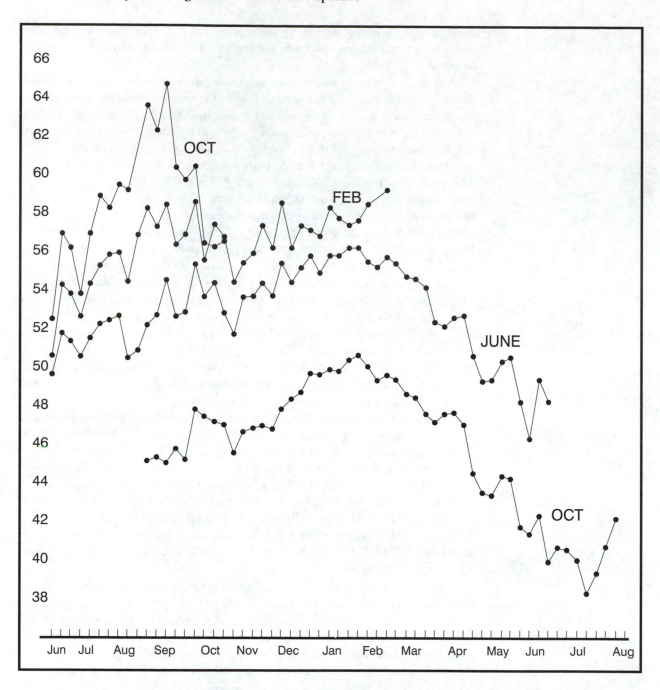

FIGURE 9–4: Relationship of Various Futures Contract Months, Lean Hog Contracts.

To exit the hedge, and lift the contracts, producers can follow one of two methods. As production equal to one contract is delivered to the cash market, one contract is lifted and not replaced. This procedure satisfies the definition of hedging, because contracted amounts are kept in balance with production amounts. Eventually, all the production will be sold and all the contracts lifted. An alternative procedure is to hold all the contracts and lift all of them when the market analysis signals that the market has bottomed. By lifting one at a time, the hedger may run out of positions before the market finishes sinking. Lifting them all at once, near the market bottom, allows the producer to capture more of the total move on the full year's worth of production. Remember, there is another batch of hogs to be finished in 18 months. There is also no law that says farmers cannot sell more than one year's worth of production when the market screams **"sell"** by offering exceptional prices. The hogs will be there eventually and they smell much better when sold at $65.00 rather than $35.00.

Bear spread hedging permits a producer to establish hedges at higher price levels than the deferred contracts offer. However, it does not guarantee that the hedges will capture the entire price move that eventually occurs. Rolling contracts month-to-month has its hazards and challenges.

Figure 9–4 shows the October option dropping to the level of the February option. Rolling from October to February could have been done at the same price. But during February, the roll to June might have been done at a $4.00 discount (February $59.00 versus June $55.00). Such slippage cannot be avoided. There is no way to capture the contract-to-contract price discounts; one can only try to roll them at the opportune times with minimum discounts (not an easy task). However, slippage is not the real issue; the real issue in January is whether June hog price will be $56.00 or $46.00. The proverbial optimist anticipates $56.00 forevermore. The realist sells a year's (or more) worth of production when the market offers the opportunity. Few producers have gone bankrupt selling $75.00 hogs.

In due time, the producer will have sold the planned production and lifted the hedge. The producer should have achieved a greater degree of price protection than if he or she had initially established hedges in lower-priced distant contract months. If the producer is lucky, he or she might even sell contracts on 12 months of production near the top of the market and capture most of the price move.

Timing Cash Sales

Within a restricted time frame, a producer can read the market for signals on when to sell cash livestock. He or she can shorten or lengthen the feeding period slightly to alter selling weight in response to changing price and changing production cost to maximize feeding profits. Hogs can be sold as light as 190 pounds or fed longer and sold at 260 pounds. Cattle can be sold early or late by adjusting feeding period and selling weight. Present prices, cost of gain, weight discounts, and the probability of price change can be combined into a decision framework to determine when to move livestock to market. Proper evaluation of added costs and added returns can give the producer some measure of control in selecting relatively more profitable selling weights and market dates.

At heavier weights, livestock are less efficient in converting feed into weight gain. Cost of gain goes up at an increasing rate as animals put on additional pounds. Table 9–4 depicts the amount of feed required for each 30 pounds of additional gain and the added cost of gain at various corn prices. As long as hog market price per hundredweight exceeds the cost of adding a pound of gain, it is profitable to feed to heavier weights. For example, if market hogs were selling for $40.00 per hundredweight and corn price was $3.30 per bushel, the added cost of feeding hogs from 220 to 250 pounds would be profitable because the added cost would be $39.49 per hundredweight. The profit-maximizing goal is to equate *added* cost of each pound of production to the *added* returns from that extra production. As long as added returns exceed added costs, net returns continue to increase. The principle applies to all livestock.

During periods of either rising or declining price trends, the analysis must be modified in response to changing price. With a changing price trend, the total value of the animal is changing, not just the additional pounds. Rising price increases the value of the first 220 pounds on the market hog, not just the value of the 30 pounds added by going to 250 pounds. Conversely, a declining price means that value loss occurs on all pounds, not just the added weight.

To decide when to sell in either up trending or down trending markets, the producer must calculate the change in total value of the animal against the cost of adding the additional weight.* If price is expected to increase to $42.00 per hundredweight over the next two weeks, the value of the animal increases by $17.00. The original 220 pounds increases in value by 2 cents per pound and the extra 30 pounds are worth 42 cents each.

TABLE 9–4: Cost of Adding 30 Pounds of Gain on Hogs Weighing 190 and 220 Pounds at Various Corn Prices

| Price of Corn per Bushel | Total Cost of Gain[1] | | | |
| | 190-Lb. Hog Fed to 220 Lb. | | 220-Lb. Hog Fed to 250 Lb. | |
	Cost of 30 Lb.	Cost per 100 Lb.	Cost of 30 Lb.	Cost per 100 Lb.
	----------------------- ($) -----------------------			
1.50	5.22	17.40	5.44	18.13
1.60	5.57	18.56	5.80	19.33
1.70	5.92	19.72	6.16	20.54
1.80	6.26	20.88	6.53	21.75
1.90	6.61	22.04	6.89	22.96
2.00	6.96	23.20	7.25	24.17
2.10	7.31	24.36	7.61	25.38
2.20	7.66	25.52	7.97	26.58
2.30	8.00	26.67	8.34	27.79
2.40	8.35	27.83	8.70	29.00
2.50	8.70	29.00	9.06	30.21
2.60	9.05	30.17	9.43	31.42
3.00	10.34	34.45	10.62	35.39
3.30	11.37	37.90	11.85	39.49
3.60	12.40	41.35	12.92	43.07

Source: Derived from data in John T. Larsen and Robert L. Rizek, *The Cost of Feeding to Heavier Weights,* ERS-371 (Washington, D.C.: USDA, February, 1968), 27-28.

[1]Calculated on the basis that 2.4 bushels of corn or its equivalent is required to raise a hog from 190 to 220 pounds, and 2.5 bushels from 220 to 250 pounds. An extra 45 percent is added to the cost of feed as the approximate costs other than feed.

Depending upon added cost and the extent of the price increase expected, additional weight increments might be profitable in rising markets. One risk is a price discount for overweight hogs that reduces the selling price. Assume that by increasing selling weight from 250 to 260 pounds, a $1.00 per hundredweight price discount is

---

*This assumes that the cost of weight gain is the only variable. This assumption is invalid in limited-space operations where overcrowding results from delayed marketing. Space and time costs are thus also important in calculating the optimum selling weight, but are ignored here to simplify the analysis and explain the procedure.

applied (from $42.00 to $41.00). The original 250 pounds would be discounted $2.50. The extra 10 pounds would be worth $4.10. Added cost of feed would be $4.00. Thus, the net result would be a reduction in profit of $2.40 per head ($4.10 value of weight gain, minus $4.00 added cost, minus a $2.50 discount on the original 250 pounds).

This method of analysis also pinpoints the profitable sale weight in a declining market. If the price is expected to decline $2.00 (from $42.00 to $40.00) per hundred-weight in the time required to grow from 220 to 250 pounds, the total value of the animal will increase only $7.60 (220 lb. × 42¢ = $92.40 versus 250 lb. × 40¢ = $100.00). As costs increase by $11.85, the result is a $4.25 reduction in profit ($7.60 – $11.85).

This analysis suggests feeding to heavier weights in up trending markets until overweight discounts are encountered and selling at lighter weights in down trending markets.

The detection of market price up trends or down trends is critically important in making cash sale decisions. Price charts of both futures and cash prices identify the general trend of market price for the analysis. The producer cannot predict exactly how much price will change over any time period, but if the producer has a feel for the trend, he or she can capture more profit than by just being an "every Monday" seller. A cattle feeder who watches price trends and reacts to them will certainly be better off than neighbors holding 1,600-pound steers, hoping for a price recovery.

Selection of "the" selling day can be further refined by watching the trend on a daily cash price chart and selling on up-price days. Buyers often raise their bids on days of light runs. The day to sell is when nobody else is selling.

If all producers utilized an **added value versus added cost** analysis to select the weight for marketing cash livestock, price declines in down trending markets would be halted earlier by reductions in the total tonnage that the market would otherwise have to absorb. In up trending markets, the price rise would be stopped short of the extreme by increased tonnage from heavier animals. Some short-term price variability would be removed from the market as producers reduced tonnage in weak markets, or increased tonnage in strong markets. In practice, feeders tend to hold heavy-weight animals hoping for price recovery. The practice is self-defeating, as increased total tonnage actually depresses price further when it eventually does come to market.

A more effective alternative is to sell those $30.00 hogs when they are ready for market, go **long** on a futures contract and "*store*" them on the floor of the Chicago Merc to speculate on a price rise. It is cheaper and the risks are no greater than heavy-weight hogs held at home.

Agricultural Price and Production Cycles

Long-term agricultural price cycles refer to fluctuating price levels over time caused by periodic expansion and contraction of the supply of agricultural products. Cycles reflect farmer production decisions and the subsequent market response to move an adjusted supply into consumption. High price entices increased production, which must be sold at a low price when the increased supply is market-ready. Low price discourages production and the reduced supply fetches a higher price at a later date. Price changes are much larger than production changes due to the inelastic demand for farm products. (See page 135 for the Cobweb Theorem.)

Price cycles are based upon producers' expectations that current price levels will continue to hold into the future. Producers tend to base production plans on current profits rather than on future expectations. They expand production in response to current high prices, thinking that no other producer will do the same. When the aggregate of increased supply becomes market-ready, prices decline to move the added volume into consumption. The cattle cycle is a good reference for understanding price cycles.

Figure 9–5 illustrates seven of the ten major beef cattle production cycles over the past 100 years. Since 1930, there have been seven major cattle cycles. Between 1930 and 1980, five cycles characterized by successively higher highs and higher lows culminated in an explosion in cattle numbers that more than doubled the U.S. herd size in less than 50 years–from 61 million head to 132 million head. All of this was geared to a burgeoning population, more disposable income, and an increased appetite for protein. The last 25 years of the 20th Century witnessed three dramatically different cattle cycles with lower highs and lower lows. As a consequence, the cattle herd size shrank from 132 to end the century with 98 million head.

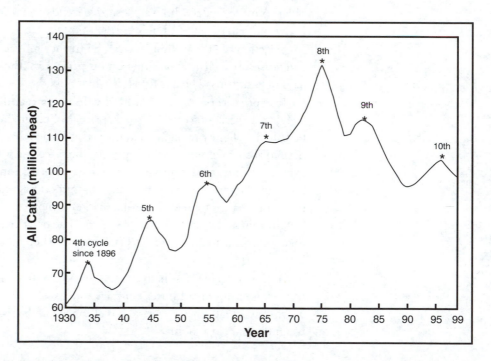

FIGURE 9–5: Ten Major Cattle Cycles in 100 Years. Source: USDA-NASS.

Much of this change can be attributed to a dramatic shift in meat consumption patterns, from beef and pork to broilers. Per capita beef consumption was at 100 pounds in 1965. It peaked at 129 pounds in 1977. Per capita beef consumption has drifted steadily lower to close the century at 96 pounds (including bones in retail cuts). Per capita pork consumption drifted between 60 and 70 pounds per capita. In the interim, broiler consumption skyrocketed from 33 pounds in 1965 to 65 pounds in 1998.

Per capita broiler consumption first exceeded per capita pork consumption in 1988. Broiler consumption surpassed beef consumption in 1997.

The Cattle Cycle

Ranching and Cow-Calf Production.

> The cattle cycle is an economic disease. We spend millions on animal diseases. How much do we spend on the real income killer, the cattle cycle?... We view the cattle cycle like death and taxes, as inevitable. Cycles are predictable and their causes are known. Through research and education, they could essentially be eliminated.
>
> Dr. Willard Williams

Characteristics of the beef enterprise create a predictable supply pattern. All the product that is produced will be consumed, at some price. The supply of livestock on the market thus becomes the dominant force in setting price. Outside factors exert relatively minor influences. Supply is controlled by thousands of independent cow-calf producers who tend to react in unison to high prices or low prices. The result is a predictable cattle-cycle supply curve and its concomitant resulting prices.

Reduced beef supplies initiate the expansion phase of the cattle cycle. At the meat counter, prices rise to ration available supplies. High retail beef prices are reflected through the channels of trade to affect prices of slaughter cattle, feeders, and breeding stock. At relatively high price levels, heifers that normally would have been marketed as feeder calves are held back by ranchers and added to the breeding herd. Slaughter numbers are further reduced, and beef supplies become more limited. Retail beef prices increase further to intensify the production signal to producers.

In response to high prices, producers ultimately expand production. In due time, expanded production means a larger meat supply in the retail market. Larger supplies in the market lead to reduced retail meat prices, lower finished cattle prices, reduced feeder prices, and lower inventory values on breeding stock. As producers reduce inventory to maintain cash flow or eliminate unprofitable production, the market must move an abnormally large volume of production to liquidate inventories. Prices drop severely because inelastic demand causes a much larger percentage drop in prices than the percentage change in volume. After a period of liquidation, marketings decline and cattle producers' prices improve. Then the expansion phase of the cycles takes over and the process is repeated.

A key factor in the cattle cycle is the biological time lag in the production process. Quick reactions to market signals are impossible because so much time is required in cattle reproduction (Figure 9–6). Heifers are bred for their first calf when they are 14 to 18 months old. Their gestation period is 9 months. The calf in turn will reach mature slaughter weight in 17 to 19 months, depending upon the individual's rate of gain and the feeding program employed. If a heifer is retained to expand the herd rather than sent to slaughter, it could take about 5½ years from the time she is born until her offspring reach slaughter weight to increase the available beef supply. On the consumption end, however, slaughter animals move through the packing plant, arrive at food stores, and are usually consumed within two weeks after leaving the feedlot.

The biological time lag is further complicated by producer response. Outside forces, such as inflation or deflation, weather, feed supplies and feed prices, influence producer decisions to expand or contract production. The psychological lag in producer response to price signals causes production cycles to extend beyond the length of time involved in the biological lag. The combined effect has caused the cattle numbers cycle to run about 10 years in length from peak to peak in numbers during the past 70 years.

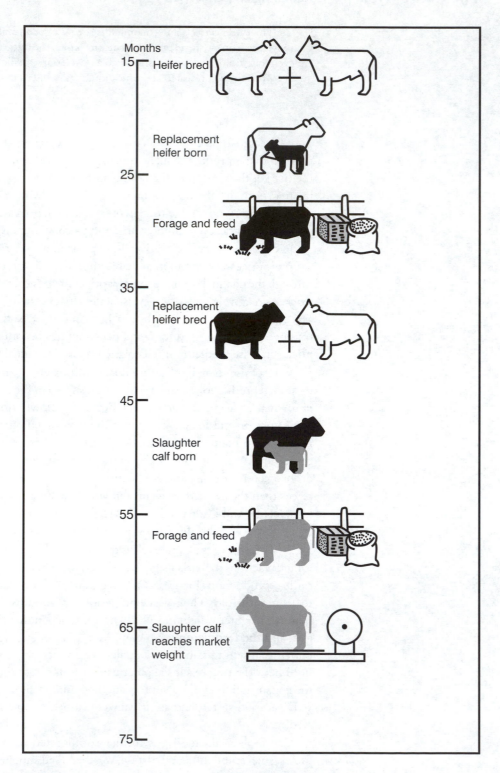

Months

15 — Heifer bred

Replacement heifer born

25

Forage and feed

35

Replacement heifer bred

45

Slaughter calf born

55

Forage and feed

65 — Slaughter calf reaches market weight

75

FIGURE 9–6: Expansion Plans Are Subject to a Biological Timetable.

The cattle roller coaster was particularly severe during the 1970's and 1980's. The early part of this period was characterized by general inflation, rising consumer income, and increasing demand for meat. During 1973, per capita supply of beef decreased for the first time in many years, putting increased upward pressure on prices. Meat boycotts, price ceilings, and panic buying distorted already high prices. Producers reacted to high prices by diverting heifers from slaughter to breeding to increase the size of their cattle herds. Producers expanded the U.S. total cattle herd to a record 132 million head by 1975. A severe drought in 1975 reduced feed and forage production. Production adjustments caused high slaughter rates and a severe price decline. In response, cattle numbers declined 25 percent (32 million head) over the next three decades from 1975's record high. By 1999, the U.S. cattle herd slumped to 98 million head; the lowest number since 1962.

Prices of beef calves, as one measure of changing prices, gyrated in response to the changing supply-demand conditions. They rose to about $70.00 per hundredweight in 1973, only to plunge into the $20's and $30's late in 1974 and 1985. After gradual improvement, feeder calf prices skyrocketed to more than $100 per hundredweight in the spring of 1979. They slipped to $65 in 1986 and began recovering to $76 in 1987. Prices topped $90 in 1990 and 1993, only to plunge below $50 in 1996 and then yo-yo between $65 and $85 in the last five years.

Decision Making by Cow-Calf Producers. Fluctuations in supplies and prices could be reduced considerably if production were geared to consumer demand.

One way for a rancher to gear production to demand is to calculate the ratio of the number of cattle (cows only) per 100 people in the U.S. population.[8] This ratio provides an indication of the supply-demand balance. It averaged 24.5 cows per 100 people from 1959 to 1975 (Figure 9–7). The ratio was higher before 1959 because of more dairy animals on farms. In the 20 years from 1955 to 1975, a ratio greater than 24.5 suggested an over supply relative to demand and therefore lower prices; a lower ratio signaled an under supply and higher prices. In 1975, the number of cows per 100 people rose to 26, a 20-year high, and the price of feeder cattle in constant (1967) dollars fell to the lowest level in the 40-year period from 1940 to 1980. The cow herd was then liquidated, and the number of cows per 100 people declined to 22 by the start of 1980. In response, feeder calf prices rose to the level of the early 1970s. Since 1982, cow population continued a dramatic and unprecedented decline, but price level also declined to pre-1940 levels in constant dollars. Something changed.

Prior to the 1980's, it was thought that multiplying the average ratio by the population would give some idea of growth in herd size that would need to occur in future years.[9] The U.S. population was forecasted to rise to 241 million by 1990; it actually hit 246.5 million. Using the measuring stick of 24.5 cows per 100 indicated the need for a cow herd of about 60 million head by 1990, assuming no change in per-capita consumption (a heroic assumption, as we have seen). This herd size would provide a per-capita supply of 89 pounds of retail-weight beef. Of the total cow herd, about 10 million head would be dairy cows. From a cow herd of 50 million cows in 1982, growth to 60 million seemed reasonable. But the 44.5 million herd in 1987 presented a different story; there was no way the herd size could rebound by 1990 to 60 million head. The predicted path of previous cycles changed in 1982 because the assumption of per capita consumption proved invalid as consumers changed eating habits. Apparently due to health concerns, they switched from beef to poultry consumption, and beef consumption declined. As further evidence that the tried and true relationships

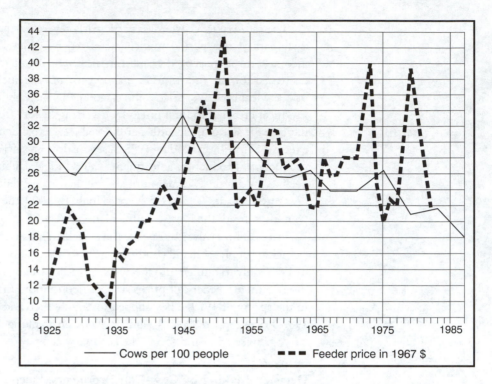

FIGURE 9–7: Cow-People Ratio and Feeder Animal Prices, 1925 to 1987.

had faltered, in 1998 there were 268 million people and 43 million cows in the United States–only 17 cows per 100 people–the world was changing and so were the fundamental relationships.

For producers to benefit from cattle cycles–and avoid economic squeezes–it is necessary to recognize potential over or under supplies, before they are realized and reflected in the market place. By monitoring herd size and the rate at which replacement heifers are entering the herd, producers can anticipate excessive expansion or contraction. To develop a cyclical marketing strategy, a rancher must take appropriate action to create a counter-cyclical reaction to what others are doing. As others are building inventory, the individual rancher should be reducing numbers and shortening the length of time he or she holds animals. The counter-cyclical rancher would cull fast, sell early, and sell light. Cows should be culled in the early stages of price decline; they will be worth less in salvage value later on. The individual should sell calves rather than hold them as yearlings. A stocker operator could buy in the spring at a lower price than the previous fall and also not incur the expense of winter feed. Purebred-cattle producers should sell bull calves rather than holding them until they are breeding age. Contracted sales of cattle should be made early in the season rather than waiting until heavy fall runs materialized at weaning time. Hedging feeder sales with futures should be used to protect price level.

When the market turns, the reverse strategy applies. The time to build inventory and hold animals longer is when others are liquidating at lower prices. Calves should be kept to sell as yearlings. Heifer calves can be held at low price levels for sale as bred heifers or cows when the market for females is more active. Stocker operators should buy in the fall and overwinter the stock for the following grass season. About halfway in the buildup phase of the cycle, lots of capital tends to flow into the cattle market. It creates a particularly active market for breeding stock. Purebred producers should hold young stock for a two-year-old bull and bred heifer sale. Contracting or futures hedging should be avoided. In the up trending phase of the cycle, the objective is to own stock as long as possible to profit from the rising price trend.

Such a counter-cyclical production strategy creates uneven cash flows, but it could improve long-run profitability. If a small number of producers practiced the counter-cyclical strategy, they would help to moderate cycle extremes. If every producer followed the strategy, the cycle would be changed. In reality, since most cattle producers have reacted the same way in the past, they can be depended upon to do so in the future. Cattle production cycles, with boom and bust prices, will continue in the future.

It May Be Fundamental, But It Is Not Simple

In our search for some relevant principles to establish a viable livestock marketing program for producers, we have approached the task from several perspectives. In Chapter 4, we explained price determination in the context of classical supply/demand fundamentals. We also demonstrated how statistical analysis is used to project future prices. In Chapter 6, we identified various pricing strategies and suggested ways to structure an analysis of a cost of production model. In the current chapter, we reported on the results of other studies analyzing additional pricing strategies. We included a time dimension emphasizing price cycles. Livestock producers should ask: "How valuable is this information in helping me to formulate a profitable marketing plan?"

Beef Cattle

To assist in answering that question, Buhr[10] recently reported on beef cattle production and price data for 1978 to 1998. These data are illustrated in Figure 9–8. A cursory review shows an apparent significant relationship between cattle price and production. Price is portrayed on the left-hand axis and by the dark line. Production is depicted on the right-hand axis and by the gray line. Economic principles suggest that as production increases, price decreases, and vice versa. But the key question is, "How useful is this knowledge in predicting whether prices will be *higher or lower next year*?"

We modified Buhr's original data in an attempt to evaluate how closely relationships in one time period match results in another time frame. Figure 9–9 modifies Figure 9–8 by using, as a matter of convenience only one data point per year, but more important, we inverted the right-hand production scale and turned the production line upside down to see how closely the lines track each other. In Figure 9–9, we observe that cattle prices and production *appear* to track each other, but not very closely. Unfortunately, these data represent annual production and yearly average prices. More important, the annual data are only finally available in the next year. Thus, 1999 production and prices are available after January 1, 2000. Consequently, we are in the unenviable position of "forecasting" 1999 prices in February 2000, which is after the fact and of little value in marketing cattle in the latter part of 1999.

Finally, we paired the annual price data points with the production data points to illustrate how closely they are related (correlated). Figure 9–10 plots the price data

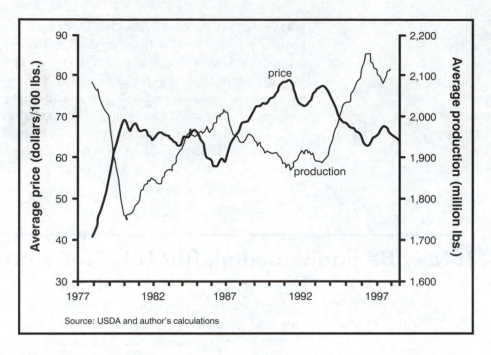

FIGURE 9–8: U.S. Beef Cattle Price (left scale) and Production (right scale) 1978 to 1998.

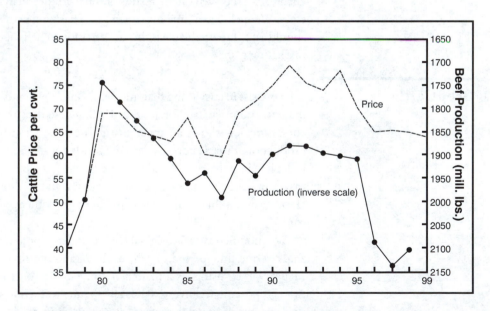

FIGURE 9–9: U.S. Beef Cattle Price (left scale) and Production (reverse right scale) 1978 to 1998.

points, noted on the vertical axis with the production data points posted on the horizontal axis.

Upon closer examination, the years 1979 through 1986 *seem* to show some relationship between beef prices and quantities. The year 1978 seems to be an outlier and belongs to a different era. The years 1985, 1987, and 1988 are inconveniently some-

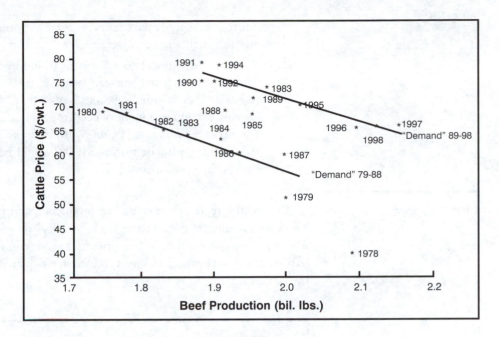

FIGURE 9–10: U.S. "Demand" for Beef, 1978 to 1998.

what removed as if in another realm, but we use them in estimating what might be described as a "demand" function. The statistical parameters of this function and the other three are summarized in the appendix at the end of this chapter. The years 1989 through 1998 are also in a different realm and thus we have estimated a separate "demand" equation for them. Unfortunately, this division of data is purely arbitrary and is only "self-evident" after the 20 years of data accumulation. There were no data to predict a shift in "demand" and it took between 1989 and 1999 to realize that perhaps the 1979-88 "demand" curve was perhaps inappropriate. In this instance, 20/20 hindsight is an enviable tool.

As noted in the appendix, 39 percent of the variation in price was explained by the pounds of cattle produced in the 1979 to 1988 "demand" era. Fully 86 percent of the variation in price was explained by production in the 1989 to 1997 "demand" era. Unfortunately, these relationships are more apparent than real and thus bring us no closer in our search for some reliable fundamental variables to aid in forecasting prices in the future. These difficulties arise from the fact that:

(1.) Ten observations a reliable demand equation do not make. Because of the paucity of data, we do not have a statistically significant relationship.

(2.) The data points are dependent. This year's cattle numbers have an impact on next year's cattle numbers.

(3.) It takes ten years **(or more)** worth of data to even begin to identify a relationship.

(4.) Quarterly data would be helpful, but, in our experience, only marginally.

(5.) The needed data is not available until well after the production period. Understandably, it takes the USDA several months to publish annual data. Quarterly estimates are available sooner, but there are severe handicaps in using last quarter's data to forecast this quarter's prices. If one is trading futures, one needs data today to forecast tomorrow's, next week's, next month's cattle futures prices.

(6.) This simplistic model does not assist us in forecasting the apparent shift in demand that occurred between 1988 and 1989. The difficulty is that it may take a number of years to become aware that "demand" has shifted. In the interim, 2.0 billion pounds of beef are generating a $57.00 price forecast in a $71.00 cattle market (see Figure 9–10). A forecast that continually predicted $65.00 cattle, **plus or minus $10.00**, would be just as useful (*useless*) and equally financially painful.

Hogs

Buhr [11] also reported on some hog production and price data for 1978 to 1998. These data are illustrated in Figure 9–11. A cursory review shows some apparent significant relationship between hog price and production. Price is portrayed on the left-hand axis and by the dark line. Production is depicted on the right-hand axis and by the gray line.

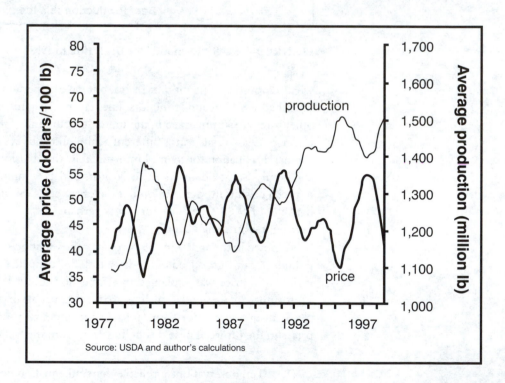

FIGURE 9–11: U.S. Hog Price (left scale) and Production (right scale) 1978 to 1998.

We modified Buhr's original data to gain some additional insight. Figure 9–12 modifies Figure 9–11 by using, as a matter of convenience only one data point per year, but more importantly inverting the right-hand production scale and turning the production line upside down to see how parallel the two lines are.

Finally, we have paired the annual price data points with the production data points to illustrate how closely they are related (correlated). Figure 9–13 plots the price data points, noted on the vertical axis with the production data points posted on the horizontal axis.

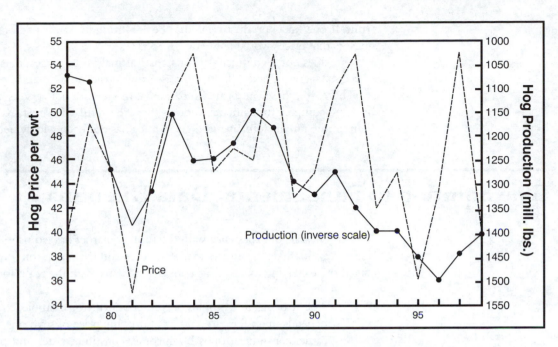

FIGURE 9–12: U.S. Hog Price (left scale) and Production (reverse right scale) 1978 to 1998.

Unlike the cattle data, the hog data appear to be scattered, at random all over the price-quantity map. One could force the issue and try to conjure up some "demand" functions but this is a real stretch. The statistical analysis, summarized in the appendix verifies that there are no significant relationships.

What is most telling is that 1984, 1988, 1992, and 1997 all generated the same $55.00 hog price. However, hog production covered the entire gamut from 1.19 to 1.44 million pounds. Similarly, there is a band of years that witnessed prices in the $44.00 to $47.00 range while production roamed across the entire production spec-

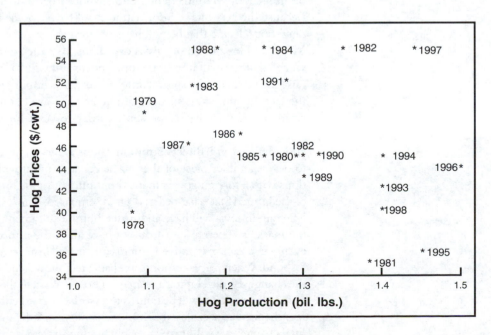

FIGURE 9–13: U.S. "Demand" for Pork, 1978 to 1998.

trum. If one knew production would be 1.44 million pounds beforehand, the best guess is that prices would fall between $35.00 and $55.00.

One does not need much data or statistical analysis to suggest that hog prices will be $45.00 – give or take $10.00. That assertion is meaningless and not useful.

Therefore, these simplistic fundamental models are not very useful. There may be some complex fundamental models that are useful. However, we have not seen them in the public domain and their efficacy has not been widely heralded.

The Solution to Fundamental Data Dilemma

Our 30 years of experience with the futures market has led us to the conclusion that we cannot trade the futures markets successfully by relying on fundamental analysis of price-making forces. This conclusion stems from the realization that:

1. The markets are complex and intertwined, for example, the price of corn depends upon the carry-over of corn, the price of soybeans, the number of cattle, consumer disposable income, consumer tastes and preferences, the cost of energy, interest rates, currency rates, consumer and producer psychology and expectations, weather and insects, corn acres planted, etc., etc., etc.

2. No one completely understands all of the interrelationships to formulate a comprehensive economic model.

3. No one has the resources to compile all of the important data in a timely manner.

4. No one has a computer large enough to compute the model and generate the needed price forecasts in a timely fashion.

Therefore, if I cannot have all of the information, I simply want today's price and nothing else. No fundamental information is better than partial information. Why? Because the very bits of information that I have may be the wrong information. My neighbor tells me that he will have no corn this year because it has not rained since May 1 and his garden in July is dry. Thus, the price of corn is going up. I am convinced that this will impact the price of corn—it will have the same effect as trying to raise the ocean tide by spitting into the surf. It is important, but far more important is the fact that there was substantial rain across Iowa last night after two weeks of dry weather and the weather person's 10-day forecast for continued dry weather was wrong for a change. Some would suggest putting rain and temperature gauges in enough fields with direct communications to record the events. But even if that were to occur, another piece of the puzzle remains unknown. Last night secret negotiations with a foreign government about human rights broke down and the President of the United States has ordered a trade embargo. The lost garden was not significant. The rain in Iowa will have an impact, but the still unannounced embargo will move the markets. Therefore, if I cannot have all of the information, I prefer none. If an event or a factor is indeed important, it will impact the market and prices will respond. When prices respond, a reliable technical trend-following model will signal the change. By developing a strategy of how to react to the market signals as prices rise or fall, the marketer will be guided to a workable marketing plan. With sufficient testing, the producer can establish the reliability, and the intricacies, of the plan. In the final chapter, we will unveil a viable pricing strategy and marketing plan.

Summary

This chapter draws attention to differences in livestock product pricing compared to storable grains. Because of these differences, futures pricing and hedging strategies also have to be different. The study cited on cattle hedging showed that no pricing strategy panacea exists. But bear spread hedging and timing cash marketings should improve profits.

The problems associated with basis complicate hedging as a means to protect prices. Basis variability does not permit accurate localizing of futures prices to project the results of hedging. Because of the biological schedule for selling livestock, the producer does not have the flexibility that grain sellers have. The delivery process and handling of live animals make delivery against contracts more complicated. Until the futures markets solve the basis problems, hedging is limited in its ability to reduce price risk.

The chapter reviewed agricultural price cycles in the context of cow-calf production. The problems and opportunities posed by cattle cycles can be extended into other livestock products. The important thing is to recognize their existence and formulate comprehensive strategies for reacting to changing production cycles.

The debate between a fundamental versus a trend-following approach was addressed.

Finally, we shared what we believe is a unique way of making the livestock futures work to your advantage. There is nothing wrong with selling several years worth of livestock production at bonanza prices. Farmers often fret when they sell $65.00 hogs. They worry that hogs will go to $70.00. We suggest that the only problem with $70.00 hogs is answering the question: "How many more years do you want to price?"

Questions for Study and Discussion:

1. What particular feature makes pricing non-storable commodities different from pricing storable commodities?

2. Explain the key fundamental factors you need to analyze in forecasting hog and live cattle prices.

3. What pricing strategies have been used for making futures pricing decisions in live cattle?

4. How do basis considerations complicate futures pricing decisions?

5. What general conclusions have been reached about the behavior of non-storable futures contract prices?

6. Explain bear spread hedging.

7. What causes agricultural price cycles?

8. Why do agricultural prices change relatively more than agricultural production?

9. Explain the biological and psychological lags in the cattle cycle.

10. Explain the system suggested as one way for farmers and ranchers to predict impending over supply and under supply of cattle.

Endnotes for Chapter 9:

1. Raymond M. Leuthold, "The Price Performance on the Futures Markets of a Non-storable Commodity: Live Beef Cattle," *American Journal of Agricultural Economics* 55(1974): 271-279.

2. Raymond M. Leuthold and Peter A. Hartman, "A Semi-strong Farm Evaluation of the Efficiency of the Hog Futures Market," *American Journal of Agricultural Economics*, Vol. 61, No. 3 (1979): 482-489.

3. Philip Garcia and Larry Martin, *Futures Markets for Selected Non-storable Commodities: Price Forecasting Revisited,* Bulletin No. 80E-148 (Champaign: University of Illinois, 1980).

4. Refer to Chapter 4 for a more detailed discussion of fundamental pricing, and to Good, Hieronymus, and Hinton, *Price Forecasting and Sales Management*, for an in-depth analysis.

5. John H. McCoy and Robert V. Price, *Cattle Hedging Strategies,* Bulletin 591 (Manhattan: Agricultural Experiment Station, Kansas State University, 1975).

6. McCoy and Price, *Cattle Hedging Strategies.*

7. Thomas A. Hieronymus, *Economics of Futures Trading* (New York: Commodity Research Bureau, 1971) 167-169.

8. USDA, "New Cattle Cycle: Can Producers Take Charge?" *Farmline* (Washington, D.C.: Economic Research Service, July, 1980) 14.

9. USDA, "New Cattle Cycle."

10. Brian Buhr, "Livestock Prices," *Minnesota Agricultural Economist* (St. Paul, Minnesota: Agricultural Experiment Station, University of Minnesota, No. 694, Fall 1998) 6-7.

11. op cit.

APPENDIX: Chapter 9

Beef Least Squares Regression Equations

1979 to 1988

Constant	148.996
Standard Error of Y	4.482
R Squared (adjusted)	**0.393**
Number of Observations	10
Degrees of Freedom	8
Coefficient	-45.021
Standard Error of Coefficient	17.245

Regression Equation: Price = 148.996 − 45.021*Quantity

1989 to 1998

Constant	170.538
Standard Error of Y	2.021
R Squared (adjusted)	0.863
Number of Observations	10
Degrees of Freedom	8
Coefficient	-49.754
Standard Error of Coefficient	6.552

Regression Equation   Price = 170.538 – 49.754*Quantity

Pork Least Squares Regression Equations

1979 to 1991

Constant	97.268
Standard Error of Y	4.556
R Squared (adjusted)	**0.296**
Number of Observations	13
Degrees of Freedom	11
Coefficient	-40.234
Standard Error of Coefficient	16.361

Regression Equation: Price = 97.268 – 40.234*Quantity

1989 to 1998

Constant	116.290
Standard Error of Y	7.496
R Squared (adjusted)	**0.000**
Number of Observations	7
Degrees of Freedom	5
Coefficient	–50.000
Standard Error of Coefficient	63.808

Regression Equation: Price = 116.290 – 50.000*Quantity

Chapter 10

730-Day Crop Marketing and Profit-Maximizing Livestock Hedging

730-Day Crop Marketing and Profit-Maximizing Livestock Hedging

Commodity marketing is not a simple process; it consists of several pieces. Each of the pieces has its unique characteristics, uses, and limitations, and each forms a part of the overall picture. The individual pieces can be combined in a number of ways. An intelligent marketer understands the pieces and fits them together into an integrated plan to best achieve his or her individual goals.

The Marketing Challenge

In the preceding chapters, we have considered a number of concepts as background for formulating a marketing plan. Let us review the pieces to see how they fit together in an integrated marketing plan.

1. Price volatility offers the opportunity to capture high profits and the challenge to avoid disastrous losses created by rising and falling commodity prices.

2. Many marketing tools are available. Each has its own characteristics and limitations. All uncouple pricing decisions from production decisions.

3. Futures markets dominate the pricing aspect of marketing. They function within a fixed business structure and play several significant roles in the marketing process.

4. Supply and demand fundamentals ultimately determine cash price in a market economy. But unknown and changing physical, economic, political, social, and psychological conditions create uncertainty about future price realizations.

5. Technical market analysis attempts to gauge the effect of human judgement and emotions in price discovery, especially in the short run. Tools of technical analysis identify price objectives and provide insights into reading market action.

6. A farmer's market strategy evolves as a method to react to market signals and to cope with risk. A strategy must be developed consistent with one's personal philosophy, attitudes, and financial condition.

7. Pricing decisions are required in two separate but related markets–futures price level and local basis level.

8. Adequate financing is vital to a marketing plan that utilizes futures contracts. Risk is increased, rather than reduced, by inadequate liquidity.

9. Storable and non-storable commodities have different pricing conditions because of the nature of the products. Pricing decisions must therefore be different.

These individual pieces must now become parts of an integrated marketing plan that addresses the issue of profitable marketing.

One-Time Selling or Continuous Marketing?

To integrate all the pieces into a successful marketing plan, let us consider a fundamental question: Is a farmer wise to be a speculative, one-time marketer, or is he or she wiser to become a continuous marketer?

One-Time Selling

On July 1, a person farming 1,000 acres could have 150,000 bushels of grain stored in bins and 150,000 bushels growing in the field. Tons of money dance through this speculator-producer's mind, as one contemplates the booty reaped from selling both the stored crop and the growing crop for $3.80 per bushel. This farmer speculates on 300,000 bushels from both crop years, hoping that another part of the country will have a drought, or that the Soviets will make a record purchase. Either or both events will reward the speculator.

Because the farmer intends to maximize profit by catching the TOP cash market price of the year, he or she stands beside the market, day by day, delaying the pricing decision. One speculates that price will rise above today's level and provide an even higher price opportunity in the future. One knows that once grain is hauled to town, or a forward contract signed, that irreversible, one-time decision locks one into a profit level for the remainder of the crop year. But when one speculates on higher price, one risks having to accept a lower price.

Lenders join in the speculative game by financing both the production of a new crop and storage of the old crop. They too have visions of the rewards from $3.80 corn. Two corn crops (300,000 bushels) would gross $1,140,000 for their borrower. That is real money!

Unfortunately, the market often fails to cooperate with these visions. Widespread rains in July and August can produce a huge crop. Hoped-for international buyers may fail to place orders. Market price typically drops to move a bountiful crop. In a $2.00 corn market, 300,000 bushels gross only $600,000, less $45,000 carrying charge on the old crop to net only $555,000. From high price expectations to low price reality, with $540,000 swings in gross receipts, a farmer's net income all too often moves from riches to rags when speculating on two crops.

Cash market speculation is *one-time, passive marketing*. Emotional selling that is based on fear, hope, and greed influences most passive marketing decisions. Farmers don't sell at bonanza prices because they firmly believe prices will go higher, but they do sell at the bottom because they fear lower prices. Others submit to "wishful thinking"–$10 soybeans, $4 corn, and $6 wheat–and wait and wait. Both types typically sell when loans are due, before, or well after, the price has peaked.

Pricing Strategies

Confronted by the challenge to capture the high of the market price and in the face of an uncertain future, farmers have tried many pricing strategies to cope with passive marketing. Some farmers sell by the calendar: equal amounts every Wednesday; or on the 15th of every month; or a third contracted in the growing year, a third from storage in December to January, and a third from storage during the June to August "summer rally." At best, they capture the average market price, minus holding costs.

Then there are the calculators who try to improve their one-time pricing decision. Some base selling decisions on the costs of production and attempt to scale up. Others analyze cycles, seasonals, fundamentals, or charts. With luck, they sometimes make high-price sales on part of their production. But usually they look back with 20/20 hindsight and say, "That's when I should have sold both crops."

An Integrated Plan of Continuous 730-Day Marketing

There is a better way. It is an entirely different marketing concept, not passive marketing, with its irreversible one-time decisions, but *active farmer pricing* through continuous 730-DAY MARKETING–a 730-day "marketing year." This approach converts a farmer from a one-time passive seller into a continuous, active crop merchandiser.

Time Framework: Marketing for 730-Days

In building an integrated marketing plan, crop producers must keep in mind the fact that pricing decisions on a *single crop span a two-year period*: the growing year and the storage year. The first stage of a crop "marketing year" begins in November as production plans are being made for the new crop and continues throughout the growing season until the end of harvest. During the second stage of the "marketing year," pricing of the harvested (*old*) crop begins at the end of the 12-month "growing" year and continues for the next 12-month storage year. Thus, the pricing of a single crop spans 730 days–the "growing year" **plus** the "storage year."

Selective Hedging/ Selective Speculating

The practice of 730-Day Marketing relies on selective hedging/selective speculating to reduce, not increase, speculation. A producer establishes short futures positions equal to planned or growing production when futures prices trend down. Price protection from short futures contracts offsets value erosion in the growing crop. When market price stops declining and turns into an up trend, the short futures positions are liquidated. The value of a growing crop is allowed to appreciate, unfettered by a futures position, as the market price rises.

Selective hedging solves some of the problems of fixed hedging. With fixed hedging, the producer sets hedges in down-trending markets but maintains the position until the product is sold in the cash market. A corn producer who sets a profitable pre-harvest hedge, say $3.00 per bushel in April, enjoys price protection as prices decline to $2.70 in June. But if the market turns around in July and rises back to $3.00 before harvest, the hedger will have seen price protection accumulate in the futures account, then disappear as the market rebounds. With selective hedging, one would lift the hedge position in July to let the value of the growing crop rise along with the market, rather than return the futures account surplus to Chicago speculators to satisfy the futures contract.

Storage Pricing for 12 Months

In up trending markets during the storage year, the crop producer benefits by owning the old crop. A rising price level increases the value of stored commodities. There are two methods for extending pricing of the old crop across the entire 12 months of the storage year. One may either own the physicals or replace the physicals with paper ownership through "Buying the Board." Once cash sales are made, it is impractical to repurchase cash grain to refill bins for further price appreciation. It is much cheaper and more convenient to replace the cash position with a paper position (long) in the futures market. The way to retain ownership throughout the storage year is by "Buying the Board." A producer must understand that holding unpriced grain in anticipation of a rise in price is pure speculation. He should realize that there is little difference in holding long futures contracts to achieve or fail to reach the same price objective. To maximize the value of farm production, the 730-Day Marketing marketer utilizes long speculative positions in futures contracts during part of the marketing year to replace a cash position with a paper one.

The practice of re-owning inventory has a major feature: It separates the basis pricing decision from the price level decision. It allows the commodity producer to capture *BOTH* a strong basis and a high price level. A high price level is typically accompanied by weak basis. Profitable basis sale opportunities are usually eliminated, therefore, in cash market speculation. The farmer-storer who is willing to make a cash sale at low price levels in response to strong basis knows full well that he or she can participate in the market price rise when it occurs by "buying the Board." One does not have to hold cash commodities to speculate. One adopts a market philosophy that permits one to react to two sets of market signals and capture advantageous prices in both. Replacing a cash position with a paper one also eliminates the physical storage risks and carrying costs of storage speculation.

In practice, when the old crop is stored after harvest and before a cash sale, it is stored unhedged in price up trends. Its value is rising as the price level rises. Should the basis market dictate a cash sale of the old crop, ownership of the physicals is replaced by buying futures contracts to continue the benefits of the price rise. When the up trend runs its course and a down trend sets in, the long futures position is liquidated. The 730-Day Marketer re-buys the old crop grain in each subsequent up trend throughout the storage year to ride its price upward on each move. One practices selective speculating to "Buy the Board" when appropriate. Thus, one can add price level increases from grain held on paper throughout the year, while having sold the physicals at the most advantageous time to do so.

Any stored old crop grain still waiting for a basis signal to sell is protected by short futures positions during price down trends. Corn stored at harvest when the price is $3.00 per bushel loses value day by day as price declines during the storage period. Elevator operators who buy cash grain from farmers at harvest cannot afford to risk a loss in value. Therefore, they hedge away their price risk on stored grain. Why do farmers, storing grain on the farm, not protect the value of inventory? When the down trend has run its course, contrary to grain warehouse practice, farmer-storers should lift their short futures position and become selective speculators by allowing the inventory value to rise in step with the market.

Across the entire 730-Day "marketing year" of pricing one crop (the growing year and the storage year) as a selective hedger, a farmer hedges in down trends but speculates in up trends. Our analysis shows that such a practice cuts cash market speculation in half. The farmer-producer is hedged about half the time. This is in direct contrast to most farmers who never hedge but speculate 100 percent of the time, all too often on 2 years' worth of production.

Flexible Hedging

Flexible hedging is required in order to execute 730-Day Marketing in a 12-month calendar year. A grain producer with 150,000 bushels of an old crop and 150,000 bushels of growing new crop must make separate decisions on the pricing of each crop. As price trends up and down during the calendar year, one seeks to protect the growing crop against declining markets with hedges, and one seeks to capture price appreciation on old crop production in rising markets by owning inventory or "Buying the Board."

The flexible marketer deals with one crop at a time in the futures market, depending on the direction of the price trend. To profitably market both the old crop and new crop in a calendar year, flexible hedging reverses futures market positions as trends change. When market down trends change to up trends, new crop hedges are lifted and an old crop is repurchased with long futures positions. When up trends change into down trends, long positions on the repurchased old crop are liquidated and new crop hedges are re-instituted. The flexible hedger holds a market position on one crop or the other at all times, but not both, as cash market speculators do. The two market positions are summarized in Figure 10–1.

Flexible hedging differs considerably from pure hedging. Textbook hedging requires owning both the physicals and an offsetting futures contract at the same time. Farmers who hedge anticipated production do not have the commodities on hand. Furthermore, when the old crop is replaced with a long position in the futures market, the "opposite" requirement of the textbook definition is again not met.

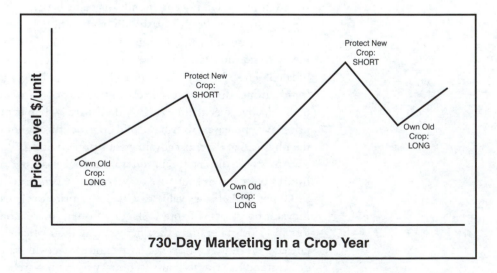

FIGURE 10–1: 730-Day Marketing Plan: Own Old Crop in Rising Markets, Protect New Crop in Declining Markets.

These differences emphasize the peculiar objectives of farmer use of the futures market in contrast to the grain trade. They further emphasize the difference between pure and flexible hedging.

Flexible hedging has one additional side benefit; it breaks the link between the biological clock and the market clock. Farmers who wait for the crop to be "knee high" or wait for it to tassel often find that the most profitable pricing opportunities have vanished. It is common for grain prices to drop 30 cents per bushel between December and May–long before the crop tassels. Flexible hedging dictates sales based on the market clock, not the biological clock.

Producers can rely on various methods for determining when market reversals occur. Our research and analysis have shown a trend-following procedure to be effective. Discussion in Chapter 5 outlined a method for calculating 10- and 40-day moving averages. The crossover definitions provided the signals for trend changes. Our discussion in Chapter 6 reviewed advantages and disadvantages of moving average analysis. A commercially available source of moving averages may be obtained from the Commodity Trend Service.[1] In addition, there are a number of private consulting firms that provide comparable services.

Basis Sales: Capturing Storage Profits

Decisions to make cash crop sales are determined by basis signals from the market, totally separate from price level decisions. The spread between local cash price and a futures price signals when to store and when to sell cash grain. Weak basis offers a storage return and suggests that a storage commitment be made. Strong basis does not offer a storage return and suggests an immediate cash sale.

Whether farmer-storers decide to sell at harvest because a strong basis offers the prospect of low storage returns, or if they decide to store to speculate on the basis, they should remember that timing the cash sale should be determined by the prospect of basis gain, *not* by hoped-for price level gain. Selling cash grain is dictated by basis signals, and "buying the Board" to speculate on a price rise is preferable to holding cash grain to speculate. Holding grain on paper is cheaper, so the net profit of a correct decision is greater, and the net loss of a wrong decision is less.

Storage, or immediate sale, depends upon the storage return offered by the market, relative to the variable costs incurred by storage. Storage profit is the difference between the basis gain and the added costs of carrying the commodity forward.[*] A picture of the relationship between changing basis and changing cost of carry is presented in Figure 10–2. It shows that as long as the *slope* of the line representing actual basis change *exceeds the slope* of the line of cost change, storage is profitable. Simply stated, actual basis gain **is rising faster** than storage costs. Storage is unprofitable when the *slope* of the actual basis line is *less than the slope of* the cost change. In this instance, actual basis gain is less than the added storage cost. Ideally, to capture all basis gain, a farmer should fill bins at A, empty them at B, refill them C, and empty them at D. Bins should be empty between B and C.

To numerically express the concept presented in Figure 10–2, rely on the methods of calculating basis gain and carrying charges presented in the earlier chapter. As long as the day-to-day amount of basis gain in the market exceeds the day-to-day increase in carrying cost, continued storage is profitable. Whenever basis gain drops

---

[*] At this point, readers should review Chapter 7 to refresh their understanding of cash-futures price convergence and the calculation of carrying charges.

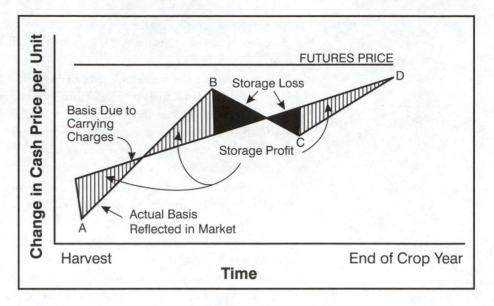

FIGURE 10–2: Basis Decisions for Cash Sale, Using Rate of Change in Basis Versus Change in Carrying Costs.

below the daily increase in cost, storage is no longer profitable and the cash sale should be made.

An analysis of daily basis change is required to utilize this concept. Gather daily cash prices from a local market and an important terminal market. Subtract both from the end-of-market-year futures option (July corn and soybeans, May wheat). The daily change in basis is the gross return to storage. It may be one cent, two cents, five cents, zero, or a negative amount on any given day. Compare the daily basis change with the daily change in the costs of storing the grain. From the earlier analysis, we calculated variable storage cost to be about four cents per month for corn and wheat, and eight cents per month for soybeans, or one cent and two cents per week, respectively. Only variable costs (interest, shrinkage, and extra drying) are important, because fixed cost-of-bin and facilities ownership is incurred whether the bins are full or empty. In practice, since the daily basis change is small (and variable) and the weekly cost change is small, we watch for a distinct change or a deteriorating basis to confirm the signal for a cash sale. A two-day basis deterioration confirms the signal to sell cash grain. This deterioration may occur in the week after harvest is completed, or it may occur after January 1 as tax-sellers begin emptying bins. A deteriorating basis says that farmers are more willing sellers than the grain trade is a willing buyer.

The daily decision as to when to sell out of storage does not answer the first question of whether to store or sell directly out of the field. In those years when harvest-time basis is weaker than the theoretical basis based upon carrying charges, the decision is easy—store. The market is saying that it does not need the grain for immediate use because the cash price being offered is less than the cost of carrying the grain forward in time. That pattern existed in most Midwest corn markets for many years. However, recent years have witnessed a dramatic shift in basis patterns. The original storage decision has become more complex and requires speculation on *expected* basis change.

The experience of the Cincinnati corn market in Figure 10–3 illustrates the shift. Over the seven years of historical basis from 1974 to 1981, harvest basis averaged 52 cents under the July futures, and it converged to zero by July. The market thus offered an average 52-cent gross storage return. That return exceeded the carrying charge of 28 cents by 24 cents. In 1983 through 1987, the historical basis pattern changed significantly (as reported in Table 7-3 in Chapter 7). Over the latter period, the October 15 basis averaged only 18 cents under July; the March basis averaged 7 cents over July; and the June basis averaged about 14 cents over July. Basis gain from October to June averaged 32 cents, barely enough to cover total carrying costs. The significant changes in basis pattern were the narrow harvest-time basis and the positive premium basis that were bid into the market, typically by January. Grain surpluses and storage programs changed the picture. Grain sellers (Ohio and Indiana farmers) had the alternative of placing grain under government loan rather than having to sell it at low cash prices, but a hungry chicken in Georgia required cash corn to maintain production. Futures prices reflected worldwide supply and demand, with huge surpluses dragging the general price level downward, but the cash market traded at a basis premium to entice cash grain into the market. Thus, the basis picture changed.

The shift in basis is not the important issue. Although it makes the original storage decision more speculative, storage profits are not ruled out by the change. As long as one expects the new basis pattern to parallel the old basis pattern, the two have the same rate of change, or slope of the basis line. Under those circumstances, the decision whether or not to store is the same–store. It is only when the slope of the basis line, or **gain** in basis, is reduced that the decision is affected. With a storage cost of 28 cents per bushel, when a harvest basis of –20 might be expected to increase to +8 or more by July 1 (gain of 28 cents or more), corn storage should be initiated. Whenever there is the expectation of less than 28 cents basis gain–sell out of the field.

The decision criterion to store or not thus reduces itself to the same as the day-to-day decision: Store and continue to hold grain as long as the change in cost of storing is less than the expected change in basis over the same time period. As soon as added costs exceed added return, sell.

Storage Decisions for 730-Day Marketers. When the 730-Day Marketer decides to store, an analysis of the trend in price level of the commodity must be made. It is at this point that price level change comes back into the storage decision in deciding whether to hedge stored grain or store it unhedged. If futures prices are in an up trend, one should delay hedging as inventory value is increasing. If the market is in a down trend when the crop is committed to storage, the inventory should be hedged immediately. If a pre-harvest production hedge had been set earlier in the season to protect the growing crop, it should be maintained during the down trend in order to protect the inventory value of the harvested crop in storage. A storage hedge, set in down trends, by selling futures contracts equal to the amount stored, protects inventory value while allowing cash and futures prices to continue to converge (basis to strengthen).

Once grain is in storage, the 730-Day Marketer analyzes the basis market every day to decide when to cease storage. One does not have to wait until the maturity of the futures contract to sell cash grain and lift the hedge. The market may offer a more profitable opportunity prior to contract maturity. The decision of whether to make a cash sale and lift the hedge or continue storing is made daily by comparing changing basis to changing costs.

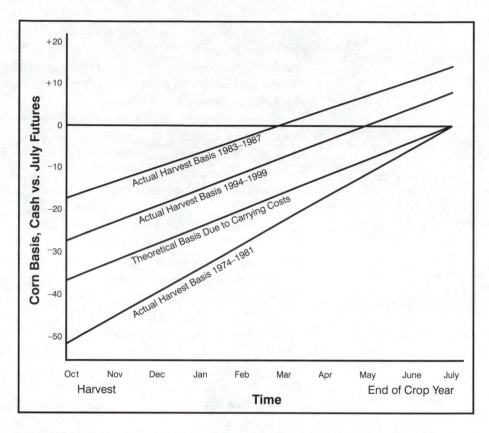

FIGURE 10–3: Cincinnati Corn Basis Patterns, 1974 to 1981, 1983 to 1987, and 1994 to 1999, as Compared to Theoretical Basis Due to Carrying Costs.

Based on 25 years of personal observation, the Cincinnati market is an extremely sensitive basis market, especially in the winter months. The Ohio River is less vulnerable to whether delays, there are several large processors in southwest Ohio, and the Cincinnati area has a complex infrastructure of rail lines and interstate highways. Pinches in feed supply in poultry production areas and empty grain vessels in New Orleans or Norfolk ring cash registers readily heard in the Cincinnati corn market. In our experience, a strong Cincinnati basis ripples across the eastern Corn Belt improving basis in other markets. Thus, Cincinnati is our choice of markets to watch for timing basis sales. Leesburg is simply ten cents removed from Cincinnati and Dayton is 15 cents away.

Furthermore, there is a sensitive time period when grain producers can expect to make basis sales. In bumper-crop years, long lines of loaded trucks wait at scales to deliver the harvest bounty. Weak basis is the charge imposed on farmers by grain merchants being begged to handle huge volumes. As soon as all the corn and soybeans are in storage, the lines disappear and basis improves dramatically, virtually overnight. By mid-November basis is usually much stronger. As Thanksgiving approaches, farmers turn their attention to the holidays and away from grain marketing. Many delay grain sales in an effort to roll tax liabilities into the next year. But hungry chickens in Georgia and empty vessels in New Orleans require a steady daily flow of grain. December 15 to January 15 is a sensitive basis period. Farmer disinterest in hauling grain, coupled with a howling snowstorm in the western Corn Belt and cackling chickens, often create lucrative basis sale opportunities for astute farmer grain merchandisers. They reward hungry markets by making cash sales or signing basis contracts. These farmers do not continue to speculate on price by holding stored grain. Their price level speculation focuses on basis contracts or by "buying the board." The December 15 to January 15 basis sale period occurs in some 8 of 10 years. The other two years are typically short crop years when strong pre-harvest or harvest basis makes storage unprofitable from the outset.

Selling $4.00 Corn in a $2.00 Market

Livestock markets occasionally offer corn farmers fantastic opportunities to sell corn for more than **$4.00** per bushel when corn buyers are otherwise offering only $2.00. To illustrate this strategy, we will draw upon the essential portions of a typical farrow-to-finish hog budget. The pigs require about 13 bushels of corn to reach a market weight of 240 pounds, including feed needed to nurture the breeding herd. The enterprise has a break-even live hog price of $45.00 per cwt. when corn costs $3.00 per bushel. At $2.00 per bushel some $10,000 in returns are generated (per 100 sows) to pay for operators' labor and management, as well as the fixed cost. If during the farrow-to-finish cycle hog prices rise to $60.00 per cwt., returns rise to the point where hog producers could afford to pay as much as $4.75 per bushel to enjoy the same returns anticipated with $45.00 hogs and $2.00 corn. Therefore, if a **corn grower** does not find a $2.00 corn market appealing and has observed a break in the up trending hog market from say $63.00 to $60.00, all she needs to do is sell lean hog futures for $60.00 and thereby capture an opportunity to "sell" corn via hogs for $4.75 per bushel.

A 240-pound live hog has a 74 percent yield and thus produces a lean carcass weight of 177.6 pounds. A 40,000 pound futures contract represents 225 hogs. Those 225 hogs require 2,928 bushels of corn. Therefore, 1.7 lean hog contracts are the equivalent of a 5,000 bushel corn contract. Thirty-four lean hog contracts sold at $60.00 and covered at $45.00 is the equivalent of selling 100,000 bushels of corn for $4.75 per bushel. If bankers have trouble with this strategy, they simply need to be reminded that $2.00 corn will make very few mortgage payments. This alternative will satisfy many mortgages. This strategy applies even if corn is $4.00 this year. There is nothing wrong with selling two years of production at $4.00 and selling a third year's crop via hogs.

Profit-Maximizing Livestock Hedging

Selective Hedging of Finished Animals

The livestock feeder can employ the selective hedging procedure to protect the value of animals on feed during the production period. The livestock marketing plan is executed in response to changes in price on the commodity one is producing. As a selective hedger, one establishes short hedges in down trending markets to prevent value erosion on growing animals. Short positions are liquidated in up trends to realize inventory appreciation, unfettered by brokerage account deficits.

Hedging Feeder Purchases

Livestock feeders who buy feeder calves or feeder pigs can use futures contracts to lock in low-cost replacement animals. Whenever the feeder cattle market enters an up trend, a selective hedger buys contracts equal to his or her planned purchases to protect against the risks of having to pay higher prices at the time feeders are placed on feed. At a market turnaround, one would liquidate the long feeder hedges.

The price relationship between fat cattle and feeder cattle essentially provides the livestock producer with a two-sided, flexible hedge in live cattle contracts. Rising live cattle prices tend to pull up feeder cattle prices; declining live cattle prices depress feeder prices. In the futures market, spread traders who execute feeder spreads keep the two futures prices aligned (see the section in Chapter 6 entitled "Profit Rather Than Price" for definition of feeder spread and reverse feeder spread trading). Prices of feeder cattle contracts rise and fall with rising and falling live cattle contracts. The spread difference between feeders and finished cattle gets wider as corn price declines. The spread narrows as corn price rises.

A cattle feeder can use price fluctuations in the futures markets to capture the best of both rising and falling prices. One sell-hedges live cattle production in price down trends, and one buy-hedges feeder animals in up trends. By using flexible hedging to deal with two markets, one effectively has a short position in down trending markets and a long position in up trending markets to protect price on both the finished product and input of feeders. The number of contracts is adjusted to reflect the relative weight of feeders bought and live cattle sold.

Feeder pig buyers could rely on the lean hog contract as a surrogate method for hedging feeder pig purchases, since no feeder pig futures contracts exist. High hog prices create strong feeder pig prices. Low hog prices drive down the value of feeder pigs. Because that economic relationship is bid into cash markets, a hog feeder can use flexible hedging to hedge-sell finished hogs in high-priced markets and hedge-buy feeder pigs in low-priced markets.* At market turns the hedger liquidates one of the positions and institutes the other.

Converting feeder pigs into lean hog contracts can be done by one of three methods. First, the hedger could buy and sell the same number of head as finished. A 1,800-head producer would sell 10 contracts to hedge 400,000 pounds of market-weight hogs. At the market turn, one could buy 10 lean hog contracts to hedge in feeders. The 400,000 pounds is too great for his feeder needs, however. As a second alternative, the feeder could buy only 2 lean hog contracts, reasoning that he or she was purchasing only 72,000 pounds (1,800 head at 40-pound purchase weight). A 10-2 combination based on weight appears out of proportion in terms of value. The third method is to hedge equivalent values. A budget analysis to calculate the value of feeder pigs relative to hog prices typically shows feeder value per head to be approximately equal to the hundredweight price of hogs. Consequently, a long feeder pig hedge matched on value appears most appropriate. The strategy is to buy half as many long futures positions to hedge feeder pigs as the farmer would sell as short positions to hedge the live hog production. The flexible hedger thus sets 10 short hog positions in down trends and 5 long feeder pig positions in up trends.

---

* Basis risk is significant in this kind of hedge.

Profit Margin Hedging

Cattle, hog, and poultry producers should also be concerned about hedging in low costs on their feed needs. Livestock producers should seek to lock in profit as the maximum difference between the cost of input items and the value of production by hedging both. Feed should also be hedged when corn prices and\or meal advance.

A livestock operation that buys all its feed, hedge-buying feed ingredients and hedge-selling production, tends to maximize profit. In up trending feed-ingredient (corn and soymeal) markets, long hedges are placed to protect against cost increases. When the market returns down, the hedges are liquidated as in a selective hedging program. They are re-instituted on the next market upturn. Each time a feed-ingredient price rises, the producer protects against the rise by hedge buying.

The need to lock in low feed prices on a farm that produces its own feed needs is less obvious. But the same procedures should be used to create an integrated marketing plan. To understand profit-maximizing hedging on an integrated farm, the enterprises should be viewed as two divisions: the crop division and the livestock division. The crop division should be concerned with maximizing profit by realizing as high a price as possible for the crops grown. The livestock division endeavors to maximize profit by buying feed inputs as cheaply as possible. The fact that one manager, rather than two vice presidents, makes pricing decisions for both divisions does not alter the basic comprehensive marketing plan. To effect the plan, the manager places short hedges in down trending markets to protect against value erosion of the growing crop and stored feed grain. In up trending markets, one lifts the crop hedge and initiates a long hedge to lock in low feed costs. The manager changes hats as the price trend changes. In down trend markets, one wears the crop producer hat; in up trend markets, one wears the livestock feeder hat. The flexible hedging strategy of the crop farmer/livestock feeder is illustrated in Figure 10–4.

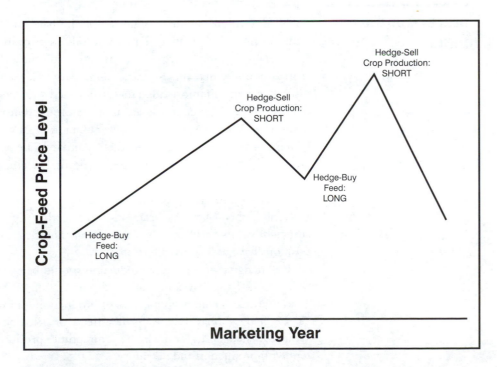

FIGURE 10–4: Flexible Hedging of Feed Needs and Crop Production by Farmer-Feeder.

By separating the decisions as to when to sell futures to hedge livestock production, when to buy futures to hedge in feeder animal costs, when to hedge-buy feed needs, and when to hedge-sell crop enterprise production, the integrated crop/livestock farmer can use market price fluctuation to widen the profit margin. It requires daily market analysis to detect price trend changes. But with an integrated hedging plan that takes different market positions at different times during the year, the integrated hedger can improve profitability.

Basis Signals for the Feed Buyer

The same technique for basis decisions used by crop producers is applicable for feed buyers in deciding when to make cash feed purchases or whether to build feed storage facilities. Feeders, however, react in a manner opposite of crop producers. When basis is weak, feeders make cash purchases; when basis is strong, they delay purchase until feed is required by the livestock.

Livestock producers who erect feed storage facilities incur ownership costs just like the grain producer with storage facilities. Buying feed ahead of needs incurs interest costs on the inventory purchases and the ownership and operating cost of the facilities. Their total costs of storage will be similar to those of any warehouser. They must capture basis advantages to pay ownership costs, if their decision is to be profitable.

In the cash market, basis discounts greater than carrying costs dictate that feeders should own storage and purchase feed ahead of needs. Their purchase of cash grain at greater discounts than storage costs would produce savings for them. On the other hand, a strong basis would not offer a sufficient discount to cover holding costs. In that case, they would be better off buying futures contracts as price protection rather than holding cash feed.

Futures Contract Months

Nearby futures contract options typically have the largest trading volume and trade on more information. Distant months, especially those at greatest distance, have relatively little activity. The increased volume of nearby contracts produces market liquidity that allows quicker and easier execution of trade orders at smaller discounts. Current information trades more actively in nearby contracts. As a consequence, the nearby contracts are much more responsive to price-making forces. Thus, the 730-Day Marketer focuses trading in nearby contract months to gain the benefits of volume trading and market sensitivity. The usual trading months for new crop production would be November soybeans, December corn and cotton, and July wheat. As the first day of each of these months arrives, contracts are rolled into the next-high-volume months, such as March, then July for fall-harvested crops and December, then May for wheat. Not every nearby month has to be used; those with sufficient volume will do. The use of fewer months requires fewer contract rolls as positions have to be moved forward.

Full hedging of the entire production should be established on each trade. Less than full hedging leaves the untraded portion subject to cash market speculation. Trading futures contracts in excess of production and inventory quantities creates "Texas hedges" that double up the effects of price changes. This too should be avoided. A hedger deals only in the commodities produced and in quantities consistent with production and inventory.

Livestock hedging should also be practiced in nearby contracts to improve responsiveness to fundamental price-making forces and to assure filling trade orders easily. The lead months move to wider extremes as they react to cash market forces. The hedger has the opportunity to benefit from wider fluctuation in futures as a consequence. Bull markets that create significant spreads between nearby and distant months can offer particularly profitable opportunities.* It should make no difference that the production is yet unborn for it to be profitably priced. Plans for production and intention to produce create a pricing requirement. It is better to set price when profitable opportunities present themselves than to have finished animals on hand that must be sold for whatever they will bring. Forward pricing of anticipated livestock production is no different from pricing anticipated crop production. Pricing ahead one year's production, or even two years, is not beyond the realm of possibilities. As long as contract obligations will be met at some time with actual production, the hedge definition of equal quantity is satisfied.

Protecting the Value of the Asset and the "Texas Hedge"

Farmers, their spouses, and their bankers tend to focus on one and only one aspect of the market – **this year's price.** To look beyond this year's price is frowned upon and those who do otherwise are denigrated as "Texas Hedgers." The mind set is to *protect the value of this year's crop and/or livestock.* Seldom does anyone plan far enough head. However, bankers lend farmers millions of dollars to assist them in their quest to acquire ownership to a piece of land and livestock facilities. With the exception of some simplistic farm budgets and cash flow analysis, no one asks the question: "How are you going to pay for it over the next 30 years?"

Late 1980 is one of several classic cases. Nearby corn futures exceeded $4.00 per bushel. Farmers were eager to buy land for $3000 per acre and their bankers were ready to lend. Midwest farmland was worth $3000 per acre because corn was selling for $4.00 per bushel in central Illinois. However, at $2.00 per bushel that same farmland is worth less than $1500 per acre. But farmers and their bankers seemed oblivious to that fact. They were betting that corn prices would stay at or near $4.00 for the next 20 years so the farm loans could be amortized (paid off). Life soon became difficult as corn prices plunged to $2.15 in the fall of 1982, rallied briefly to $3.75 with the drought of 1983 and then plummeted to $1.70 with the 1986 harvest. The tragic outcome was that farmers could not meet their loan payments and were forced into bankruptcy. Their neighbors refused to pay $3000 per acre in a $2.00 corn market. The distressed farms sold for $1200 per acre and with those horrendous losses went the bank's capital and the bank went into bankruptcy or acquisition for pennies on the dollar.

Farmers and bankers are reluctant to "speculate" in the futures market. Yet those same individuals are willing to speculate in the $3000 land market betting their fortunes that **corn will sell for $4.00 per bushel every year over the next the next 20 years**. They gambled in the cash market and lost because they did not protect the

---

* See "bear spread hedging" in the previous chapter.

value of the land asset by selling 20 years worth of corn (150,000 bushel × 20 = 600 contracts) in the nearby contract for $4.00. Five dollar corn in 1996 offers a comparable scenario. That bonanza year was followed by $2.75 corn in 1997, $2.15 in 1998, and below $2.00 in 1999. Many farmers are in grave financial circumstances and complain about low prices. The market offered $4.00 to $5.50 in 1996.

The only question that farmers and that bankers had to ask and answer was how many years worth of production do you want to sell at $4.00 or higher. Some call that a "Texas Hedge," we call it sound business management. There is nothing wrong with selling a number of years worth of production. The capacity is there, all it takes is time. The difference between $2.00 and $4.00 corn is not simply $2.00. It is the difference is between success and failure. It is the difference between paying the bills and the mortgage, putting food on the table, and sending children to college: these are not trivial issues. There is no law that says farmers cannot sell $4.00 corn. There is no law that says farmers cannot earn a decent living.

To succeed, many firms enter into long-term production contracts. They build airplanes and submarines. They acquire raw materials. They lease facilities. They enter into labor and retirement contracts. They also prosper. Long-term purchases and sales contracts work for industrial firms. Why are farmers reluctant to use similar business practices to their advantage?

Summary: The 730-Day Crop Marketing Plan and Profit-Maximizing Livestock Hedging

Farmer-producers must understand markets and marketing. They must understand themselves and their psychology. They must design a marketing plan that fits their situation. They also need to develop a financing plan, usually with their lender. They need to employ a broker who understands the plan and can effect the futures market transactions required by it. Some busy farmers may add a fourth person to their marketing team, an advisor. The advisor becomes the market watcher and analyzes the market signals. The advisor becomes the farmer's market watchdog.

A marketing program is not a simple process if it is to be rewarding. 730-DAY CROP MARKETING AND PROFIT-MAXIMIZING LIVESTOCK HEDGING are offered as integrated, profitable hedging programs.

730-Day Marketing for the Crop Farmer

730-Day Marketing for the grain producer deals with pricing decisions in three stages: the growing crop, the cash sale, and the stored crop. The marketing period for a crop starts December 1, 2000 during the winter planning stages of the growing year and continues for 730 days until the end of the storage year (November 30, 2002).

Throughout the planning, planting, and growing season, the growing crop is protected against price erosion by hedging with short futures positions.

At harvest time, the decision to sell out of the field or to store is determined by basis market analysis. A strong basis dictates sale from the field; a weak basis dictates storage. When storage is chosen, timing of the cash sale decision depends upon the daily improvement in basis as compared to the added daily variable cost of holding grain. 730-Day Marketers make cash sales to capture a strong basis when the market dictates the move. They sell to capture basis profits and do not store to speculate on price level.

Throughout the second year, the storage year, farmers own grain when market price is going up. But farmers DO NOT HAVE TO OWN GRAIN STORED IN BINS TO SPECULATE ON A PRICE RISE. They can replace their cash positions with a futures position and speculate on the Board instead of speculating with grain in bins. It is also less expensive to speculate on paper (commission is one cent per bushel) than to pay interest and storage to speculate with stored cash grain.

In summary, using the 730-Day Marketing concept, a farmer hedges the growing year crop against price erosion, owns the stored crop in bins or on paper for price appreciation in the storage year, and sells grain on a strong basis (pre- or post-harvest) to capture storage profits.

Squeezed into a calendar-year period, 730-Day Marketing constantly deals with one or the other of two crops. In price down trends, the farmer hedges the growing crop. At a market turnaround, he or she lifts the short hedge and simultaneously holds the old crop unhedged in storage, or buys it back on paper to *"refill"* empty bins. During price up trends, the old crop is owned as a long futures position, after the cash sale has been made.

When the price up trend turns into a down trend, the long position is liquidated and a short hedge on growing production is reinstated. The 730-Day Marketing plan goes a step beyond selective hedging. It employs FLEXIBLE HEDGING. At changes in price trends, market positions are reversed. The producer is almost always in an active marketing position.

Finally, the cash sale is made after an analysis of the basis market. Whenever prospective gains in basis no longer exceed added holding costs, the cash sale is made. This practice converts farmers into grain merchandisers, intent on capturing storage profits, as opposed to storage speculators.

730-Day Marketing incorporates two innovative procedures not found in other marketing strategies.

First, the farmer can speculate without holding grain in bins. This allows one to capture BOTH a strong basis AND a high price level. One seeks to capture high prices in *both* markets.

Second, pricing decisions are not irreversible one-time decisions, but a constant process. The farmer holds a futures market position almost all the time, either short or long. One is not standing outside the market as a cash market speculator wondering when to fix a one-time pricing decision. With continuous pricing, a person is always in the market, letting the market work for, not against, him or her. In practice, continuous marketing reduces cash speculation significantly. Production is hedged about half the time. Speculation is elected the other half of the time only for reasons that move the market (drought, strong demand) and is not carried out as the standard accepted practice of passive marketing.

With continuous 730-Day Marketing, farmers can capture marketing profits in all three stages:

1. price protection from growing crop hedges,

2. storage profit from basis sales, and

3. speculative profit from owning cash or paper-stored grain.

With these added together, 730-Day Marketing offers greater potential marketing profits than most other marketing plans.*

Profit-Maximizing Hedging for the Livestock Farmer

The concepts practiced in flexible hedging serve the livestock producer with an integrated PROFIT-MAXIMIZING HEDGING plan as well as 730-DAY MARKETING serves the grain farmer. Profit-Maximizing Hedging deals with pricing decisions in three important activities, the:

1. sale of finished animals,
2. purchase of feeder stock, and
3. purchase of feed ingredients.

The objective of the plan is to maximize returns over costs of major input items. A full year's production and requirements are committed to the program. Throughout the year, the value of the production to be eventually sold is protected from price erosion during market price down trends. As price trends change into up trends, selling hedges are lifted to enjoy inventory value appreciation on the anticipated market livestock. Correspondingly, at market turns, feeder animals are hedge-bought to protect against rising costs, since feeders and live cattle tend to trend upward together. At the end of the up trend, feeder hedges are liquidated and the production hedges reestablished. Whether the price level is going up or down, the livestock feeder enjoys the results, protecting either high price on production or low cost on feeders.

Feed purchases of corn and soybean meal are added into the hedging program to acquire these major cost items as cheaply as possible. When the ingredient markets turn into up trends, corn and soymeal are hedge-bought for cost protection. In down trends, feed hedges are liquidated and the buying of cash commodities proceeds hand to mouth, taking advantage of dropping cash market prices.

Farmer-feeders producing on-farm feed inputs (corn, oats, barley, or sorghum**) carry trading activities one step further. They switch to their "corn-producer cap" by establishing short positions to hedge growing crops when they detect a down trend signal that says lift the feed corn hedge. They hedge feed needs in market up trends and grain production in down trends. They let the market work at capturing both $4.00 corn "sales" and $2.00 feed cost. There is nothing illegal or immoral about making both the corn enterprise AND the hog enterprise profitable!

Profit-Maximizing Hedging works for the farmer-feeder to capture, in effect, wide corn/hog or corn/steer ratios, irrespective of price levels in the individual markets. A farmer seeks price protection in all three segments of the operation:

1. Price protection on fed livestock,
2. Purchase price of feeders,
3. Major cost feed ingredients.

---

* The 730-DAY MARKETING concept has been successfully employed by AG COM, INC., a commodity marketing consulting firm for farmer-hedgers. Ag Com, Inc., has defined the tools for making 730-Day Marketing an effective marketing program. On the internet contact: hoepner@vt.edu

**Feed commodities produced but not traded on exchanges may be converted into corn equivalent as a surrogate and traded in the corn pits, as long as the price of various feed items trade almost lock step to corn.

When these are added together, integrated Profit-Maximizing Hedging offers a greater profit potential than hit-and-miss programs.

The name of the game in the new millennium is PROFIT. Much of production is a necessary exercise; the bottom line depends on marketing. A logical, integrated plan makes profit possible–we wish you the best of 730-Day Marketing years.

Questions for Study and Discussion:

1. Over what time period should a producer plan to price his or her production?
2. How does the 730-Day Marketing Plan operate to protect price level on new and old crop production?
3. Why should basis decisions be separated from price level decisions?
4. What is the key feature of making profitable basis decisions?
5. Outline a method for calculating basis profit.
6. How is "buying the Board" speculation similar to storing cash grain?
7. How does it differ?
8. Explain how to lock in production profits in livestock.
9. What are the basis signals that the input buyer should watch for in the market?
10. Explain why nearby futures contracts offer advantages over distant contracts in placing futures hedges.

Endnote for Chapter 10:

1. Commodity Trend Service, "Futures Charts" (Palm Beach Gardens, FL 33420).

Glossary of Terms

ACREAGE ALLOTMENT: A voluntary limitation on the number of acres of a given crop farmers plant as established under a federal farm program.

ACREAGE RESERVE: An arrangement under which farmers agree to withdraw a stated acreage of cropland from production for a specified number of years. This is a conservation measure under the Conservation Reserve Program, which provides for annual compensation for any loss of income.

ACTUALS: Physical commodities, as distinguished from futures contracts.

ASK: An offer to sell a specific quantity of a commodity at a stated price.

AT THE MARKET: An order to buy or sell at the best price obtainable when the order reaches the trading floor.

AT THE MONEY: An option with a strike price equal to the current price of the futures contract the option is based on.

BAR CHARTS: Charts of futures prices drawn with vertical lines, showing the daily high, low, and closing prices.

BASIS: The difference between a cash price at a specific location and the price of a particular futures contract.

BASIS QUOTE: An offer or sale of a cash commodity in terms of the difference above or below a futures price.

BEAR: One who believes prices will move lower.

BEAR MARKET: One in which prices are declining.

BEAR SPREAD: The simultaneous purchase and sale of two futures contracts in the same or related commodities with the intention of profiting from a decline in prices; the sale of a nearby futures contract (short) while being long the distant cash.

BID: An offer to buy a specific quantity of a commodity at a stated price.

BREAK: A rapid, significant decline in price, frequently falling below a commonly recognized "support" level.

BREAKOUT: A movement of prices to outside a well-defined trading range.

BROKER: (1) A person paid a fee or commission for acting as an agent in making contracts or sales; (2) **floor broker:** in commodities futures trading, a person who actually executes orders on the trading floor of an exchange; (3) **account executive:** the person who deals with customers in commission house offices.

BULL: One who expects prices to rise.

BULL MARKET: One in which prices are rising.

BULL SPREAD: The simultaneous purchase and sale of two futures contracts in the same or related commodities with the intention of profiting from a rise in prices; being long the nearby futures and short the distant futures.

BUYING THE BOARD: Buying a futures contract to replace a long cash position (grain in a bin that has been sold) with a long futures position to speculate on a price rise.

CALL OPTION: A contract that gives the buyer/taker the right but not the obligation to buy a fixed quantity of a commodity at a stipulated striking price at any time up to the expiration of the option.

CARRYING CHARGES: Those costs incurred in owning a physical commodity, generally including interest, insurance, and storage; may also include transportation, grading, sampling, and so forth.

CARRYOVER: That portion of current supplies of a commodity comprised of excess stocks from previous production/marketing seasons.

CASH COMMODITY: Actual stocks of a commodity as distinguished from futures contracts; goods available for immediate delivery or delivery within a specified period following sale; or a commodity bought or sold with an agreement for delivery at a specified future date.

CASH FORWARD SALE: The sale of a cash commodity for delivery at a later date. Price may be fixed at the time of the agreement, or there may be an agreement to determine the price at the time of delivery on the basis of prevailing local cash price or on some future price. Also known as "deferred delivery" or "forward sale."

CASH MARKET: (1) The market for the physical commodity; (2) an organized, self-regulated central market, such as a commodity exchange; (3) a decentralized over-the-counter market; (4) a local organization, such as a grain elevator or meat processor, which provides a market for a small region.

CASH SETTLEMENT: Payment for physical commodities received.

CHARTER: A technical trader who reacts to signals read from graphs of price movement.

CHARTING: The use of graphs to view market patterns in order to plot trends and patterns of price movements, volume, and open interest, in the expectation that such graphs and charts will help one to anticipate and profit from future price movements. An aspect of technical analysis as contrasted with **fundamental analysis.**

CLEARINGHOUSE: An agency connected with commodity exchanges that provides the mechanism by which all futures contracts are made, offset, or fulfilled through delivery of the commodity and through which financial settlement is made.

CLOSING OUT: Liquidating an existing long or short futures position with an equal and opposite transaction, also known as "existing."

CLOSING PRICE: The price for a commodity futures contract generated by trading through open outcry during the closing period of a trading session.

COMMISSION: The charge that a commodity broker must make to the client when the broker buys or sells contracts. Traditionally this charge is levied at the time of liquidation; however, there is a shift toward charging one-half when the position is entered.

COMMODITY CREDIT CORPORATION (CCC): A government-owned corporation established in 1933 to assist U.S. agriculture by providing price-support programs in which the CCC purchases excess supplies of commodities and by assisting in the export of agriculture commodities.

COMMODITY FUTURES TRADING COMMISSION (CFTC): A federal regulatory agency empowered under the Commodity Futures Trading Commission Act of 1974 with regulation of all commodities trading on all domestic contract markets. The CFTC consists of five commissioners, one of whom is chairperson. All are appointed by the President subject to Senate confirmation. The CFTC replaced and assumed all powers of the Commodity Exchange Authority.

CONGESTION: In technical analysis, a price range within which buying power and selling pressure are about equal, resulting in a sideways movement of prices.

CONSOLIDATION: A price pattern represented by a prolonged period of narrowing price movement.

CONTRACT: An agreement to buy or sell a specified amount of a particular commodity. The contract details the amount and grade of the product and the date on which the contract will mature and become deliverable, if it is not liquidated earlier.

CONTRACT MONTH: The month in which a given contract becomes deliverable, if not liquidated or offset before the date specified.

CONTRACT UNIT: The specific amount of a commodity represented by the futures contract.

CONTRARIAN: A trader who takes positions opposite to the market direction anticipated by the majority of traders.

CONVERGENCE: The tendency for prices for physical and futures to approach one another, usually during the delivery month. Also called a "narrowing of the basis."

CORNER: The securing of significant control of a commodity, enabling price manipulation, often called a "squeeze"; in its extreme, the control of more contracts requiring delivery than the available supply.

COVERING: Offsetting a previously established futures short position by a PURCHASE. Also "evening up" or "liquidating."

CROP YEAR: The period beginning with the harvesting of a crop to the corresponding time the following year.

CYCLE: The tendency for an entity to move through phases of high and low with sufficient regularity to be at least partially predictable, e.g., the hog cycle.

DAY TRADE: A trade that is entered and liquidated on the same trading day, as different from an **overnight trade**.

DEFERRED DELIVERY: The more distant months in which futures trading is taking place, as distinguished from the nearby futures delivery months. See also **Cash forward sale**.

DELIVERY: Tender of a commodity by (1) issuing a warehouse receipt or a bill of lading, (2) issuing a shipping certificate for some commodities, or (3) delivering actuals against a futures delivery contract.

DELIVERY DATE: Date on which the commodity must be delivered to fulfill the terms of the contract.

DELIVERY MONTH: The calendar month in which the futures contract matures and within which delivery of the physical commodity can be made.

DELIVERY NOTICE: Written notice from a seller through the clearinghouse, stating intention to make delivery on a short futures position on a particular date. This notice also specifies quantity, grade, and place of delivery. A notice precedes and is distinct from a warehouse receipt of shipping certificate, which are instruments representing transfer of ownership.

DELIVERY POINTS: The places specified by a commodity exchange to which delivery of a physical commodity can be made in fulfillment of a contract.

DELIVERY PRICE: The price fixed by the clearinghouse where futures deliveries are invoiced. Also, the price at which a commodities futures contract is settled when deliveries are made.

DELTA: The amount of change in the option premium compared with the change in price of its underlying futures contract.

DEMAND: The quantity of a commodity that consumer (users) are willing to buy at various market prices.

EFFICIENT MARKET: A market in which new information is immediately and costlessly available to all investors and potential investors; a market in which all information is instantaneously assimilated and therefore has no distortions.

ELASTICITY: The percentage change in quantity relative to the percentage change in price. This concept applies to both supply and demand.

ELEVATOR: A commercial facility for grain storage.

EQUILIBRIUM: A state of rest from which there is no tendency to change. Where the quantity of a commodity supplied equals the quantity of that good demanded at a price that is acceptable to all successful buyers and sellers.

EXCESS: The amount of money that can be withdrawn from a futures account without liquidating the positions; the amount by which equity exceeds margin requirements.

EXCHANGE (COMMODITY): An organized marketplace where particular commodities are bought and sold.

FEED/VALUE RATIO: The relationship of the cost of feed to market-weight sales prices of livestock, expressed as a ratio, such as the *hog/corn ratio*. These ratios serve as indicators of the profit margin or lack of profit in feeding animals to market weight.

FIRST NOTICE DAY: The first day on which notices of intention to deliver actual or physical commodities against a short futures contract can be presented by sellers through the clearinghouse.

FLOOR BROKER: A member of the commodity exchange or representative of a member firm of the commodity market who is "on the floor" (i.e., in the market, ready to do business) at all times when the market is open.

FORWARD CONTRACT: A cash market transaction in which two parties agree to the purchase and sale of a commodity at some future time under such conditions as the two agree–in contrast to **futures contracts**.

FORWARD SALE (OR PURCHASE): See **Forward contract**.

FUNDAMENTAL ANALYSIS: The forecasting of price movement traditionally using factors such as income, acreage, inventory, yields, government programs, and techniques of econometrics, as contrasted with **technical analysis** or **charting**.

FUTURES: Contracts for the purchase and sale of commodities for delivery some time in the future on an organized exchange and subject to all terms and conditions included in the rules of that exchange.

FUTURES COMMISSION MERCHANT (FCM): An individual, association, partnership, or corporation engaging in soliciting or in accepting orders for the purchase or sale of any commodity for future delivery (on and subject to the rules of any contact market) and registered with and regulated by the CFTC.

FUTURES CONTRACT: An agreement to make or take delivery of a standardized amount of a commodity, of standardized quality grade, during a specific month, under terms and conditions established by an authorized exchange.

FUTURES EXCHANGE: A membership organization whose activity involves trading sessions between members in the specific commodities listed for trading on the exchange.

GOOD TILL CANCELED ORDER (GTC): An order that is valid at any time during market hours until executed or canceled by the client. Also known as an "open order."

GRADES: Standards set for the quality of a commodity.

HEDGER: One who sells futures contracts in anticipation of future sales of cash commodities as a protection against possible price declines, and who purchases futures contracts in anticipation of future purchases of cash commodities as a protection against the possibility of increasing costs.

HEDGING: The establishment of an opposite position in the futures market from that held in the physicals.

HISTORICAL VOLATILITY: The actual amount that prices increased or decreased in the past. Usually expressed as a percentage.

IMPLIED VOLATILITY: The amount prices are expected to go up or down in the future. Usually expressed as a percentage.

IN THE MONEY: A term used to describe an option that has intrinsic value. A call at $4.00 on wheat trading at $4.10 is in the money 10 cents.

INTRINSIC VALUE: A measure of the value of an option if immediately exercised. Amount by which the option is "in the money," i.e., has real value.

INVERTED MARKET: A futures market in which distant-month contracts are selling below near-month contracts.

LAST TRADING DAY: The day on which trading ceases for a particular delivery month. All contracts that have not been liquidated by an offsetting purchase or sale by the end of trading on that day must therefore be settled by delivery of the actual physical commodity.

LEVERAGE: Buying or selling with deposits, collateral, or margin less than the face value of the item being traded. Commodity futures markets usually require about 10 percent of the face value of the contract, giving 90 percent leverage.

LICENSED WAREHOUSE: A warehouse designated for delivery by the exchange on which a commodity is traded. Only such designated warehouses may be used to store a commodity for delivery.

LIFE OF CONTRACT: The period of time from the first trading day in a futures month through the last trading day.

LIMIT (UP OR DOWN): The maximum price advance (or decline) that the rules of the exchange permit a commodity to trade at from the previous day's trading session.

LIMIT ORDER: An order that has some restriction of execution, such as price or time.

LIMIT PRICE: The highest or lowest price that a commodity may trade at during a trading session, as fixed by the contract market's rules. See also **Limit (up or down)**.

LIQUIDATION: The closing out of a long position. The term is sometimes used to denote closing out a short position, but this is more often referred to as "covering."

LIQUID MARKET, LIQIUDITY: A market where selling and buying can be accomplished with ease, because of the presence of a large number of interested buyers and seller willing and able to trade substantial quantities at small price differences. Often measured by volume and open interest.

LOAN PRICES: Prices at which the U.S. government will lend producers (farmers) money for their crops placed in storage and under loan.

LOAN PROGRAM: The U.S. government agricultural price support operation. A program under which farmers commit their crops to the government with the assurance that they will receive a certain minimum loan price. If the price of the commodity rises above the loan price, the farmers may sell their crops in the open market and repay the loan. Even if the price of the commodity falls below the low level, the *non-recourse* character of the loan makes it possible for farmers to deliver their crops to the CCC (Commodity Credit Corporation), discharging their obligation in full.

LOCALS: Members of the U.S. exchange who trade for their own account and/or fill orders for customers and whose activities provide market liquidity.

LONG: One who has bought a cash commodity or a commodity futures contract, in contrast to a **short**, who has sold a cash commodity or futures contract.

LONG HEDGE: Purchase of futures to lock in a low (favorable) price in anticipation of acquiring the resource (feed) at some future date.

LONG LIQUIDATION: Closing of long positions.

LONG THE BASIS: A hedge consisting of ownership of actuals against short futures in a particular commodity. See also **Hedging**.

MARGIN: (1) An amount of money deposited by both buyers and sellers of futures contracts to insure performance against the contract, i.e., to deliver or take delivery of the commodity (not an equity or down payment for the goods represented by the futures contract); (2) **profit margin**: the difference between the price that one pays for goods and the price at which the goods or products of them are resold.

MARGIN CALL: A call from a brokerage firm to a customer to bring margin deposits back up to minimum levels required by exchange regulations; similarly, a request by the clearinghouse to a member firm to make additional deposits to bring clearing margins back to minimum levels required by clearinghouse rules.

MARKET: A region or place where buyers and sellers come together to transact business.

MARKET IF TOUCHED (MIT) ORDER: An order to buy or sell when the market reaches a specified price. An order to buy becomes a market order when the commodity sells (or is offered) at or below the order price; an order to sell becomes a market order when the commodity sells (or is bid) at or above the order price.

MARKETING: A process whereby buyers and sellers transact business in the purchase and sale of commodities.

MARKET ORDER: An order to buy or sell at the best price available and as soon as possible after the order reaches the trading floor of the exchange.

MAXIMUM PRICE FLUCTUATION: The limit of fluctuation in the price of a futures contract during any one trading session, as established by the exchange.

MINIMUM PRICE FLUCTUATION: The minimum amount by which the price of a commodity can fluctuate during one trading session, as established by the exchange.

MOMENTUM: In technical analysis, the relative change in price over a specific time interval.

MOVING AVERAGE: In technical analysis, a method of averaging prices of the most recent days in an effort to determine and/or forecast the direction of movement.

NEARBY DELIVERY (MONTH): The futures contract closest to maturity. Also called "spot month."

NON-MEMBER FCM: A brokerage firm (FCM) that does not own its own seats on futures exchanges; a firm that executes its orders through a member FCM.

NOTICE OF INTENTION TO DELIVER: A notice that must be presented by the seller to the clearinghouse. The clearinghouse then assigns the notice and the subsequent delivery instrument to the longest-standing buyer on record.

OFFER: A price at which a trader is willing to sell; the asking price.

OFFSET: The liquidation of a purchase of futures (longs) through the sale of an equal number on contracts of the same delivery month, or the liquidation of a short sale of futures contracts through the purchase of an equal number of contracts of the same delivery month. Either action transfers to other persons the obligation to make or take delivery of the actual commodity.

OPEN CONTRACTS: Contracts bought or sold and not offset by an opposite trade.

OPENING RANGE: The range of prices at which transactions took place during the period designated as "the opening" of trading by an exchange.

OPEN INTEREST: The total number of futures contracts of a given commodity that have not yet been offset by opposite futures transactions nor fulfilled by delivery of the actual commodity; the total number of open transactions, with each transaction having a buyer and a seller.

OPEN OUTCRY: A method of public auction for making bids and offers in the trading pits of a commodity exchange.

OPEN POSITION: A market position that has not been closed out.

OPTION: An agreement that gives the taker (purchaser) the right, but not the obligation, to buy from or sell to the grantor (seller) of the option at any time before its expiration a specified quantity of the commodity concerned at an agreed price.

OPTION BUYER: The person who buys calls, puts, or any combination of calls and puts.

OPTION SELLER: Also, **option writer**. The person who originates an option contract by promising to perform a certain obligation in return for the price of the option.

ORIGINAL MARGIN: See **Margin**.

OUT OF THE MONEY: A term used to describe an option that has no intrinsic value. A call at $4.00 on wheat trading at $3.90 is out of the money 10 cents. The right to sell $4.00 wheat in a $3.90 market has no real value.

OVER-SOLD: In technical analysis, an opinion that the market price has declined too steeply and too fast in relation to underlying fundamental factors.

PARALLEL PRICES: The constant relationship between cash and futures markets. The difference between the two being the basis.

PARITY: A theoretically equal relationship between farm product prices and all other prices. In farm program legislation, parity is defined in such a manner that the purchasing power of a unit of an agricultural commodity is maintained at its level during an earlier historical base period.

PHYSICAL: See **Actuals**.

PIT: The place where futures are traded on the floor of a commodity exchange.

POINT: A unit expressing minimum price change. Synonymous with "minimum price fluctuation."

POINT-AND –FIGURE: A method of charting that uses prices to form a pattern of movement without regard to time. It defines a price trend as continued movement in one direction until a reversal of a predetermined criterion is met.

POSITION: A market commitment. A buyer of futures contracts is said to have a long position, and conversely, a seller of futures contracts is said to have a short position.

PREMIUM: (1) The amount by which a cash commodity sells over a futures price or another cash commodity price; (2) the excess of one futures contract price over another; (3) in options, the price paid by the purchaser of the option to the grantor (seller).

PRICE LIMIT: The maximum price fluctuation on a futures contract during one trading session, as determined by the exchange. Also known as "limit."

PRICE TREND: See **Trend**.

PRICING: The act of fixing a price.

PUBLIC ELEVATORS: Storage facilities for grain licensed and regulated by federal and state agencies. Grain of the same grade but owned by different persons is usually commingled, or mixed together. Some elevators are designated as depositories for delivery by exchanges dealing in commodities stored at these elevators.

PUT OPTION: A contract that gives the purchaser the right, but not the obligation, to sell a fixed quantity of a commodity at a stipulated strike price at any time during the life of the option.

RALLY: An upward movement of price after a decline.

RANGE: The difference between the highest and lowest prices recorded during a given trading session, week, month, life of contract, or any given period.

RESISTANCE: In technical trading, a price area where new selling will emerge to dampen a continued rise. See also **Support**.

RETRACEMENT: A reversal within a major price trend.

REVERSAL: A change of direction in prices.

ROLL, ROLLING FORWARD: See **Switching**.

SCALPER: A speculator on an exchange floor who trades in and out of the market on very small price fluctuations. The scalper, trading in this manner, provides market liquidity, but seldom carries a position overnight.

SELLER: A risk taker who takes a short future position or grants (sells) a commodity option.

SELLER'S OPTION: The right of the seller to select commodity quality, time, and place of delivery within the limits prescribed by the exchange upon which futures contracts are traded in making physical delivery.

SETTLEMENT PRICE: The price on the close of the day, or a price determined to be in line with actual closing values. Settlement prices are used to set the next trading day's price fluctuation limits and are used as the basis for adjusting margins and determining equity by the clearinghouse for contracts traded on each exchange.

730-DAY MARKETING PLAN: To hold a position in the futures market at virtually all times during the 365-day growing year and the subsequent 365-day storage year. A short futures position is held in down trending markets to protect the value of the growing crop and the value of grain in the bin. Cash grain is sold when basis calculations signal that the added cost of holding grain longer exceeds the added return from basis improvement. Cash grain is replaced by a long futures position when the market is in an up trend to participate in price appreciation.

SHORT: Starting a transaction by the sale of a futures contract. An open position on a futures market that was initiated by a sale.

SHORT COVERING: See **Covering**.

SHORT HEDGES: The sale of futures against holdings of the physical commodity.

SHORT THE BASIS: The purchase of futures as a hedge against a commitment to sell in the cash or spot market.

SPECULATOR: A person using the futures market for a purpose other than hedging and willing to assume varying degrees of risk in exchange for potential profit.

SPOT: (1) The characteristic of being available for immediate (or nearly immediate) delivery; usually refers to a cash market price for the physical commodity available for immediate delivery; (2) sometimes used in reference to the futures contract of the current month, in which delivery is possible at any time.

SPOT COMMODITY: An actual or physical commodity.

SPOT MARKET: The market in which cash or spot transactions occur.

SPOT MONTH: The first deliverable month for which a quotation is available on the futures market.

SPOT PRICE: The commodity cash sale price, as opposed to a futures price.

SPREAD (STRADDLE): The purchase of one futures delivery month against the sale of another futures delivery month of the same commodity; the purchase of one delivery month of one commodity against the sale of that same delivery month of a different commodity; or the purchase of one commodity in one market against the sale of that commodity in another market. The purpose of any one of these transactions is to take advantage of distortions in normal price relationships. The term "spread" is also used to refer to the difference between the price of one futures month and the price of another month of the same commodity. There are four basic types of spreads: (a) **Interdelivery spread**: The purchase and sale of the same commodity, in the same market, in different delivery months. Also called an **intramural spread**. (b) **Intermarket spread**: The purchase and sale of the same commodity, in the same or different delivery months, in two different markets. (c) **Intercommodity spread**: The purchase and sale of different but related commodities, in the same or different markets, in the same or different delivery months. (d) **Commodity product spread**: The purchase of futures raw material and the sale of the derived processed products futures, or vice versa.

STOP ORDER, STOP LOSS ORDER: An order that becomes a market order to buy only if the market advances to a specified level, or to sell only if the market declines to a specified level. As soon as this specified level is traded, the order is executed for the client at the next obtainable price. There is no guarantee the order will be executed at the price specified. A stop loss order is used to prevent or minimize losses in either a short or long position.

STRIKE PRICE: The price at which an option can be exercised irrespective of current market conditions.

SUPPLY: The quantity of a commodity that producers are willing to provide at various market prices.

SUPPORT: In technical analysis, a price area where new buying is likely to come in and stem any decline.

SWITCHING: Liquidation of a position in one delivery month of a commodity and simultaneous initiation of a similar position in another delivery month of the same commodity. When used by hedges, this tactic is referred to as "rolling forward the hedge."

TECHNICAL ANALYSIS: An approach to forecasting commodity prices based on the study of price movement itself without regard to underlying fundamental market factors. Contrasted with **fundamental analysis**.

TECHNICAL RALLY: Price variations arising from factors other than those affecting supply and demand for a commodity.

TECHNICAL TRADING: Systematic trading based on charts or technical analysis that indicate buy and sell signals. See also **Technical analysis**.

TERMINAL ELEVATOR: In the movement of grains, a storage facility located at a point of accumulation or distribution.

TIME VALUE: The part of an option's value that cannot be realized by immediately exercising the option.

TRADER: A person who takes positions in the futures market, usually without the intention of making or taking delivery.

TREND: The general direction of price movement, within a specified time interval.

TRENDLINE: In charting, a line drawn across the bottom or top of a price chart indicating the direction or trend of price movement. If up, the trendline is called **bullish**; if down, it is called **bearish**.

UNDERLYING FUTURES CONTRACT: The futures contract that a particular option gives a person the right to buy or sell.

VARIABLE PRICE LIMIT: A price limit schedule, determined by an exchange, that permits variations other than the normally allowable price movements for any one trading day.

VOLUME: The quantity of business or transactions done.

WAREHOUSE RECEIPT: A document guaranteeing the existence and ability of a given quantity and quality of a commodity in storage.

Index